Theory and Advances of Tribology

Theory and Advances of Tribology

Edited by **Bernice Wyong**

NY RESEARCH PRESS
P R E S S

New York

Published by NY Research Press,
23 West, 55th Street, Suite 816,
New York, NY 10019, USA
www.nyresearchpress.com

Theory and Advances of Tribology
Edited by Bernice Wyong

International Standard Book Number: 978-1-63238-445-4 (Hardback)

Printed in the United States of America.

Contents

Preface

Every book is a source of knowledge and this one is no exception. The idea that led to the conceptualization of this book was the fact that the world is advancing rapidly; which makes it crucial to document the progress in every field. I am aware that a lot of data is already available, yet, there is a lot more to learn. Hence, I accepted the responsibility of editing this book and contributing my knowledge to the community.

The science of interacting surfaces in relative motion is referred to as tribology. The field of tribology consists of lubrication, friction and wear of contact elements relevant to practical operations. Latest results and advancements are discussed and analyzed in the context of established information. The book combines fundamental concepts with novel findings. It discusses topics such as lubrication, the characteristics of lubricants, boundary lubrication applications, testing and modeling, and sustainability of tribosystems. The book includes contributions by prominent experts and attempts to convey latest research trends in tribology to students, scientists and practitioners.

While editing this book, I had multiple visions for it. Then I finally narrowed down to make every chapter a sole standing text explaining a particular topic, so that they can be used independently. However, the umbrella subject sinews them into a common theme. This makes the book a unique platform of knowledge.

I would like to give the major credit of this book to the experts from every corner of the world, who took the time to share their expertise with us. Also, I owe the completion of this book to the never-ending support of my family, who supported me throughout the project.

Editor

Lubrication and Properties of Lubricants

Lubrication and Lubricants

Nehal S. Ahmed and
Amal M. Nassar

Additional information is available at the end of the chapter

1. Introduction

1.1. Lubrication

The primary purpose of lubrication is to reduce wear and heat between contacting surfaces in relative motion. While wear and heat cannot be completely eliminated, they can be reduced to negligible or acceptable levels. Because heat and wear are associated with friction, both effects can be minimized by reducing the coefficient of friction between the contacting surfaces. Lubrication is also used to reduce oxidation and prevent rust; to provide insulation in transformer applications; to transmit mechanical power in hydraulic fluid power applications; and to seal against dust, dirt, and water.

1.1.1. The lubrication regimes

The modern period of lubrication began with the work of Osborne Reynolds (1842-1912). Reynold's research was concerned with shafts rotating in bearings and cases this show in Fig. 1. When a lubricant was applied to the shaft, Reynolds found that a rotating shaft pulled a converging wedge of lubricant between the shaft and the bearing. He also noted that as the shaft gained velocity, the liquid flowed between the two surfaces at a greater rate. This, because the lubricant is viscous, produces a liquid pressure in the lubricant wedge that is sufficient to keep the two surfaces separated. Under ideal conditions, Reynolds showed that this liquid pressure was great enough to prevent direct contact between the metal surfaces. Fig.2 taking a plain journal bearing as example, Fig.3 which is known as Stribeck curve summarizes the lubrication regimes by describing the relationship between speed, load, oil viscosity, oil film thickness, and friction.

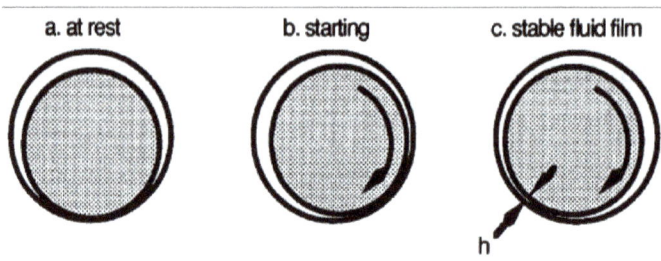

Figure 1. Three positions of shaft in a bearing

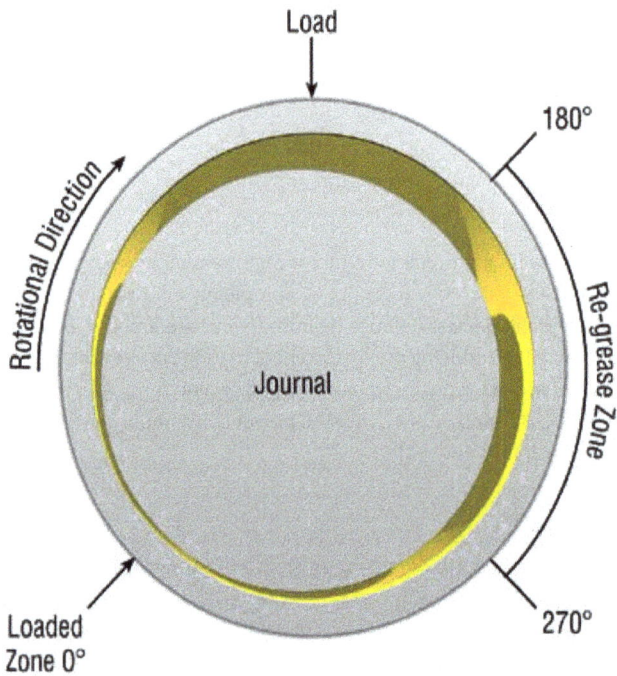

Figure 2. Plain Journal bearing

In this graph, the coefficient of friction is plotted against the expression ZN/P (sometimes referred to as the Hersey number)

$$\text{Where } ZN/P = \frac{\text{oil viscosity} \times \text{shaft speed}}{\text{bearing pressure}} \tag{1}$$

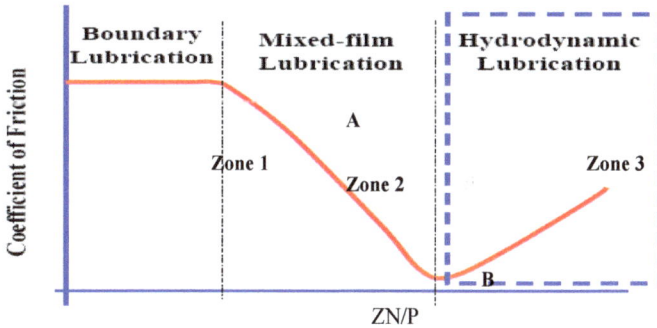

Figure 3. Stribeck curve

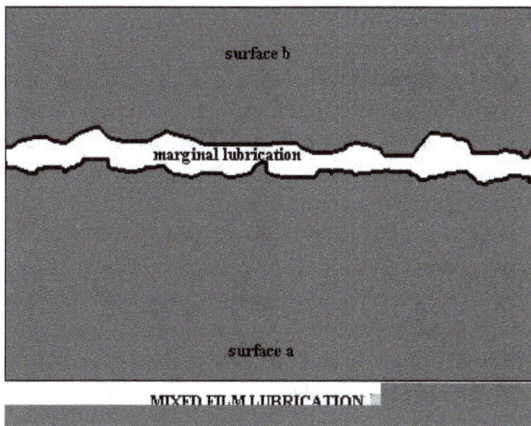

Scheme 1. Mixed-film lubrication

As shown there are three distinct zones separated by points A and B. At B the oil film is just thick enough to ensure that there is no contact between asperities on the shaft and bearing surfaces. Smoother surfaces shift B to the left, while at point A the oil film thickness reduces virtually to nil. Zone 2, between A and B is known as the zone of mixed lubrication. Mixed-film lubrication is unstable at which increase in lubrication temperature causes further increases in lubrication temperature.

1.1.2. Hydrodynamic lubrication

Basically, lubrication is governed by one of two principles: hydrodynamic lubrication and boundary lubrication. In the former, a continuous full-fluid film separates the sliding surfaces. In the latter, the oil film is not sufficient to prevent metal-to-metal contact. Hydrodynamic lubrication is the more common, and it is applicable to nearly all types of continuous sliding action where extreme pressures are not involved. Whether the sliding occurs on flat surfaces, as it does in most thrust bearings, or whether the surfaces are cylindrical, as in the case of journal (plain or sleeve) bearings, the principle is essentially the same.

It would be reasonable to suppose that, when one part slides on another, the protective oil film between them would be scraped away. Except under some conditions of reciprocating motion, this is not necessarily true at all. With the proper design, in fact, this very sliding motion constitutes the means of creating and maintaining that film.

In zone 3 is the zone of hydrodynamic or fluid film lubrication where there is no wear because there is no contact between the surfaces. Hydrodynamic Lubrication is often referred to as stable lubrication. There are four essential elements in hydrodynamic lubrication, a liquid, relative motion, the viscous properties of the liquid, and the geometry of the surfaces between which the convergent wedge of fluid is produced. Only friction present in a hydrodynamic lubrication system is the friction of the lubricant itself, it would make sense to have a less viscous fluid in order to minimize friction: the less viscous a liquid the lower the friction. Too low of a viscosity jeopardizes our system though. We have to be very careful that the distance between the two surfaces is greater than the largest surface defect. The distance between the two surfaces decreases with higher loads on the bearing, less viscous fluids, and lower speeds. The surface geometry is also very important. The surfaces have to be such that a converging wedge of fluid can develop between the surfaces, allowing the hydrodynamic pressure of the lubricant to support the load of the shaft or moving surface. Hydrodynamic lubrication is an excellent method of lubrication since it is possible to achieve coefficients of friction as low as 0.001, and there is no wear between the moving parts. Special attention must be paid to the heating of the lubricant by the frictional force since viscosity is temperature dependent. One method of accomplishing this is to cycle the lubricant through a cooling reservoir in order to maintain the desired viscosity of the fluid. Another way of handling the heat dissipation is to use commercially available additives to decrease the viscosity's temperature dependence which are known as viscosity index improvers.

The formation of fluid film is influenced by the following factors:

• The contact surfaces must meet at a slight angle to allow formation of the lubricant wedge.

• The fluid viscosity must be high to maintain adequate film thickness to separate the contacting surfaces at operating speeds.

• The fluid must be adhering to the contact surfaces for conveyance into the pressure area to support the load.

• The fluid must be distributing itself completely within the bearing clearance area.

• The operating speed must be sufficient to allow formation and maintenance of the fluid film.

• The contact surfaces of bearings and journals must be smooth and free from sharp surfaces that will disrupt the fluid film.

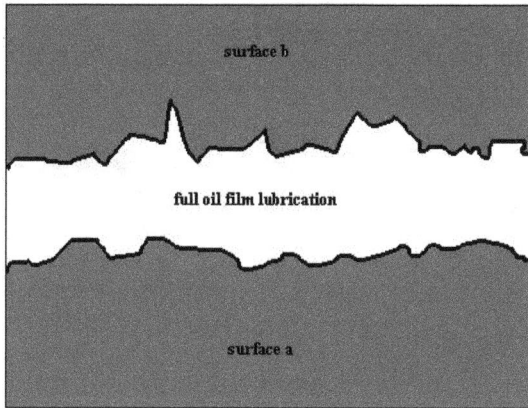

HYDRODYNAMIC LUBRICATION

Scheme 2. Hydrodynamic lubrication

1.1.3. Boundary lubrication

The oil film has become so thin in Zone 1 that there is no hydrodynamic contribution and only boundary lubrication which is defined by Campbell in 1969 as the lubrication by a liquid under conditions where the solid surfaces are so close together that appreciable contact between opposing asperities is possible. The friction and wear in boundary lubrication are determined predominantly by interaction between the solids and between the solids and the liquid. The bulk flow properties of the liquid play little or no part in the friction and wear behavior.

As mentioned, boundary lubrication is effective when a complete fluid film does not develop between potentially rubbing surfaces, the film thickness may be reduced to permit momentary dry contact between wear surface high points or asperities. Boundary lubrication occurs whenever any of the essential factors that influence formation of a full fluid film are missing. The most common example of boundary lubrication includes bearings, which normally operate with fluid film lubrication but experience boundary lubricating conditions during routine starting and stopping of equipment. Other examples include gear tooth contacts and reciprocating equipment.

A brief explanation of what needs to be added to basic mineral oil in order to create an effective boundary lubricant. Generally, the best additives are active organic compounds with long chain molecules and active end groups. These compounds bind tightly and intricately with each other, forming a film that builds up on the surface of the metal itself. This results in a thin film that is very difficult to penetrate. When two surfaces, each covered with a boundary layer, come in contact with each other they tend to slide along their outermost surfaces, with the actual faces of the surfaces rarely making contact with each other. Liquids are rarely good boundary lubricants. The best boundary lubricants are solids with long chains of high inter-chain attraction, low shear resistance so as to slip easily, and a high temperature tolerance. The

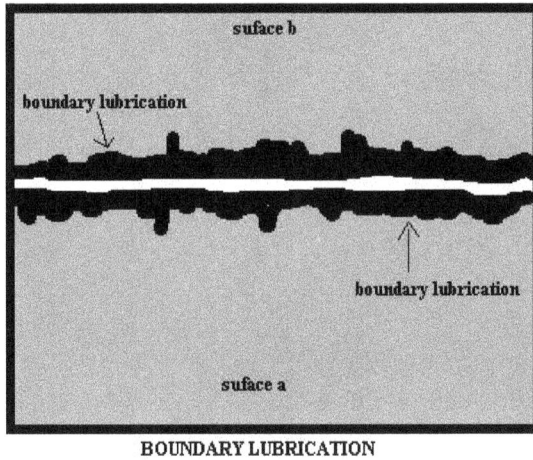

BOUNDARY LUBRICATION

Scheme 3. Boundary lubrication

boundary lubricant should also, obviously, be able to maintain a strong attachment to the surfaces under high temperatures and load pressures.

The most common boundary lubricants are probably greases. Greases are so widely used because they have the most desirable properties of a boundary lubricant. They not only shear easily, they flow. They also dissipate heat easily; form a protective barrier for the surfaces, preventing dust, dirt, and corrosive agents from harming the surfaces

1.2. Base stock

Petroleum is one of the naturally occurring hydrocarbons that frequently include natural gas, natural bitumen, and natural wax. The name "petroleum" is derived from the Latin *petra* (rock) and *oleum* (oil). According to the most generally accepted theory today, petroleum was formed by the decomposition of organic refuse, aided by high temperatures and pressures, over a vast period of geological time.

Although petroleum occurs, as its name indicates, among rocks in the earth, it sometimes seeps to the surface through fissures or is exposed by erosion. The existence of petroleum was known to primitive man, since surface seepage, often sticky and thick, was obvious to anyone passing by prehistoric animals were sometimes mired in it, but few human bones have been recovered from these tar pits. Early man evidently knew enough about the danger of surface seepage to avoid it.

The petroleum remaining from the distillation is thick like pitch; if the distillation has been pushed far, the residuum will flow only languidly in the retort, and in cold weather it becomes a soft solid, resembling much the maltha or mineral pitch Fig. 4 shows that the distillation of crude oil.

Crude Distillation	Crude Oil Fractions			Refinery Processing	Refined Product Categories
	Common Name	Carbon No.	Temp. (°F)		
	Light gases	C1 to C4	< 60		LPG
	SR naphthas	C5 to C9	60 – 175		Gasoline, petrochemicals
	SR naphthas	C5 to C10	175 – 350		Gasoline, jet fuel
Crude oil	SR kerosene	C10 to C16	350 – 500		Jet fuel, kerosene
	SR distillates	C14 to C20	500 – 625		Diesel fuel, heating oil
	SR gas oils	C20 to C50	500 – 850		Lubricating oil, waxes
	SR gas oils	C20 to C70	625 – 1050		Fuel oil
	Residual oil	> C70	> 1050		Bunker fuel, asphalt

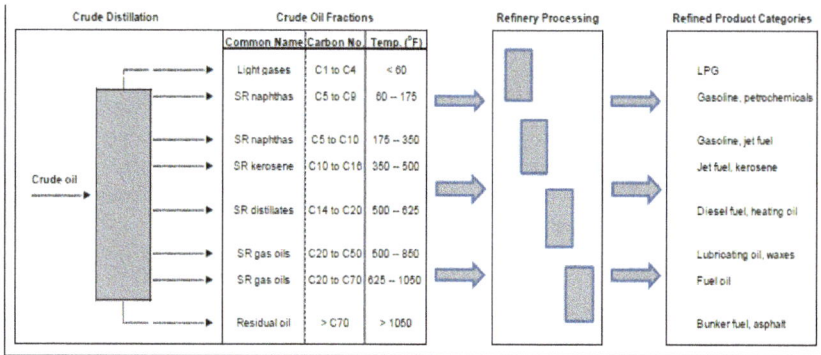

Figure 4. Schematic View of Crude Oil Distillation

Base stocks are refined from crude oil to obtain products with the best lubricating properties. Base stocks generally make up 80-95% of a typical engine oil and 5% additives [1]. Base stock is used to describe plain mineral oil. The physical properties of an oil depend on its base stock. In most cases it is chemically inert there are three sources of base stock: biological, mineral and synthetic. The oils manufactured from these sources exhibit different properties and they are suitable for different applications. For example:

a. Biological oils are suitable in applications where the risk of contamination must be reduced to a minimum, for example, in the food or pharmaceutical industry. They are usually applied to lubricate kilns, bakery ovens, etc. There can be two sources of this type of oil: vegetable and animal. Examples of vegetable oils are: castor, palm and rape-seed oils while the examples of animal oils are: sperm, fish and wool oils from sheep (lanolin).

b. Mineral oils are the most commonly used lubricants throughout industry. They are petroleum based and are used in applications where temperature requirements are moderate. Typical applications of mineral oils are to gears, bearings, engines, turbines, etc.

c. Synthetic oils are artificially developed substitutes for mineral oils. They are specifically developed to provide lubricants with superior properties to mineral oils. For example, temperature resistant synthetic oils are used in high performance machinery operating at high temperatures. Synthetic oils for very low temperature applications are also available [2].

1.3. Lubricants

All liquids will provide lubrication of a sort, but some do it a great deal bettor than others. The difference between one lubricating material and another is often the difference between successful operation of a machine and failure. For almost every situation, petroleum products have been found to excel as lubricants. Petroleum lubricants stand high in metal-wetting ability, and they possess the body, or viscosity characteristics, that a substantial film requires, these oils have many additional properties that are essential to modern lubrication, such as good water resistance, inherent rust-preventive characteristics, natural adhesiveness, relatively good thermal stability, and the ability to transfer frictional heat away from lubricated parts. What is more, nearly all of these properties can be modified during manufacture to produce a suitable lubricant for each of a wide variety of applications. Oils have been developed hand-in- hand with the modern machinery that they lubricate; indeed, the efficiency, if not the existence, of many of today's industries and transportation facilities is dependent upon petroleum lubricants as well as petroleum fuels.

The basic petroleum lubricant is lubricating oil, which is often referred to simply as "oil." This complex mixture of hydrocarbon molecules represents one of the important classifications of products derived from the refining of crude petroleum oils, and is readily available in a great variety of types and grades.

Any description of lubricating oils would be incomplete without consideration of oils for vehicle engines. These oils are used in greater quantity than all other lubricants combined, and are of interest to more people than any other lubricants. Engine oils are generally recommended by automotive builder according to the Society American of Automotive Engineers (SAE) viscosity classification.

Engine oil lubricants make up nearly one half of the lubricant market and therefore attract a lot of interest. The principal function of the engine oil lubricant is to extend the life of moving parts operating under many different conditions of speed, temperature, and pressure. At low temperatures the lubricant is expected to flow sufficiently in order that moving parts are not starved of oil. At higher temperatures they are expected to keep the moving parts apart to minimize wear. The lubricants reduce friction and removing heat from moving parts.

1.3.1. General classification of the lubricating oils

The term lubricating oil is generally used to include all those classes of lubricating materials that are applied as fluids [3]. Lubricating oils are made from the more viscous portion of the crude oil which remains after removal by distillation of the gas oil and lighter fraction [4-8]. Although crude oils from various parts of the world differ widely in properties and appearance, there is relatively little difference in their elemental analysis. Thus, crude oil samples will generally show carbon content ranging from 83% to 87 %, and hydrogen content from 11% to 14%. The remainder is composed of elements such as oxygen, nitrogen, and sulfur, and various metallic compounds. An elemental analysis, therefore, gives little indication of the extreme range of physical and chemical properties that actually exists, or of the nature of the lubricating base stocks that can be produced from a particular crude oil.

An idea of the complexity of the lubricating oil-refining problem can obtained from a consideration of the variations that can exist in a single hydrocarbon molecule with a specific number of carbon atoms. For example, the paraffinic molecule containing 25 carbon atoms has 52 hydrogen atoms. This compound can have about 37,000,000 different molecular arrangements [3]. The hydrocarbons of the crude oils are:

1.3.1.1. Paraffinic components

The paraffinic components, show in Fig. 5 (a, b), which determine the pour point, contain not only linear but also branched paraffins. The straight chain paraffins of high molecular weights raise the pour point of oils (waxy compounds) and should be removed by dewaxing processes.

The branched paraffins are chemically interesting hydrocarbons and they are found in large quantities in lubricating oil fractions from paraffinic crudes. Oil rich in paraffinic hydrocarbons have relatively low density and viscosity for their molecular weight and boiling range. Also, they have good viscosity/ temperature characteristics. In general, paraffinic components are reasonably resistant to oxidation and have particularly good response to oxidation inhibitors [9, 10].

1.3.1.2. Naphthenic components

They have rather higher density and viscosity for their molecular weight compared to the paraffinic components. An advantage which naphthenic components have over the paraffinic ones is that they tend to have low pour point and so do not contribute to wax. However, one disadvantage is that they have inferior viscosity/ temperature characteristics. Single ring alicyclics with long paraffinic side chains, however, share many properties with branched paraffins and can in fact be highly desirable components for lubricant base oils. Naphthenic components, Fig. 5 (c), tend to have better solvency power for additives than paraffinic components but their stability to oxidative processes is inferior [9, 10].

1.3.1.3. Aromatic components

They have densities and viscosities which are still higher viscosity/ temperature characteristics are in general poor but pour point is low, although they have the best solvency power for additives, their stability to oxidation is poor. As for alicyclics, single ring aromatics with long paraffinic side chain may be very desirable base oil components, Fig. 5 (d). The classifying of hydrocarbon as paraffinic, naphthenic and aromatic groups which are generally used for characterizing the base oil should not be taken as absolute but as an expression of the pre-dominating chemical tendencies of the base stocks [11].

1.3.1.4. Non hydrocarbon components

The non hydrocarbons in lubricating oil are analogous in many ways to the hydrocarbons. Sulfur and nitrogen compounds are found almost entirely in ring structures such as sulfides, thiophene, pyridine and pyrrol types. More complex molecules are also thought to exist in lubricating oil in which nitrogen and sulfur atoms are found in the same molecule. As in the case of hydrocarbons, these compounds will probably also have paraffinic side chains and possibly be condensed with naphthenic and aromatic ring structures [11]

Although these non hydrocarbons may be present in only trace amounts, they often play a major role in controlling the properties of lubricating oils. In general they are chemically more active than the hydrocarbon, and hence they may markedly affect properties such as oxidation

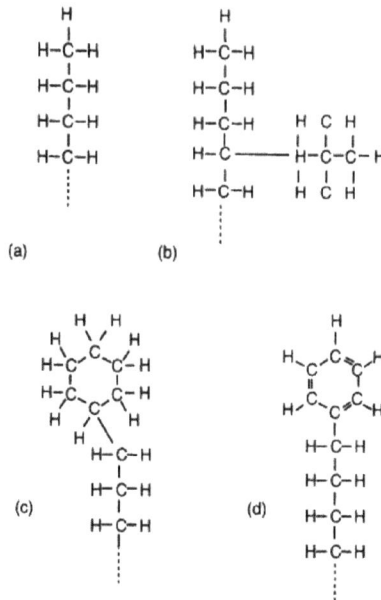

Figure 5. Chemical Structure of Lubricating oil

stability, thermal stability and deposit forming tendencies. In refining the general tendency is to reduce the non hydrocarbons content to a minimum.

Naphthenic acid account for most of the oxygenated compounds found in petroleum. These are removed in the refining processes by neutralization and distillation. The naphthenates are retained in the residue from the distillation and can be removed by deasphalting process. Modern refining methods generally remove most of resins, asphaltenes, polycyclic aromatic, di aromatic and their analogous non hydrocarbons, so that the final lubricant consists chiefly of saturated and monocyclic aromatic fraction [12].

1.3.2. Main properties of lubricating oils

The main properties which a lubricating oil must posses to full performance are :

1.3.2.1. Physical properties of lubricating oil

a.   Viscosity

Viscosity is the measure of the internal friction within a liquid; the way the molecules interact to resist motion. It is a vital property of a lubricant because it influences the ability of the oil to form a lubricating film or to minimize friction [8]. Newton defined the absolute viscosity of a liquid as the ratio between the applied shear stress and the resulting shear rate.

b. Viscosity index

The most frequently used method for comparing the variation of viscosity with temperature between different oils by calculation of dimensionless numbers, known as the viscosity index (VI). The kinematic viscosity of the sample is measured at two different temperatures (40°C, 100°C) and the viscosity compared with an empirical reference scale. VI is used as a convenient measure of the degree of aromatics removal during the base oil manufacturing process, but comparison of VI of different oil samples is only realistic if they are derived from the same distillate feedstock [8].

c. Low temperature properties.

When a sample of oil is cooled, its viscosity increases in a predictable manner until wax crystals start to form. The matrix of wax crystals becomes sufficiently dense with further cooling to cause an apparent solidification of the oil. Although the solidified oil does not pour under the influence of gravity, it can move if sufficient force is applied. Further decrease in temperature cause more wax to form, increasing the complexity of the wax/oil matrix. Many lubricating oils have to be capable of flow at low temperatures and a number of properties should be measured.

* **cloud point**

It is the temperature at which the first sign of wax formation can be detected. A sample of oil is warmed sufficiently to be fluid and clear. It is then cooled at a specified rate. The temperature at which haziness is first observed is recorded as the cloud point, the ASTM D 2500/IP 219 test. The oil sample must be free of water because it interferers with the test.

* **pour point**

It is the lowest temperature at which the sample of the sample of oil can make to flow by gravity alone. The oil is warmed and then cooled at a specified rate. The test jar is removed from the cooling bath at intervals to see if the sample is still mobile. The procedure is repeated until movement of the oil doesnot occur, ASTM D 97/IP 15. the pour point is the last temperature before the movement ceases, not the temperature at which solidification occurs. This is an important property of diesel fuels as well as lubricant base oils. High- Viscosity oils may cease to flow at low temperatures because their viscosity becomes too high rather than because of wax formation. In these cases, the pour point will be higher than the cloud point.

d. High temperature properties.

The high temperature properties of oil are governed by distillation or boiling range characteristics of the oil.

* **volatility**

It is important because it is an indication of the tendency of oil to be lost in service by vaporization.

* **flash point**

It is important for oil from a safety point of view because it is the lowest temperature at which auto-ignition of the vapour occur above the heated oil sample. Different methods are used, ASTM D 92, D93, and it is essential to know which equipment has been used when comparing results.

e.   Other physical properties

Various other physical properties may be measured, most of them relating to specialized lubricant applications. Some of the more important measurements are:

• **density**

Important, because oils may be formulated by weight, but measured by volume.

• **demulsification**

Ability of oil and water to separate.

• **foam characteristics**

Tendency to foam formation and stability of the foam that results.

• **pressure/viscosity characteristics**

• **thermal conductivity**

Important for heat transfer fluid.

• **electrical properties**

Resistively, dielectric constant.

• **surface properties**

As surface tension, air separation.

1.3.2.2. Chemical properties of lubricating oils

a.   Ease of starting rapidity of warming up.

The ease of starting depends chiefly on the cranking speed which is influenced by oil viscosity at the temperature of the crankcase. The major factor in the usage of a lubricant is its viscosity. It's not enough that the lubricants should have the proper viscosity but also they should maintain the little viscosity change within the temperature range during and after the appertain. So, viscosity controls not only frictional and thermal effect but also oil flow as a function of the load speed, temperature and design of the device lubricated. In other words, if the equipment will often have no make a cold start, it's also important that the viscosity at starting temperature is not so high that the machine can not be started. The rapidity with which an engine can be put to work is dependent on the speed of circulation and supply of oil to vital components, all forms of wear and even the safety of the engine are influenced by rapidity of circulation of the lubricants.

b.   Low Carbon Forming Tendency.

This property is important for high compression ratio petro engine where carbon deposit will adversely affect combustion quality. The extent and also the composition of such formed deposits are causing noisy and rough burning which subjects the engine to high thermal and mechanical stresses resulting in lowering of performance and reduction of engine life. The typical symptoms will be knocking, preignition and surface ignition. These call higher octane fuels which are more expensive and do not eliminate the need for ultimate decarbonizing.

Carbon residue test methods.

Provide with some indication about the relative coke forming tendency of the oil in some application and quality-controlled lubricants. So, the test can be helpful in selecting oils for certain industrial applications such as heat treating, lubrication of bearing subjected to high temperature and air compressors. It is claimed that the presence of viscous oil (bright stock) in the base oils plays an important role in the formation of carbon deposits.

c.   High Oxidation Stability.

One of the most important requirement of the lubricant is that its properties are not changed during use [5-10]. The lubricant is often subjected to several oxidizing conditions which are primarily due to the oxidative changes of the oil. While the temperature of the oil, engine parts presence of oxygen, nature by products of fuel composition contribute to the oxidative change the properties of the lubricant during use. Therefore, It's essential that the lubricating oil; when exposed to high temperature; doesn't contribute to the forming of deposits even after a long period of continuous engine running. So, the lubricant resistance to the oxidative depends mainly on the nature of the lubricant and the presence of anti-oxidant additives.

d.   Wear Reduction.

Wear occurs in lubricated systems by three mechanisms (abrasion, corrosion and metal-to-metal contact. i.e adhesion). The lubricant play an important role in combating each type of wear.

i.   Abrasive wear

It is caused by solid particles entering into the area between the lubricated surfaces and physically eroding these surfaces and may contaminate wear fragments. To cause wear, the solid particles must be larger than the oil-film thickness and harder than the lubricated surfaces. The flushing action of the lubricant, especially in forced feed or once through systems, severs to remove potentially harmful solid particles from the area of lubricated surfaces.

ii.   Corrosive wear

Corrosive wear is generally caused by the products of oxidation of lubricants. The high sulfur content of the fuel helps the corrosive attack. In other words, corrosion is the principal cause of wear in the internal combustion engines because the products of combustion are highly acidic and contaminate the lubrication oil, lubricants function to minimize corrosive wear is in two ways: proper refinement plus the use of oxidation inhibitors which reduces lubricant deterioration and keeps the level of corrosive oxidation products low.

iii.   Adhesive wear

This type of wear can significaly affect certain parts of the engine where metal-to-metal contact takes place. Adhesive wear takes place also if power was increased without corresponding modification is design, finishing and composition of the metal parts. Wear of this type also results form breakdown of lubricant film. It can also be the result of excessive surface roughness or interruption of the lubricant supply. A plentiful supply of the proper viscosity of oil is often the best way to avoid these conditions. The composition of the base oil and addition of certain chemical additives are also the important factors in protection of engine parts components against adhesive wear.

e.   Detergency and Dispersancy.

With the exception of detergency and dispersancy in the combustion chamber, deposit in the oil are controlled by its detergent power. The source of the deposits found in engines are many and their volume depends mainly on the used, the quality of combustion, the temperature of lubricating oil and coolant, and on the gas sealing of the ring in the cylinder. It these deposits are not removed with the oil when it is drained, their accumulation in the engine would drastically shorten the engine life. The role of the detergent additives is to reduce the amount of deposits formed and their removal easy. The detergent property imparted to oils by additives seems to perform differently depending upon whether deposits result from high low temperature, low temperature deposits are mainly yielded from the fuel combustion and the detergency function is to keep them in suspension or solution in the lubricating oil. However, high temperature deposits are mainly related to the oxidized fraction of the oil.

The role of detergency here is not only to maintain these products in suspension, but also to stop the development of those chain reaction which promote the formation of varnishes and lacquers. The physical and functional properties of the lube oil will depend on the properties of carbon atoms in the various ring structures and aliphatic side chain

f.   Seal compatibility

Lubricants are often used in machines where they come into contact with rubber or plastic seal. The strength and degree of swell of these seals may be affected by interaction with the oil. Various tests have been devised to measure the effect of base oils different seals and under different test conditions [13]. The strength and degree of swell of these seals may be affected by interaction with the oil. Various tests measure the effects of base oils on different seals and under different test conditions.

1.3.3. Required performance characteristics for lubricating oils

Selection and application of lubricating oil are determined by the functions which are expected for performance. In one application, such as delicate instrument bearing, the reduction friction is paramount and in another, such as metal cutting, the temperature control may be most important. A lubricating oil performance or requirement for a modern high speed engine should fulfill the following five important functions:

1.   Reduction of the frictional resistance:

The reduction of engine resistance to minimum is necessary to ensure maximum mechanical efficiency (running costs of a vehicle or engines are influenced by the lubricant viscosity)

2.   Protection of the engine against all types of wear:

All users wants minimum maintenance costs, longer engine life and increased usefulness. Modern oil has allowed longer intervals between engines over hauls.

3.   Reduction of gas and oil leakages:

The reduction of gas and oil leakages in an efficient and lasting manner is necessary to maintain engine performance and to prevent the combustions products from adulterating the oil.

4.   Contributing the thermal equilibrium of the engine:

In modern engines, the oil functions and more as a heat exchange medium, dissipating the heat is not converted into work. This is often associated with the first function in this list where the viscous oil give greater frictional resistance and its slow internal circulation leads to a rapid temperature raise of some vital part of the engine to cool efficiency, the oil must be able to circulated quickly.

5.   Removal of all injurious impurities:

The lubricant give the function of protecting the engine against corrosive and mechanical wear which caused by all injurious impurities. So, the removal of these impurities by lubricants is very important for engine. The function and the corresponding qualities required for engine lubricating oils are summarized in Table (1).

1.3.4. Types of lubricants

1.3.4.1. Gaseous lubricants

Gaseous lubricants belong to the simplest, lowest viscosity lubricants known and include air, nitrogen, oxygen, and helium. They are applied in aerodynamic and aerostatic bearings. Since the chemical properties and the aggregate state of most gases remain unchanged over a wide temperature range, gaseous lubricants offer several advantages over liquid lubricants. First, they can be applied at both very high and very low temperatures. Their chemical stability eliminates any risk of contamination of the bearing by the lubricant, important for the machinery used in many branches of industry, primarily in the food, pharmaceutical and electronic industries.

A useful property of gases is that their viscosities increase with temperature, wheras the opposite is true of liquids, resulting in load – carrying capacity of gas – lubricated bearings

Main functions required	Qualities required
Reduce frictional resistance	• Viscosity not too high to provide good pumpability or to cause undue cracking resistance. • Minimum viscosity without risk of metal to metal contact under the varying condition of temperatures, speed and load. • Sufficiently high viscosity a high temperature; good lubrication property outside the hydrodynamic condition. • Anti-seizure properties, especially during the run-in period.
Protect against corrosion and wear	• Must protect metallic surface against corrosive action of fuel decomposition product (wear, So_2, HBr, HCl, etec.) • Must resist degradation (resist oxidation and have a good thermal stability). • Must counteract action of fuel and lubricant decomposition product at high temperatures, especially on non-ferrous metals. • By intervention in the friction mechanism, must reduce the consequences of unavoidable metal-to-metal contact. • Must resist deposit formations which would affect lubrication (detergency or dispersancy action). • Must contribute to the elimination of dust and other pollutants (dispersancy action).
Assist sealing	• Must have sufficient viscosity at high temperature and low volatility. • Must limit wear. • Must not contribute to formation of deposits and fight against such formation.
Contribute to cooling	• Must good and thermal stability and oxidation resistance. • Must have low volatility. • Viscosity must not be too high.
Facilitate the suspension and eliminate undesirable products	• Must be able to maintain in fine solid material whatever the temperature and physical and chemical condition.

Table 1. Function and qualities required for engine oils.

increasing with temperature. However, the relatively low viscosity of gases generally limits the load-carrying capacity of self-acting, aerodynamic bearings to 15-20kPa. It is possible to achieve better bearing performance with gaseous lubricants than with liquid lubricants due to the very low viscosity of the gases which results in smaller heat generation by internal friction. In some cases, such as in foil air bearings, sliding contact occurs during stops and starts [14], therefore solid lubricants such as PTFE are used to reduce friction.

1.3.4.2. Liquid lubricants

Mineral oils: As the hydrodynamic behaviour of plain bearings of plain bearings is totally dependent on the viscosity characteristics of the lubricant, typical liquid bearing lubricants are straight mineral oil raffinates of various viscosity grades. The viscosity grade required is dependent upon bearing speed, oil temperature and load. table (2) provides a general guideline to selecting the correct ISO viscosity grade. The ISO grade number indicated is the preferred grade for the speed and temperature range. ISO 68- and 100- Grade oils are commonly used in indoor, heated applications, with32- grade oils being used for high-speed, 10.000 rpm, units and some outdoor low temperature applications. The higher the bearing speed, the lower the oil viscosity required and also that the higher the unit operating temperature, the higher the oil viscosity required. If vibration or minor shock loading is possible, a higher grade of oil than the one indicated in table (2) should be considered.

Bearing Speed (rpm)	Bearing / Oil Temperature (°C)			
0-50	60	75	90	
300-1,500	-	68	100-150	-
1,800	32	32 - 46	68 -100	100
3,600	32	32	46 - 68	68 -100
10,000	32	32	32	32-46

Table 2. Plain Bearing ISO viscosity grade selection

Other methods for determining the viscosity grade required in an application are to apply minimum and optimum viscosity criteria to a viscosity – temperature plot. A third and more complex method is to calculate the oil viscosity needed to obtain a satisfactory oil film thickness.

The lubrication of bearings for machine tools usually requires mineral oils of ISO VG 46 or 68. for fast – running grinding spindles with plain bearings, mineral oils of ISO VG 5 or 7 are required, dependent on bearing clearance and speed. Bearings operating under high loads need lubricants of ISO VG 68 or 100. the service life of the bearing can be increased if the viscosity of the selected liquid lubricant at operating temperature exceeds the calculated optimum viscosity.

On the other hand, increased viscosity also increases operating temperature. In practice, therefore, the extent to which lubrication can be improved in this way is often limited. The chemical compositions of these oils differ from typical base oils in that they contain somewhat more aromatic hydrocarbons and heterocyclic compounds, which act as natural oxidation inhibitors. An increased viscosity for oils derived from the same crude oil does not significantly change their chemical composition; the difference generally lies with the increasing chain length of the paraffinic hydrocarbons, mostly isoparaffins, and in the aliphatic substituents of naphthenic and aromatic rings, together with a slight increase in the number of naphthenic and aromatic rings. More highly refined mineral oils and oxidation inhibitors are used for applications where higher temperatures or longer service periods require better ageing stabilites.

Synthetic lubricants: in practice, every synthetic oil of adequate viscosity and good viscosity-temperature behavior can be used as a bearing lubricant, e.g. polyglycols are very good bearing lubricants for mills and calenders in the rubber, plastics, textile and paper industries. However, in most cases the synthetic oils specifically developed for lubricating particular equipment are also used to lubricate its bearings. Although synthetic oils do not form a lubricant film under pressure as well as mineral oils and may not be effective bearing lubricants despite their higher temperature viscosity.

Biodegradable products: Biodegradable products of vegetable or animal origin are also considered for liquid lubrication, e.g. the effects of sunflower oil added to base oil on the performance of journal bearings. The use of vegetable oils as lubricants is likely to increase due to environmental and government requirements and is becoming increasingly important.

1.3.4.3. Solid lubricants

General description: bearings used under vacuum, at very high temperatures or under very high radiation cannot be lubricated by liquid lubricants or greases. For these and many other cases, solid lubricants are used, deemed to be any solid material used to reduce friction and wear between two moving surfaces.

In general, the solid material is interposed as a film between sliding and /or rolling surfaces. Simply stated, an adequate solid material is required for the special lubrication requirements of extreme operating conditions, such as very high or very low temperatures over a wide range, e.g. -200 to 850°C, and corrosive atmospheres. Such materials normally have a layered crystalline structure which ensures low shear strength, thereby minimizing friction. The shear strength between the crystalline layers is weak and sets up a low and sets up a low friction mechanism by slippage of the crystalline layers under low shearing forces. Examples of layer-lattice solids are molybdenum disulphide, graphite, boron nitride, cadmium iodide and borax. Solid lubricants are used mainly in the form of powders or as bonded solid films.

A good solid film lubricant has strong adhesion to the bearing substrate material, full surface coverage and good malleability. It should also be chemically stable and prevent corrosion, taking into account operational and environmental conditions. Many solid film lubricants have poor wear resistance, since any breaks in the film are not self-healing, in contrast to the surface coating formed by a liquid lubricant. Advanced solid film lubricants perform reliably in many specific applications and much experience has been gained to better understand their limitations. The most commonly used disulphide, graphite, polytetrafluroethylene propylene.

Another group of materials, the self-lubricated materials, are related to solid lubricants and are particulary important for bearings. Their self-lubricating characteristics eliminate the need of grease or other lubrication and gove improved performance under high temperature conditions. Graphalloy (Graphite/matal) alloys make use of special properties of graphite, the structure of which can be compared to a deck of cards with individual layers able to easily slide off. This phenomenon gives the material a self lubricating ability matched by few other materials and allows for the elimination of grease or oil that would evaporate, congeal or solidify, causing premature failure. The graphite matrix can be filled with a variety of embedded lubricants to enhance chemical, mechanical and tribological properties to give a constant, low friction coefficient rather than just a surface layer, helping to protect against catastrophic failure. Lubrication is maintained during linear motion where lubricant is not frawn out and dust is not pulled in.

A recent development in solid bearing lubricants is micro –porous polymeric lubricants, MPL, where a polymer containing a continous microporous network has oil contained within the pores, which may include appropriate additives [14]. The oil content in the polymer can be more than 50% by weight and the microporous polymer acts as a sponage, releasing and absorbing oil when necessary.

1.3.5. Lubricant impurities and contaminants

• **Water Content**

Water content (ASTM D95, D1744, D1533, and D96) is the amount of water present in the lubricant. It can be expressed as parts per million, percent by volume or percent by weight. It can be measured by centrifuging, distillation and voltametry. The most popular, although least accurate, method of water content assessment is the centrifuge test. In this method a 50% mixture of oil and solvent is centrifuged at a specified speed until the volumes of water and sediment observed are stable. Apart from water, solids and other solubles are also separated and the results obtained do not correlate well with those obtained by the other two methods. The distillation method is a little more accurate and involves distillation of oil mixed with xylene. Any water present in the sample condenses in a graduated receiver. Voltametry method is the most accurate. It employs electrometric titration, giving the water concentration in parts per million.

Corrosion and oxidation behaviour of lubricants is critically related to water content. An oil mixed with water gives an emulsion. An emulsion has a much lower load carrying capacity than pure oil and lubricant failure followed by damage to the operating surfaces can result. In general, in applications such as turbine oil systems, the limit on water content is 0.2% and for hydraulic systems 0.1%. In dielectric systems excessive water content has a significant effect on dielectric breakdown. Usually the water content in such systems should be kept below 35 [ppm].

• **Sulphur Content**

Sulphur content (ASTM D1266, D129, D1662) is the amount of sulphur present in an oil. It can have some beneficial, as well as some detrimental, effects on operating machinery. Sulphur is a very good boundary agent, which can effectively operate under extreme conditions of pressure and temperature. On the other hand, it is very corrosive. A commonly used technique for the determination of sulphur content is the bomb oxidation technique. It involves the ignition and combustion of a small oil sample under pressurised oxygen. The sulphur from the products of combustion is extracted and weighed.

• **Ash Content**

There is some quantity of noncombustible material present in a lubricant which can be determined by measuring the amount of ash remaining after combustion of the oil (ASTM D482, D874). The contaminants may be wear products, solid decomposition products from a fuel or lubricant, atmospheric dust entering through a filter, etc. Some of these contaminants are removed by an oil filter but some settle into the oil. To determine the amount of contaminant, the oil sample is burned in a specially designed vessel. The residue that remains is then ashed in a high temperature muffle furnace and the result displayed as a percentage of the original sample. The ash content is used as a means of monitoring oils for undesirable impurities and sometimes additives. In used oils it can also indicate contaminants such as dirt, wear products, etc.

• **Chlorine Content**

The amount of chlorine in a lubricant should be at an optimum level. Excess chlorine causes corrosion whereas an insufficient amount of chlorine may cause wear and frictional losses to increase. Chlorine content (ASTM D808, D1317) can be determined either by a bomb test which provides the gravimetric evaluation or by a volumetric test which gives chlorine content, after reacting with sodium metal to produce sodium chloride, then titrating with silver nitride [14].

2. Conclusion

1. The technology of lubrication has been used from the ancient times, from the pyramid building where massive rock slabs are moved, up to present modern times.

2. The main purpose of lubrication is to reduce friction and wear in bearings or sliding components to prevent premature failure.

3. Adequate lubrication also helps to prevent foreign material from entering the bearings and guards against corrosion and rusting. Satisfactory bearing performance can be achieved by adopting the lubricating method that is most suitable for the particular application and operating conditions

4. A lubricant prevents the direct contact of rubbing surfaces and thus reduces wear. It keeps the surface of metals clean.Lubricants can also act as coolants by removing heat effects and also prevent rusting and deposition of solids on close fitting parts.

5. lubricant is consisting of either oil or grease. Most grease is from animal fats or vegetable lard.

6. Lubricating oils are made from the more viscous portion of the crude oil which remains after removal by distillation of the gas oil and lighter fraction

7. There are three major types of lubricants: Gaseous lubricants e.g. air, helium, Liquid lubricants e.g. oils, water and Solid lubricants e.g. graphite, grease, teflon, molybdenum disulphide etc. Liquid lubricant is the most commonly used lubricant because of its wide

range of possible applications while gaseous and solid lubricants are recommended in special applications.

8.   Lubricants do not persist working without additives.

9.   Additives are chemical compounds added to lubricating oils to impart specific properties to the finished oils. Some additives impart new and useful properties to the lubricant; some enhance properties already present, while some act to reduce the rate at which undesirable changes take place in product during its service.

Author details

Nehal S. Ahmed and Amal M. Nassar

Additives lab., Department of Petroleum Applications, Egyptian Petroleum Research Institute, Nasr City, Cairo, Egypt

References

[1] Rudnick Leslie R., Ewa A. Bardasz, and Gordon D. Lamb; "Lubricant Additives: Chemistry and Applications", Marcel Dekker, pages 387-427, (2003).

[2] Stachowiak Gwidon W.,and Andrew W. Batchelor; Engineering Tribology", third edition, Amsterdam: Elsevier, pages 2,12,-22,52,62-67,77, (2005).

[3] Geore J.W.; "Lubrication Fundamentals", (1980).

[4] Dowson D.; History of Tribology, 2nd Edition, Professional Engineering Publishing, London (1998).

[5] Pirro D.M. and Wessol A.A.; "Lubrication fundamentals"; Marcel Dekker, Inc. New York and Basel 3, 37 (2001).

[6] Spikes H.; Tribology International 34, 789 (2001).

[7] Stachowiak G. W. and Batchelor A.w.; Engineering Tribology, 2nd Edition, Butterworth-Heinemann, Boston,(2001).

[8] Pawlak Z., "Tribochemistry of Lubricating Oils"; Elsevier, UK, 45, 17 (2003).

[9] Avilino S.Jr.,"Lubricant Base Oil and Wax Processing", Morcel Dekker,Inc.,New York, Chapter (2),pp.17-36,(1994).

[10] Mortier R.M. and Orszulik S.T.; "Chemistry and Technology of Lubricant", Blackie Academic and Professional Publications, Chapter (1), pp.2-12,(1993).

[11] O'Connar J.J., Boyd J. and Auallane E.A.; "Standard Hand Book of Lubrication Engineering", McGrow Hill, New York, 14-2 (1968).

[12] Allyson M., Keith D., Vincent R. and Thibon A.; Tribology International 34, 389-395 (2001).

[13] Anon, Machinery and Production, 19 July 24 (1996).

[14] Roy M. M., Malcolm F. F., and Stefan T. Orszulik; Chemistry and Technology of Lubricants, 3rd Edition, 12-13,(2010).

Chapter 2

Fundamentals of Lubricants and Lubrication

Walter Holweger

Additional information is available at the end of the chapter

1. Introduction

Literature about lubricants is available in all public domains. Readers should search at those platforms in the case of special interests. Citations given here do not represent the full scale but reflect an overview from a today's perspective. [1-7]

Part of this chapter will be the basic chemical structure of lubricants including some property descriptions. Since literature in tribology is innumerous, the reader should check his special area of interest.

Lubricants play a key role in machinery element safety. Their main tasks are

- to keep moving parts apart from each other,

- to take heat out of the contact by their through pass,

- to keep surfaces clean,

- to transport functional additives toward the surface and

- to transfer power in the application (hydraulic, automatic transmission, breaks). [6, 8]

Functionality of lubricants is defined by their chemical structure and their physical properties. Basics of lubrication are covered by organic chemistry to a major and inorganic chemistry to a minor extent. [2, 3]

Lubricants are regulated internationally and locally, e. g. by ASTM (American Standard of Testing Materials) or DIN (Deutsche Industrienorm). Regulation covers the physical, chemical and toxicological description of lubricants including safety guide lines and others. [2, 3]

2. Some basics

The spatial structure of carbon chemistry defines all activities of the lubricants derived from them. The spatial structure of organic carbon chemicals is given by the binding state of carbon. [10]

Three main types are discussed. Two are essential for lubrication: single and double bonds.

2.1. Single bonds: Tetrahedral binding

In the tetrahedral binding state, reflecting the status of single bonds, carbon is placed in the centre of a pyramid with bindings into space from the centre to the corner (Figure 1).

Figure 1. Tetrahedral binding of carbon

Carbon is placed in the centre of the tetrahedral with four attached valences. Within chemical convention in order to abbreviate the structure denotation the atom symbols are neglected.

Carbon may bind to another one by corner to corner. (Figure 2)

Figure 2. Corner to corner binding state

Corner to corner binding leads to zigzag chains, where the angle of carbon to the hydrogen atoms is 108°. In general the hydrogen is neglected, leading to a skeleton drawing of the structure.

Beyond the fixed angle of 108° and the zigzag shape of such hydrocarbon structures, a high variety of structures arise due to the fact that those bindings may branch or bind to cyclic structures. (Figure 3 and Figure 4)

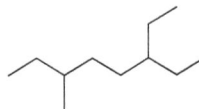

Figure 3. Branched structures by carbon to carbon binding

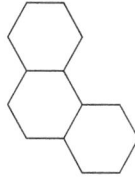

Figure 4. Cyclic Structures by carbon-carbon binding

Single bonds in hydrocarbons are free to rotate (Figure 5). Rotation leads to the situation that hydrogen atoms within the chain get close to each other. As a consequence the energy of the molecule rises.

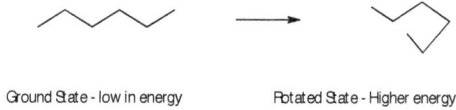

Ground State - low in energy   Rotated State - Higher energy

Figure 5. Energy rise in rotated structures

Similar to internal rotation, molecular energy rises if molecules get under stress by moving them closely together without giving time to relax. (Figure 6)

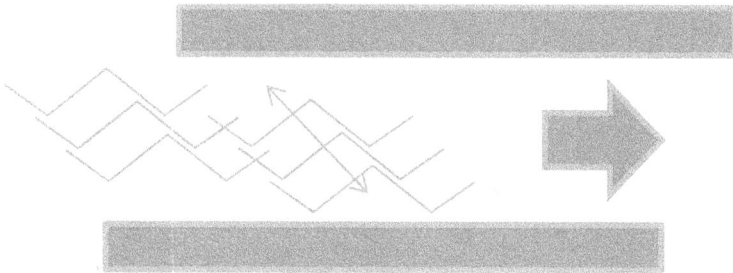

Figure 6. Excitation by pushing molecules to one another by shear stress

Also the fact of putting or pressing molecules toward a surface may lead to a steep increase in internal molecular energy, sometimes high enough to cut them.

2.2. Double bonds

Carbon may also bind to others by double bonds, such that two of the four bindings attach to the other as double, whereas the remaining bonding stays single. (Figure 7)

Figure 7. Double bonding

Double bond shows a 120° neighborhood angle to the carbon. This angle is kept constant and will lead toward different structures in the double bond chemistry. (Figure 8)

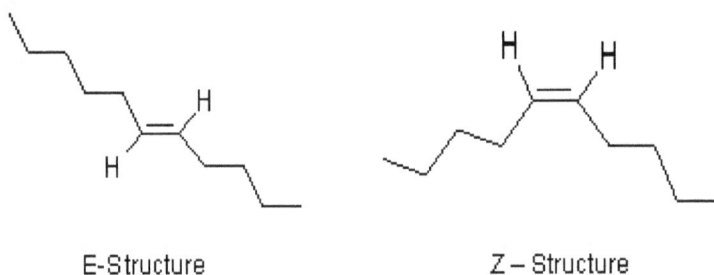

E-Structure   Z – Structure

Figure 8. E and Z structures in double bond

Both structures differ in their energy. Double bonds are part of biodegradable additives (native oils) but also additives and thickeners in the case of greases. Z-Structures are dominant in native oils.

2.3. Triple bonds

Triple bonds are seldom found in tribology. They represent a high energy state in molecules with very high reactivity. As such, they are part of catalytic degradation processes in lubricants. Within a triple bond carbons attach to each other by a linear principle (Figure 9):

Figure 9. Triple bond present in a hydrocarbon

3. Base oils in lubrication: General comments about specie and groups

Hydrocarbon Base Oils for Lubrication derive from organic chemistry. Different categories are given by their chemical composition and structure. [2]

Hydrocarbons, e.g. Structures that contain solely Hydrogen and Carbon (H, C)

Ester Oils, e.g. Structures containing Hydrogen, Carbon and Oxygen. Some Esters are derived from other precursors, such as phosphoric acid esters.

Polyglycoles: Structure containing Hydrogen, Carbon and Oxygen but being different in binding state compared to Esters.

Within a general scheme, base oils are identified as Groups.

Group I: Those lubricants are built from saturated hydrocarbons, e.g. hydrocarbons without alkenes (hydrocarbons with double bonds)) (> 90%), obtained by solvent extraction processes and catalytic hydrogenation. Sulfur may part in amount of > 0.03%. Viscosity index (VI) is in between 80 and 120.

Group I+: Oils that are in a VI range of 103-108.

Group II: Hydrogenated (saturated) hydrocarbons (> 90%) and sulfur below 0.03% per weight with viscosity index (VI) of 80 till 120.

Group II+: Oils in the VI range of 113-119.

The base oil within this group is manufactured by hydrocracking, solvent extraction or catalytic dewaxing processes. Those oils are pale or water like colored.

Group III: Oils with a saturation > 90%, sulfur < 0.03% and a viscosity index > 120. Those oils are produced by catalytic procedures with a concurrent rearrangement of the carbon backbone during hydrogenation.

Group III+: Oils providing a VI at least of 140.

Group IV: Poly-α-Olefins with sulfur content approximately 0, viscosity index 140–170, being produced by catalytic polymerization of low molecular weight end terminated olefins.

Group V: All other oils, e.g. esters, polyglycoles, phosphate esters.

North America states Group III, IV and V as synthesized hydrocarbons (SHC) while in Europe Group IV and V is declared as synthetic oil.

4. Saturated natural hydrocarbons

Saturated hydrocarbons are those who do not contain double bonds in their structure. They derive from the tetrahedral binding of carbon (bindings that point into corners of tetrahedron). The simplest structure is given by methane, ethane, propane, butane with carbons attached at the corners of the tetrahedral. These representatives are present in the natural gases, while methane is found in enormous quantities as methane-ice cluster. The gases themselves are not in use as lubricants but are components of fuels. (Figure 10)

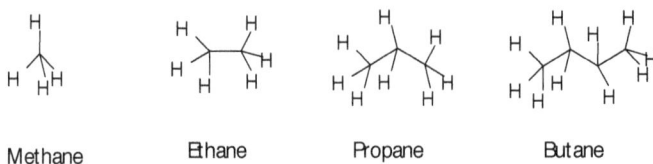

Methane   Ethane   Propane   Butane

Figure 10. Methane, Ethane, Propane, Butane

Starting from pentane the hydrocarbons get liquid and are the principal components of fuels, solvents, and raw materials for the chemical industry. To facilitate reading and drawing only the carbon backbone is drawn without explicitly showing hydrogen. (Figure 11)

Figure 11. Pentane, Hexane, Heptane

Binding of carbon to carbon may be realized in chains, but also in branched chains and different cycles (Figure 12).

Figure 12. Methylbutane (Isopentane), 2,2-Dimethylpropane (Neopentane), Cyclohexane

Hydrocarbons from C10 on till C14 are in use as solvents for cleaning (C11-C13 isoparaffines) (Figure 13):

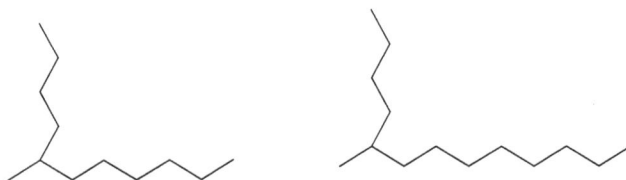

Figure 13. C11-C13 Iso paraffines

From C16 on, hydrocarbons represent typical structures present in lubricants. As the linear hydrocarbons, beginning at C18 are solids, they are common in waxes and thickeners for liquid hydrocarbons. Due to their high solidification point they are a threat if present in Diesel fuels by blocking filters.

Apart from their function as hydrocarbon waxes they are not suitable as lubricants for machine oil circuits.

Suitable lubricants are derived from C16–C70 hydrocarbons with branched chains. Branching leads to low pour points (the point where the lubricant starts to get solid). Machine oils with low pour point, suitable for low temperature applications are branched in their carbon chain. (Figure 14)

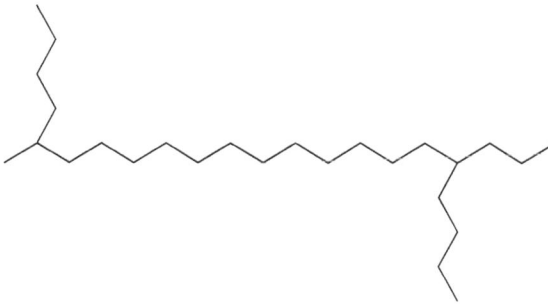

Figure 14. Representatives of saturated hydrocarbons as typical lubricants

In general, the viscosity of a lubricant - as a measure for the ability to move across - increases with the molecular weight, e.g. the number of carbon atoms attached. Viscosity is measured by different techniques. Basically the lubricant is pushed or moved in between plates or by moving it in the gravity field. International convention states 16 classes of viscosity as an ISO Standard (ISO VG classes): ISO VG 5, 7, 10, 15, 22, 32, 46, 68, 100, 150, 220, 320, 460, 680, 1000 and 1500. Low numbers indicate low viscosity, higher numbers high viscosity. Since viscosity is strictly related to temperature, the ISO VG classification refers to 40°C as a standard temperature. The nature of measuring the viscosity leads to the physical value of an area per time: mm^2/s. Hence, ISO VG 68 for example denotes a viscosity of the lubricant, measured at 40°C within 68 mm^2/s within a range of roughly 10% below and 10% beyond the given 68mm^2/s.

Low molecular weight, branched hydrocarbons are often used in *pneumatic spraying*, due to their viscosity range, starting at 2 (water-like), 5, 10 and 15.

Low viscous hydrocarbons from ISO VG 10, 15, 22, 32, 46, 68 and 100 are in use as *hydraulic oils*. Common hydraulic oil viscosity is around ISO VG 32, 46 and 68.

Hydrocarbons with higher viscosities are part of *machine oils*, carrying out the ordinary lubrication functions. Machine oil viscosities are in the range of ISO VG 68, 100, 150, 220, 320, 460. The number of carbons is in the range of 30–80 in the chain.

Some applications in heavy duty processes demand viscosities even higher in the range of ISO VG 680, 1000 and 1500.

4.1. Cyclic hydrocarbons (Naphtenes)

Naphtenic hydrocarbons are derived from hydrocarbon cycles with more or less long chains attached to the cycle. Due to their high branching they are very common in low temperature applications (below -30°C) for hydraulics; low temperature greases. (Figure 15)

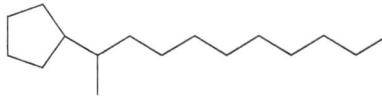

Figure 15. Principal Structure of Napthenic Hydrocarbons

4.2. Aromatic hydrocarbons (Alkyl aromats)

Aromatic Hydrocarbons (Alkyl Aromats) derive from the six-membered benzene ring system, attached by hydrocarbon chains. Aromatic hydrocarbons are in use for low temperature applications.

Alkyl Naphthalenes are a modern group of aromatic hydrocarbons. They may act as solvent improvers for synthetic oils, facilitators in generating greases, low temperature applications and much more (Figure 16):

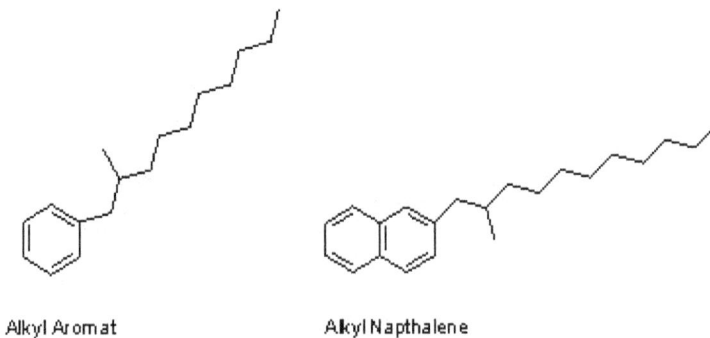

Alkyl Aromat   Alkyl Napthalene

Figure 16. Alkyl Aromats and Naphtalenes

Aromats and aromat-containing hydrocarbons are very vulnerable toward oxygen. Oxidation of aromats starts at the attached hydrocarbon chain, proximate to the aromat nucleus by a radical attack. This position is always very sensitive in similar structure, due to the fact, that the intermediate carbon radical is stabilized by the aromat and thus starts to stay persistent. As a fact, aromats may strongly boost oxidation of hydrocarbons if present in the mixture due to the mentioned persistency of the reactive intermediates. (Figure 17)

Aromats and naphtenics (containing unsaturated hydrocarbons and aromats) should be stabilized against oxidation.

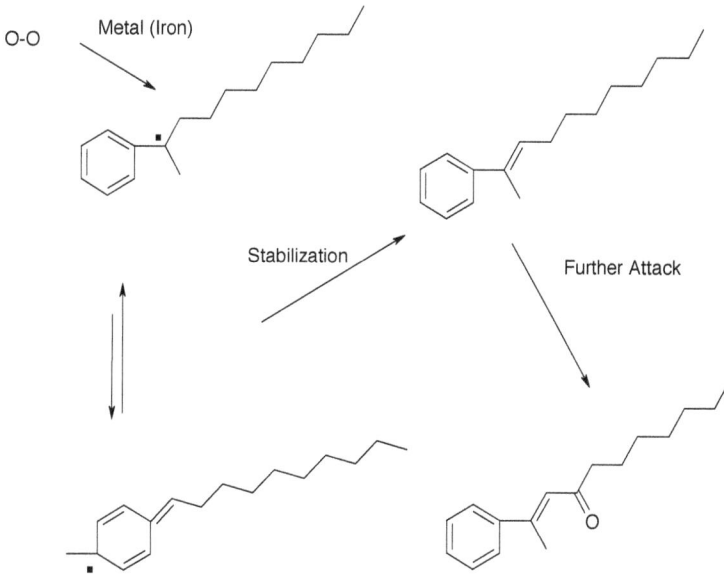

Figure 17. Oxygen attack in the oxidation mechanism of Alkylaromats

5. Synthetic hydrocarbons

5.1. Poly-α-Olefins (PAO)

PAO is dominating all synthetic hydrocarbons by amount of production and worldwide turnover. Syntheses start from Dec-1-ene, a linear C10 hydrocarbon with a double bond at the beginning of the molecule. Polymerization and hydrogenation leads to PAO, as a highly branched and fully saturated hydrocarbon (Figure 18). [2, 4]

Modern PAO may also start from a variety of hydrocarbons (C8-C12) by the same processes. PAO are the most prominent worldwide used hydrocarbons and found in all important applications, e.g. gear oils, circuit oil, hydraulic oil, base stock for automotive applications and others. [2, 3, 6, 7]

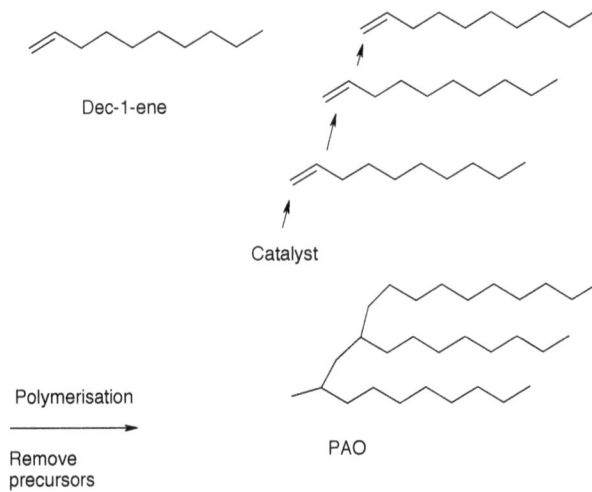

Figure 18. Principle of PAO formation

The extraordinary importance of PAO is due to its applicability at very low temperatures (pour points below –30°C) and, in the case of suitable antioxidant prevention also at higher temperatures (> 120°C). While PAO is, by its structure, very common in low temperature applications, it is very poor in the contact with metal surfaces beyond 120°C if not properly additivated by antioxidants.

Principal antioxidants for PAO are Phenyl-α-Naphtylamine (PAN) and octyldiphenylamines (see antioxidants (AO)).

5.2. Polyisobutenes (PIB)

PIB are a sub class of polymerized olefins. They are widely used to boost low viscous oils to higher ISO VG grades or as functional additives to improve the viscosity index (VI): the attitude of the lubricant to lower its viscosity strongly by temperature is reduced by addition of PIB. Synthesis is carried out starting from isobutene by catalytic oxidation processes (Figure 19):

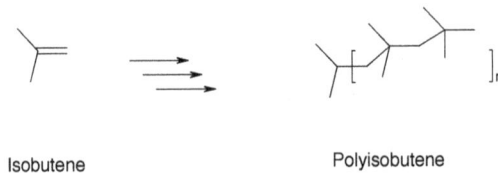

Isobutene   Polyisobutene

Figure 19. PIB formation by catalytic polymerization of Isobutene

Sulfurization with activated sulfur precursors lead toward sulfurized isobutenes (SIB) widely used as extreme pressure (see also section about EP/AW additives).

6. Ester oils

6.1. General

Esters are in general reaction products between alcohols and acids. Their formation is also possible by means of other techniques, e.g. specific oxidation reactions, rearrangements in organic molecules or different reactions. [10]

Carboxylic Acid esters are created by the reaction of alcohols and carboxylic acids [A] and their derivatives, by trans–esterification (B), or catalytic reactions, e.g. epoxides with carbon dioxide (C) (Figure 20). [10]

6.2. Esters in lubrication technology

Despite their high variety in structure esters are used in different categories: [1, 2, 4]

6.3. Mono-esters

Mono-Esters derive from a monocarboxylic acid (Carboxylic Acid that contains only one acidic centre) and monofunctional alcohols (Alcohole with only one OH group). [10]

Esters derived from this structure are seldom used as pure lubricants, more as solvents or dispersants. For example alcohol ethoxylates, formed by addition of alcohols to epoxides may be esterified by a monocarboxylic acid leading toward a dispersant or self-emulsifying solvent. (Figure 21)

6.4. Di-esters

Di-Esters are synthesized by use of dicarboxylic acids, mainly adipaic or sebacaic acid and two molecules of an alcohol. 2-Ethylhexylalcohole (Iso Octanole), leading to Di-isooctyladipate (DOA) or Di-isooctylsebacate (DOS, DEHS = Di-ethylhexylsebacate). (Figure 22)

They constitute an important group of oils, with either the function of base oil by themselves but also as adjuvant to mineral oil or PAO formulations.

For Di-Esters the reaction of alcohols (A) with two hydroxyl groups and a monofunctional carboxylic acid (B, B') is also applicable. (Figure 23)

For technical purposes the reaction product of Neo Pentylgylcole (3.3 Dimethyl-propane-1.4-diol) with oleic acid is important in lubrication technologies for use as a friction reducer and in minimal lubrication systems. (Figure 24)

A

Formation of Carboxylic Acid Esters (C) by Reaction of Acid (A) and Alcohole (B)

B

Formation of Carboxylic Acid Esters by Transesterification

C

Formation of Carboxylic Acid Esters by reaction of epoxides toward Lactones
(Cyclic Esters)

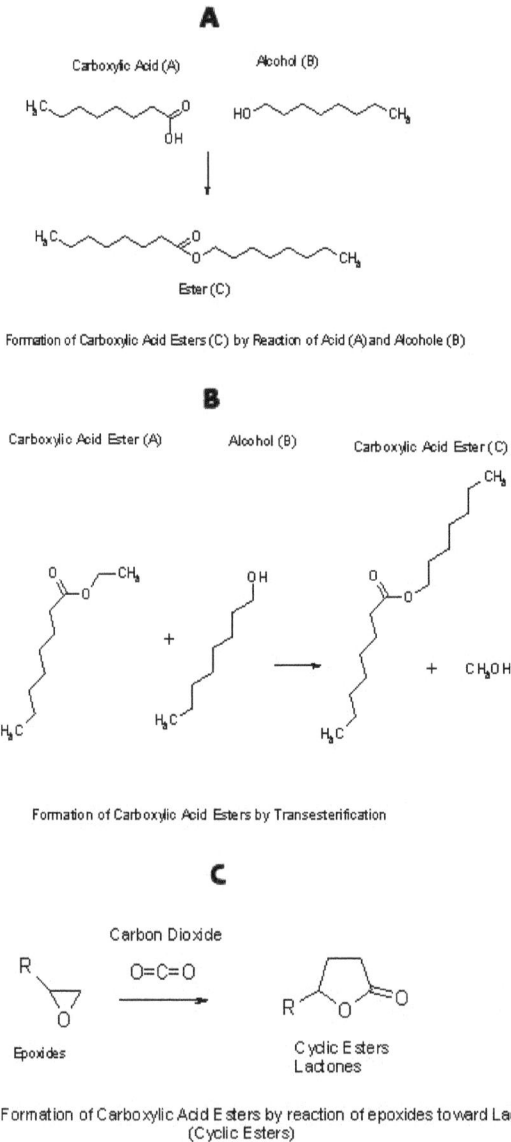

Figure 20. Examples for creation of esters. (A) Reaction of Carboxylic Acids with Alcohols, (B) Transesterfication, (C) Reaction of Expoxide to cyclic Esters.

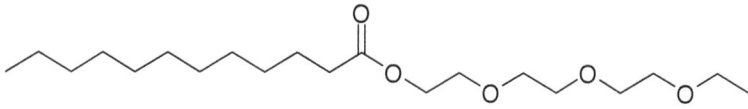

Figure 21. Mono Ester Formation with the specialty of esterified alcohole ethoxilates

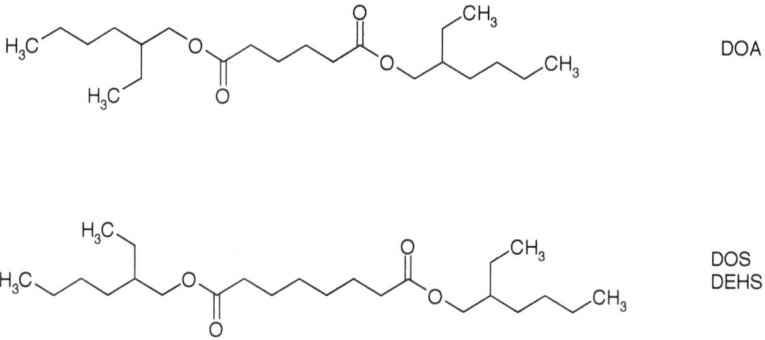

DOA

DOS
DEHS

Figure 22. DOA and DOS

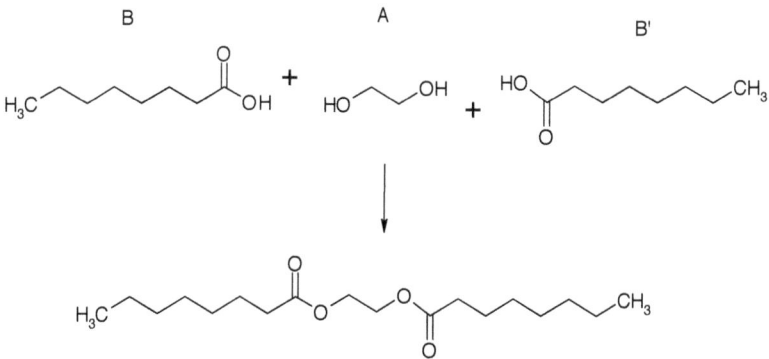

Figure 23. Formation of Di-Esters by Di-Alcoholes (A) reacted with Monocarboxylic Acids (B, B')

6.5. Tri-esters

Tri-Esters are mainly created by the reaction of trivalent alcohols with monocarboxylic acids. They are mainly represented by to major groups:

6.5.1. Glycerol esters

Esters derived from glycerol as a trivalent alcohole leads to tri-Esters. (Figure 25)

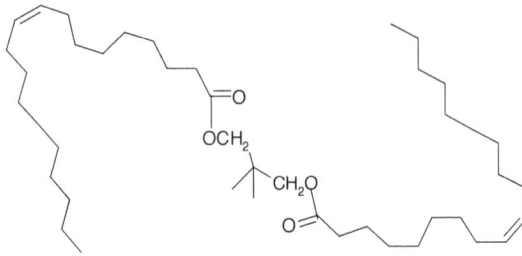

Figure 24. Neopentylgylcoledioleate (NPG-Dioleate)

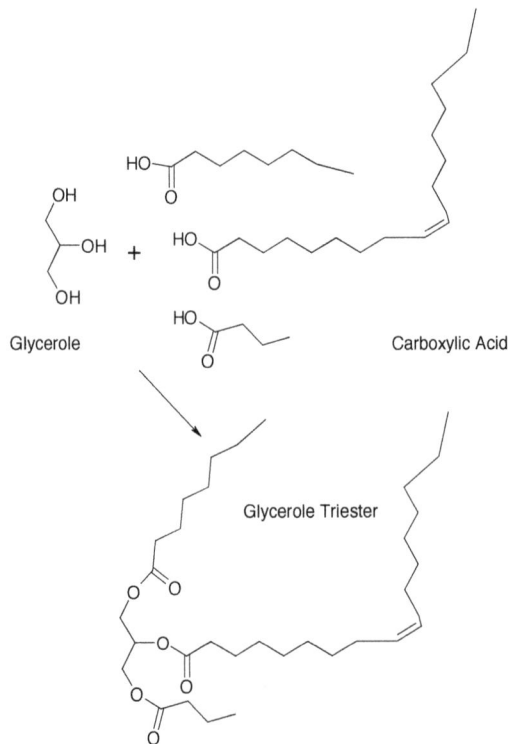

Figure 25. Esterification of Glycerole to tri-Esters

Glycerol Tri-Esters represent the huge group of natural oils. Sunflower, rapseed oil are prominent representatives. A mixture of short chain carboxylic acids with unsaturated long chain acids is used.

As a fact of the presence of short chain carboxylic acids those esters are nutrients, biological degradable and widely used as natural, biodegradable oils.

As a special glycerol ester, important for lubrication, ricinoleic acid esters have to be mentioned.

Within this group ricinoleic acid represents the group of 12-hydroxy substituted C18 carboxylic acids.

Hence, alkaline cleavage of ricinoleic acid glycerol esters lead to 12-Hydroxi-oleic acid on the one hand and to sebacaic acid on the other hand by degradation of the double bond.

Catalytic hydrogenation of 12-Hydroxioleic acid results in the formation of 12-Hydroxistearic acid, which is important for modern grease concepts. Sebacaic Acid on the other hand is a raw material for DOS (see above) but also for the production of complex greases. (Figure 26) [1, 2, 4]

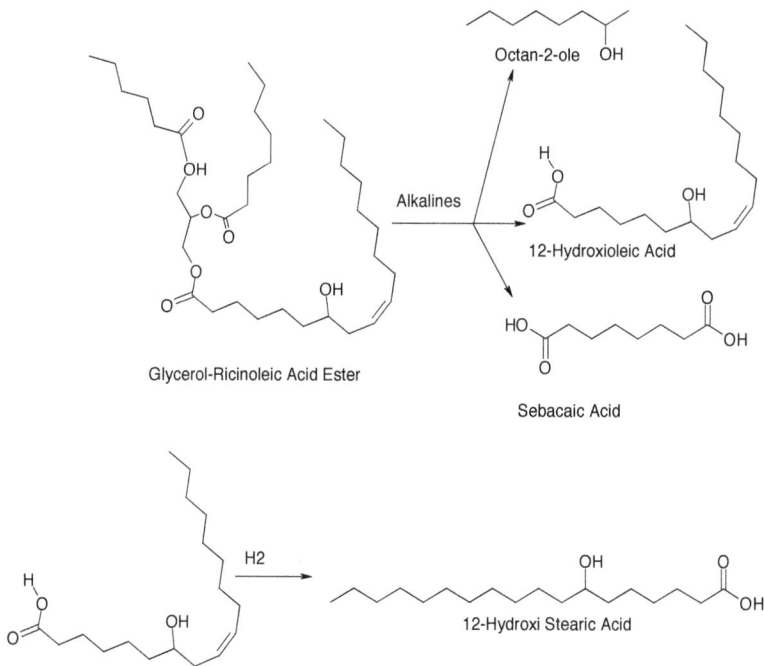

Figure 26. Cleavage of Glycerol – Ricinioleic Acid and hydrogenation to 12-Hydroxistearic Acid, Sebacaic Acid and Octan-2-ole [10].

Glycerole Esters with long chain carboxylic acids only, e.g. Glycerole Tristearate, are no longer nutrients and sparingly biodegradable. They are used as emulsifiers, consistency givers.

Glycerole Trioleate is a powerful friction reducer in tribological applications.

6.5.2. Triesters, derived from alcohols else than glycerole

6.5.2.1. Trimethylolpropane esters (TMP-esters)

TMP-Esters are created out of Trimethylolpropane (TMP) by reaction with short chain carboxylic acids, e.g. the range from C6 to C10. (Figure 27)

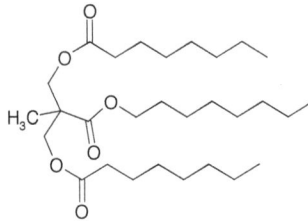

Figure 27. TMP Esters

TMP Trioleate is created by reaction of TMP with oleic acid or by trans-esterification.and commonly used as lubricant in minimal lubrication.

6.5.2.2. Trimellitic esters (TM-esters)

Apart from the described structures where trivalent alcohols get reacted with monocarboxylic acids, trimellitic Esters (TM-Esters) are products from Trimellitic Acid Anhydride with Mono alcohols. (Figure 28)

Due to the aromatic core those esters are high in thermal stability and widely used in high temperature applications.

6.6. Tetra esters (Pentaerythrolesters, PE-esters)

Pentaerythrole acts as a four-valent alcohole which may be esterified by four carboxylic acids. (Figure 29)

Carboxylic acids are in the range from C6 to C10.

Dipentaerythrol Esters (Di PE Esters) are formed starting from Dipentaerythrole as a six- valent alcohole reacted by six monocarboxylic acids in a Carbon Chain length from 6 to 10. (Figure 30)

6.7. Polyesters

In the past 20 years new groups of esters have been created by reaction of polycarboxylic polymers with alcohols. Those are reaction products of maleic acid anhydride (MSA) by Ene-Reaction with PAO precursors, leading to the PAO-backbone MSA addition product that might be esterified by butanole, leading to carboxylic complex esters (Figure 31).

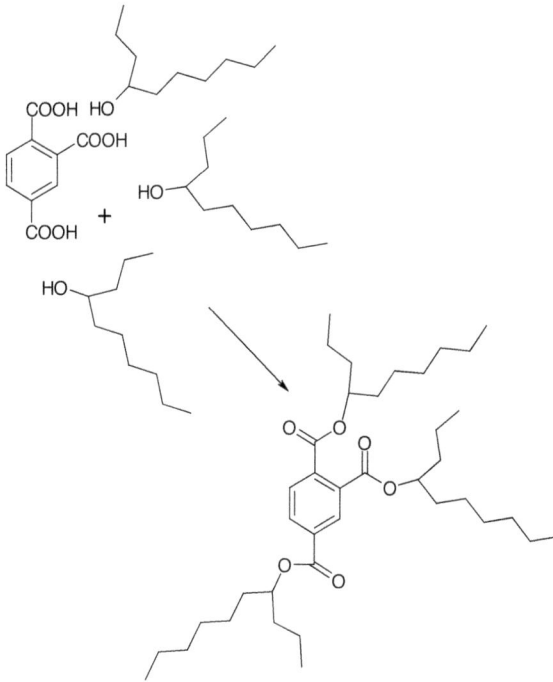

Figure 28. TM Esters by reaction of trimellitic acid with branched alcohole

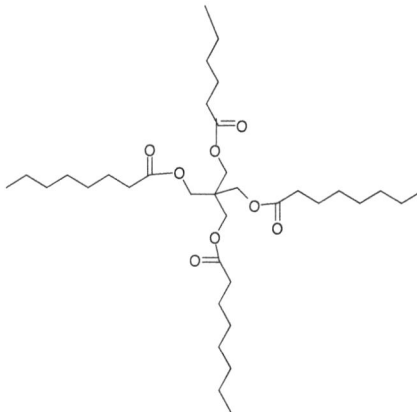

Figure 29. PE Esters

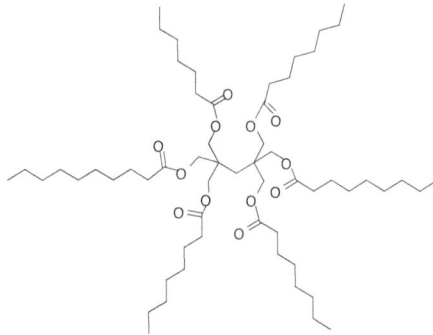

Figure 30. Di PE Esters

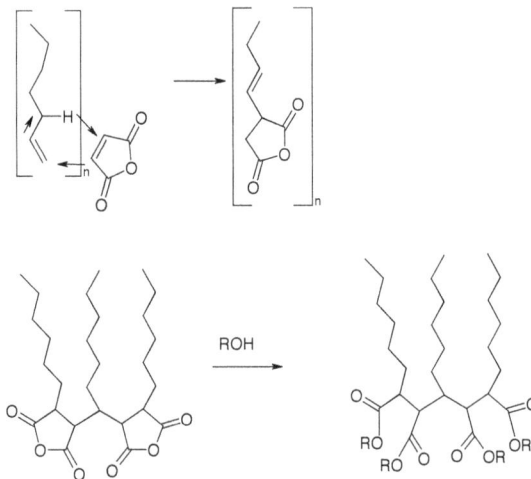

Figure 31. Complex Ester Formation by Ene-Reaction Sequences

Complex Esters from those structures are widely used to improve the additive solubility and performance. Their structure with shielding the carboxylic groups causes less aggressiveness toward sealings.

7. Structure activity relationship in esters

Esters are prominent representatives of lubricants where the chemical structure promptly leads to a specific tribological activity. However, if a tribological acitivity is demanded, the specific construction of esters may offer the solution.

7.1. Polar acitivity

Esters are polar by their nature due to the central element where a carboxylic acid tail binds toward an alcohol. Polarity gives some advantage but also disadvantage in the case esters are used. In general, esters enhance the solubility of functional additives and keep them away from fall-out. Esters also enhance the cleaning of metal surfaces in operation, preventing a formulation by creation of sludge. Esters are, as a fact of their polarity, aggressive toward sealings with a general tendency to shrink them. Plastics and elastomers under bending are susceptible toward stress corrosion cracking if attacked by esters. Hence, stress-corrosion cracking has to be considered explicitly in the case if esters are used. Since hydrocarbons, like PAO have a tendency to swell elastomers, the addition of esters may counteract such that the effect is neutralized. As a fact synthetic oils based on PAO are additivated by addition of 10 or 20 % esters per weight to create this effect.

7.2. Low temperature (Pour point) properties

Di-Esters, e.g. DOA, DOS are very useful in temperature ranges that undergo -40°C. This effect might be explained by the lack of hydroxyl groups that might associate at low temperature via hydrogen bridging, but also as a consequence of the crystallization hindrance due to the spatial structure of esters which does not allow a dense crystal packing. In contrast, esters may be designed such that their low temperature properties are lost, just by changing their structure. Also, if the number of polar groups increase the tendency to molecular association increases, and hence the pour point rises.

7.3. High temperature properties

High temperature applications in the use of esters are achieved by

• Sterical hindrance of the ß-Position in the Alcohol

• Use of Aromatic Nuclei in the Ester structure

As a specialty esters may rearrange within their structure via a preferred six-membered cyclic intermediate that creates an alkene on one side and a carboxylic acid on the other side (Figure 32).

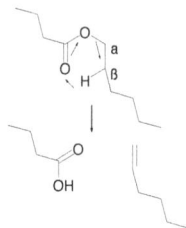

Figure 32. Decomposition of esters via cyclic rearrangement

Degradation of esters via such mechanisms takes place at ambient temperatures, e.g. by copper activation even at 70°C. The formation of carboxylic acids and alkenes may lead to corrosion and unfavorable deposits on metals. In the case of blocking the ß-position, as in the NPG and TMP esters, the cyclic rearrangement is blocked and the ester does not undergo the thermal degradation. Such oils are commonly used as turbine oils.

7.4. Side reactions

7.4.1. Hydrolysis

Ester Oils generally hydrolyze by interaction with water. The hydrolytic process is somehow the reverse reaction how esters form. The attack of water is enhanced if alkalinity is present but also acids may catalyze the hydrolysis. Common understanding states the attack of so called nucleophiles, like water at the carbonyl C-atom, followed by rearrangement sequences, leading to carboxylic acid and alcohols. (Figure 33)

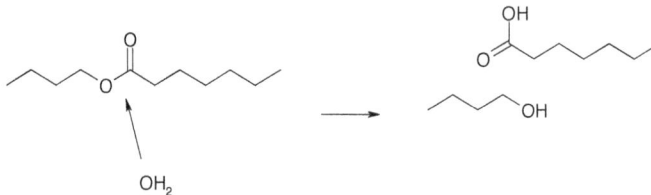

Figure 33. water-based cleavage of Esters toward carboxylic acids and alcohols.

Catalytic hydrolysis of ester oils also take place at metal surfaces, e.g. under tribological conditions. Formation of carboxylic acids may lead to corrosion as a consequence.

7.4.2. Biodegradation

Esters may decline under the interaction of bacteria and combust. Biodegradation is observed in the case of vegetable oils, e.g. glycerol esters, seldom on technical esters. In principal biodegradation cleaves esters, like water does to carboxylic acids. Biodegradation as a complex process does not stop there but lead to further products. Esters may oxidize as described in mineral oils and PAO at the organic tail. As a specialty they may undergo hydroxylation at a side position followed by trans-esterification to lactones. The lactone sequence is described already in the mineral oil section. Lactones are observed if esters, but also PAO are decomposed on iron at higher temperature. Infrared Spectra show absorption at 1800 -1760 cm^{-1} caused by lactone formation (see also chapter of antioxidants). (Figure 34)

7.5. Other esters

Esters may be created, as already mentioned by reaction of acids, in a different way as carboxylic ones. Prominent representatives are esters derived from phosphoric acid. Phos-

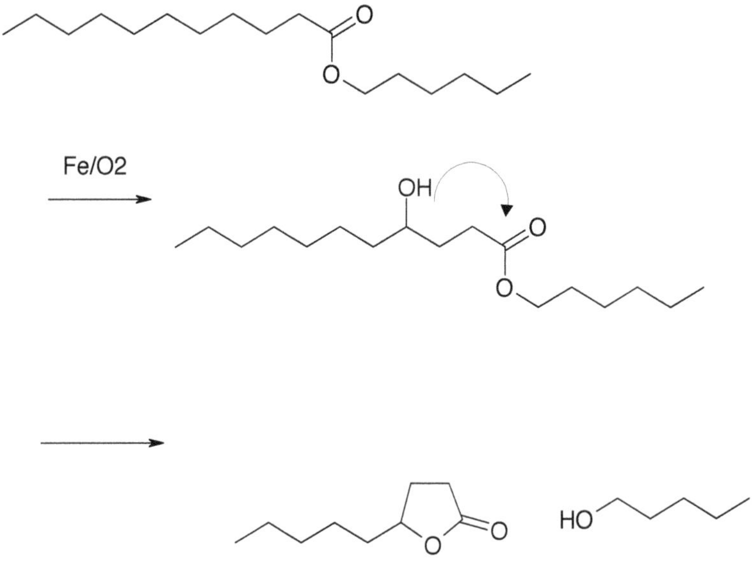

Figure 34. Lactone formation by side-chain oxidation of esters

phorous offers two main oxidation states (+III and +V) from which acids are derived. Depending on the oxidation state and the alcohols, phosphoric esters are different in use. Also phosphorous overtakes the role of anti-wear activity in such substances. [9]

A common representative is Trilaurylphosphite (Figure 35).

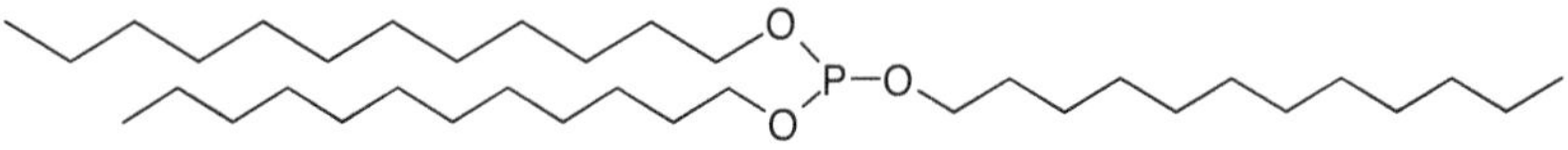

Figure 35. Trilaurylphosphite as a representative of Phosphinic Acid Esters

7.5.1. *Phosphoric acid esters*

Phosphorus in the oxidation state (+V) creates a plenty of variant Acids, such as Orthophosphoric Acid, Diphosphoric Acid, Triphosphoric Acids switching into each other. (Figure 36)

Phosphoric Acids are created by reaction of either phosphoric acid anhydride with alcohols or phosphoric acid derivatives, e.g. POCl₃ (Phosphorous-Oxi-Chloride) with alcohols. Aliphatic alcohols are in use, but also Phenols [10].

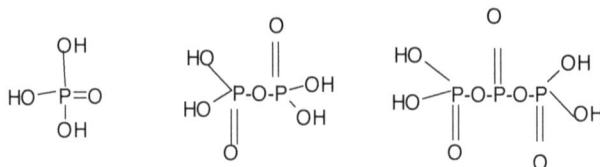

Figure 36. Representation of Phosphoric Acid Ester

7.5.1. Formation of phosphoric acid ester

Reaction of aliphatic alcohols, e.g. hexanole, with phosphorous pentoxide leads to hexylphosphate. In general some acidity remains due to insufficient esterifications. As a consequence those esters are often neutralized with amines to give amine phosphates. [10] (Figure 37)

Amine phosphates are widely used as anti-wear and anti-corrosion additives in all kinds of applications. Phosphoric Acid Esters derived from Phenoles are shown below. (Figure 38).

In a different reaction Scheme Phosphoroxichloride reacts with alcohols. Those reactions are convenient to come to aryl phosphoric acid esters. Arylphosphates are somehow used to come to non-flammable high temperature lubricants at temperatures beyond 200°C.

Apart from the use as base oil phosphoric esters like Tricresylphosphates, based on the reaction from Phosphorous Oxichloride with Cresol (Methylphenols) are common additives for lubricants in bearing industry.

Whilst TCP with the methyl group in the para-position is seen as hazardous, TCP isomers in the ortho is registered to be highly toxic. Also mixtures of TCP isomers, due to the content of the highly toxic ortho isomer are registered as highly toxic. TCP, despite its superior behavior as AW additive for bearing lubrication is restricted for use (Figure 39).

Use of Thiophosphorylchloride as precursor, the reaction with Phenole leads to EP/AW additives like Triphenylphosphorothionate (TPPT) and its derivatives. (Figure 40)

TPPT is widely used as non metal EP Additive as a substitute for Zn and Molybdenum Dithiophosphates. Due to its thermal stability, TPPT undergoes reactions at higher temperatures (>100°C). As to the fact that TPPT is ashless and starts to react at higher temperatures, it is a preferred additive in high temperature lubrication in combination with sterically hindered esters and PAO. In contrast to TCP, TPPT is not registered to be toxic, even more, the use of TPPT is allowed at level of 0.5% per weight for incidental food contact.

8. Polyglycoles (PG)

8.1. General

Synthesis of Polyglycoles starts from Epoxides, obtained by catalytic oxidation of Alkenes from Petrol- or hydrocarbon chemistry. Polymerization catalysed either by acids or alkaline result

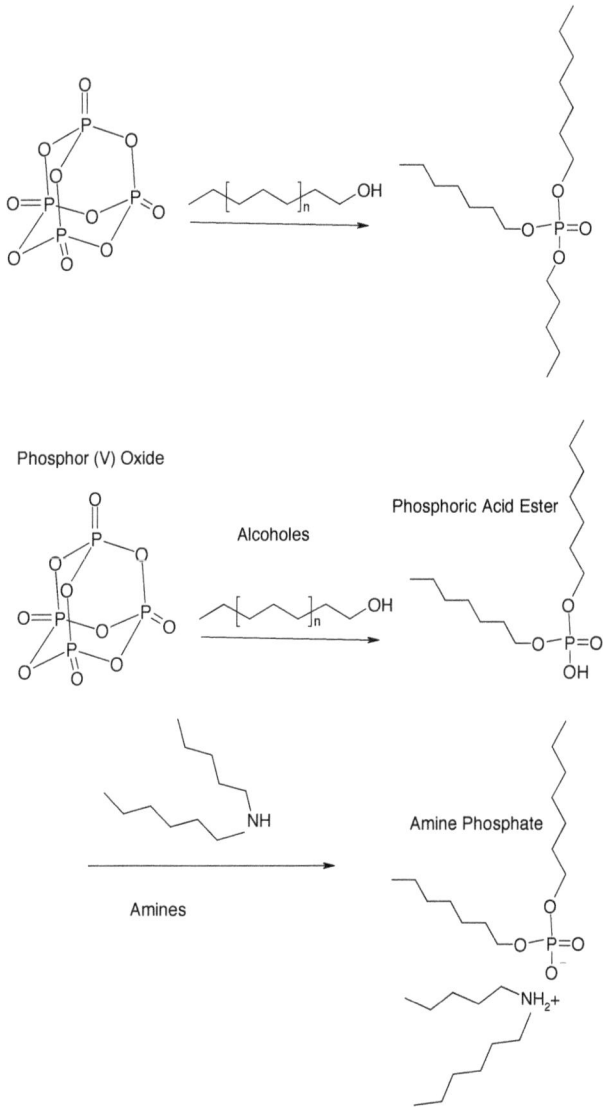

Phosphor (V) Oxide

Alcoholes

Phosphoric Acid Ester

Amines

Amine Phosphate

Figure 37. Phosphate Esters and Amine phosphates

in the formation of polyglycoles. In the case of alkaline catalyst, e.g. alkoxides on half of the PG contains a hydroxyl group while the end is capped by an ether function. (Figure 41) [1, 2, 7, 10]

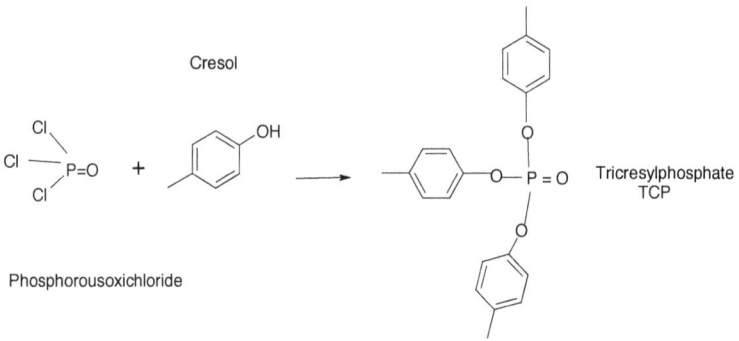

Figure 38. Arylphosphates derived from Phosphor Oxide Chloride Reactions

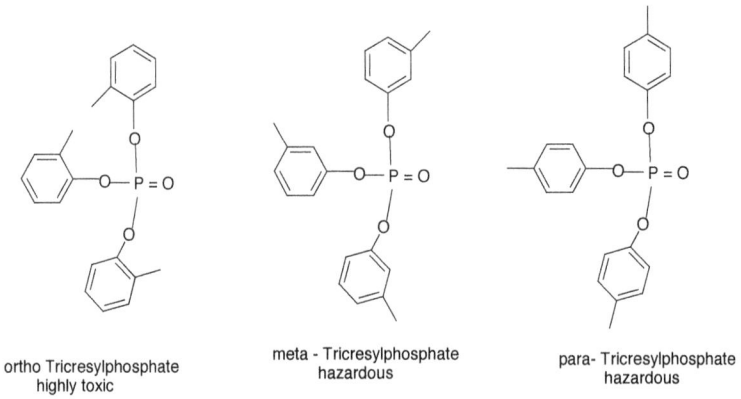

Figure 39. TCP and some isomers

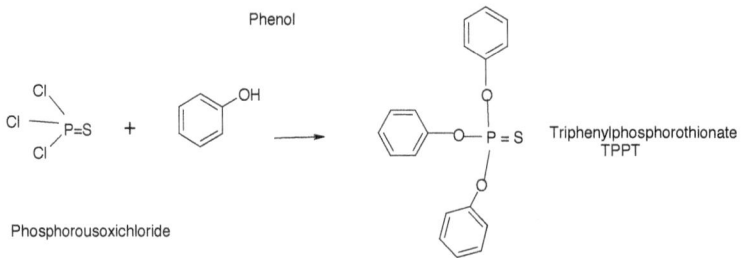

Figure 40. Synthesis of TPPT

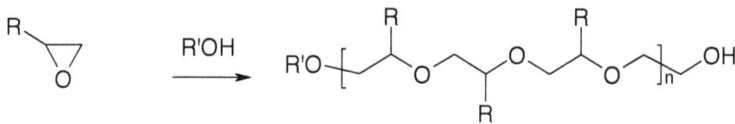

Figure 41. General Formation of Polyglycoles by alkaline catalytic polymerization of Alkene Epoxides

The choice of either different alkenes (Group R) or alkoxides (R') leads toward a huge variety of PG, all of them with different chemical and physical properties.

Polyglycoles			PEG	PPG-PEG	PPG	PBG
Description	Ethyleneoxide Polymer			Mixed Polymers Ethylene/Propylene Oxide	Proyleneoxide Polymers	Butyleneoxide Polymers
Chemical Data						
Physical Data	Density	approx.	1	0.95-0.98	0.95-0.98	0.95-0.98
	Flashpoint	approx.				
	Pourpoint	approx.				
	Water miscible(%)		100	partially		
	Hydrocarbons		non miscible	partially	partially	partially
	Ester Oils		partially-full	partially-full	partially-full	partially-full
	Other PG	PPG	partially-full	partially-full	partially-full	partially-full
		PBG	partially-full	partially-full	partially-full	partially-full
Tribological Data	Viscosity, 40°C	Range	32-46	32-100	32-100	32-100
	VT Coefficient	Range	180-..>200	180-..>200	180-..>200	180-..>200
	VP Coefficient	Range				
Others	Seals	NBR	compatible			
		ABS	compatible			
	Paintings		not compatible			

Table 1. PG and their data and applicability

Technically only a couple of variances are produced in a larger scale, such as:

- *Polyethylenglycoles (PEG)* where Ethylene Oxide is the starter
- *Polypropyleneglycoles (PPG)* where Propylene Oxide is the starter
- *Polyethylene- Polypropylene Oxide Mixtures* started from mixtures or Ethylene and Propylene Oxide

Table 1 offers an overview across the most common PG their data and applicability.

Single addition of long chain alcohols lead to the formation of fatty alcohol ethoxilates, for use as non-ionic detergants and dispersants in lubricant formulations, as silicone free defoaming and emulsifiers for lubricant formulae.

In general PG are not thermally stable by themselves and tend to decompose by emission of volatile degradation products, e.g. low boiling compounds, such as aldehydes, ketones, acids and others. Due to this behavior PG are used in high temperature applications where the formation of polymers and lacquers due to heat induced degradation of lubricants is not convenient, for example high temperature chain lubrication.

Presence of alkalines, such as overbased sulphonates, widely used in motor oils, as corrosion inhibitor lead to multiple cross-reactions with the decomposition products of PG (aldol reactions): Results of the aldole reaction are tars, sludge and slurries in the system. In consequence corrosion resistance of PG should always be carried out by acidic corrosion inhibitors, such as succinic-esters, Zinc-Naphtenates or Phosphoric partial esters. (Figure 42)

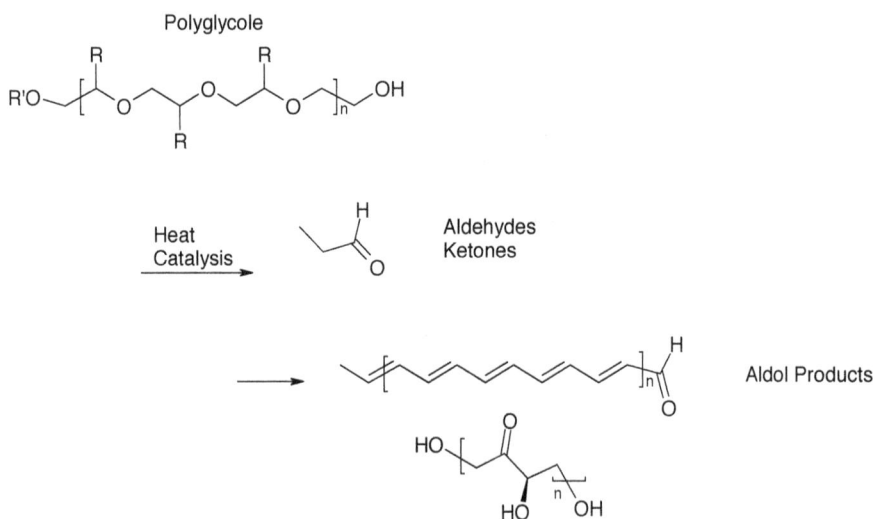

Figure 42. Aldole sludge formation in PG by use of alkaline

It has to be considered that PG are poorly soluble even amongst themselves and should be carefully checked. In general their solubility in mineral oils is poor, better in esters (depending on the structure). However, PG needs to be stabilized by antioxidants in order to prevent the early thermal degradation. By doing so, the application of PG are enhanced significantly, such, that even applications temperatures > 160°C are approached.

Convenient stabilizers are Phenyl-α-Naphtylamine, Phenothiazines or Alkyldiphenylamines. The amount should be adapted to the application.

In general PG offer very high viscosity indices, mainly above 160 (compared to mineraloils at ranges from 20 (alkylnaphtalenes), napthenics (70), paraffine base solvates (110), Poly-α-Olefines (140).

This high VI allows reducing the calculated viscosity in a given tribological application down to one or two levels. For example, if in a given application ISO VG 320 (320 mm²/s, at 40°C) is calculated for a mineral oil with a VI of 100, this viscosity maybe reduced by use of a PG down to 220 mm²/s or even 150 mm²/s. Pour points are low in the case of PG, very often in the range of -30.. – 40 °C even. Reaching the Pour point, PG tend to form highly viscous liquid, however, crystallization-inhibited. As a fact of this huge increase, PG is not for use even at temperatures above the pour point. Realistically PG is not suitable in the vicinity of their pour point. Therefore they are not very good low temperature base oils compared, for instance, with esters or PAO fluids. Due to their chemical structure PG are somehow strong solvents, e.g. paintings. In the case PG is used, the system has to be checked whether the paintings of the tank, the machine housing or others are affected. Dissolved painting from the tank may cause severe problems in the oil circuit by blocking filters. Additive response, known from standard applications, may change seriously by use of PG due to their different solvent capability. Extreme Pressure Additives have to be checked in their performance if used in PG. Normally anti-wear and anti-friction additives may be decreased in their content.

As a fact of the presence of epoxides in PG and due to their cancerogenic potential, the use of PG formulations drops down.

8.2. Polyethylene Glycols (PEG)

Polyethylene Glycols are made from ethylene oxide by polymerization (Figure 43).

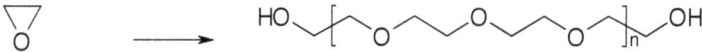

Figure 43. PEG formation and structure

PEG, apart from its wide use in cosmetic industry is completely water miscible. Due to its water miscibility PEG is only or sparingly soluble in hydrocarbons. Compatibility of the PEG with a given fluid has to be checked before use. PEG, as facts of its water miscibility will uptake water without separation. In case of the use of PEG in applications within water environment the water ingress should be checked carefully. Effects of water ingress are increasing threat of corrosion and thinning due to the mixture.

8.2.1. Use of PEG

Water miscibility is of use in non-flammable hydraulics in coal mining industries, but also in applications of pharmacy and food processing. In general PEG is allowed within the FDA regulation to be safe for incidental food contact. Due to the positive effect of sliding especially in worm gears PEG is somehow recommended for use in such applications. [2, 6, 7, 9]

8.3. Polypropylene Glycoles (PPG)

Polypropylene Glycols are made from Proplyene Oxide by polymerization by use of butoxides leading to a half ether structure (Figure 44):

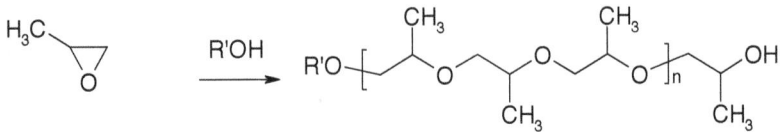

Figure 44. PPG Structure

Due to the additional methyl group in the structure water miscibility drops down (contrast to PEG) while the oil miscibility promotes. Also by choosing longer alkyl chain butoxides, PPG structures may be obtained with enhanced oil solubility.

While PEG is highly dissolved in water, PPG forms droplets immersed in the water. Due to this fact water separation out of PPG is difficult to achieve.

The partial solubility and immersion of PPG in water causes a very high fish toxicity. PPG should never be used in the case of its break out-in free lands or water

8.3.1. Use of PPG

PPG is commonly used as high temperature circuit oil, e.g. calandars, compressors, high temperature chain lubrication. All over PPG has to be stabilized by acid corrosion inhibitors, e.g. phosphoric partial esters and antioxidants like Phenyl-α-Naphtylamine [2] [3] [5] [8].

8.4. Polybutylene Glycoles (PBG)

PBG are seldom in use and made consequentially from Butylene Oxide polymerization by use of alkoxides, leading to half esters (Figure 45)

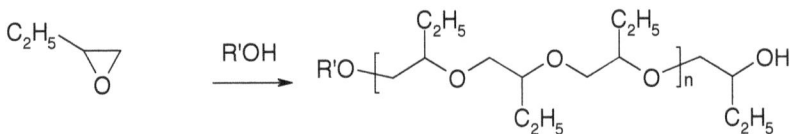

Figure 45. PBG Structure

8.4.1. Use of PBG

PBG is useful for enhancing the solubility of additives, boosting the viscosity index of mineral oil variants.

8.5. Alcohole Ethoxilates

Alcohole Ethoxilates are formed by a cross reaction of epoxides with alcohols. (Figure 46)

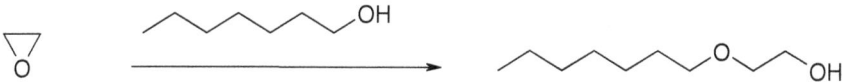

Figure 46. Alcohole Ethoxilates

The use of long chain alcohols leads to alcohole ethoxilates being used as non-ionic surfactants, emulsifiers and dispersants in multiple applications, e.g. hydraulic oils with dispersant capability, cutting fluids, and dispersants for applications where sludge is expected.

Due to their non ionic nature alcohole ethoxilates are widely compatible in lubricant formulations.

9. Siloxanes

Siloxanes (Silicone Oils) are common lubricants in multiple applications, such as food and pharmacy, but also in applications where special low friction properties are demanded [2, 3, 8]

In general silicones are the result of alkylchlorosilane hydrolyses [10] (Figure 47).

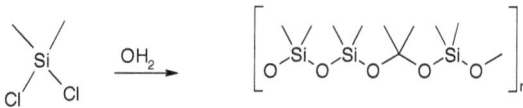

Figure 47. Scheme of Silicone Oil formation

Side groups are methyl, but also phenyl groups leading to polydimethylsiloxanes or polyarylsiloxanes. Mixtures of methyl and arylsiloxanes are in use with different spreading between in the side chain (Figure 48).

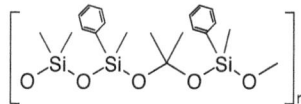

Figure 48. Polymethyl-Aryl Siloxanes

Siloxanes with different structures are not generally miscible amongst each other. Miscibility has to be checked carefully. As mentioned, siloxanes are widely used in lubrication technology due their exceptional properties concerning low friction capability, high temperature stability and low toxicity in various applications. Prominent applications are starter components in cars, valves in food industry, slow speed bearings and high temperature applications where arylsiloxanes are in use. Siloxanes creep widely across surfaces and may cause problems in coatings, lacquering and paintings.

10. Polyfluorinated Polyether (PFPE) base oil

PFPE Base Oil is created by polymerization of Perfluoroepoxids. Structure of PFPE is similar to polyglycoles but with overall substitution of hydrogen by fluorine [2] [3] (Figure 49)

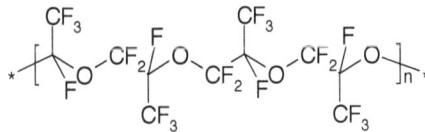

Figure 49. PFPE Base Oil

Due to the effective shielding of the C-O-C backbone in the structure of PFPE by the trifluoromethyl side chain group PFPE are completely insoluble in water, inert toward alkaline and acids and even oxygen.

PFPE Base oil is used for high temperature purposes and in the presence of aggressive media, mentioned above in junction with PTFE thickener. (Figure 50)

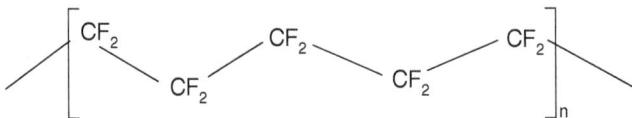

Figure 50. PTFE as thickener for PFPE

PFPE sparingly adheres to metal surfaces due to droplet formation. The low adhesion causes creeping across surfaces and mal-lubrication if the surfaces are not cleaned thoroughly. Creeping and low adhesion may cause low friction in certain applications. PFPE is insoluble in most of the common base oils. Use of PFPE hence is restricted to the fluorine group of base oil.

Inertness of PFPE and PTFE make greases suitable for incidental food contact lubrication.

High temperature combustion of PFPE may cause the emission of hydrogen fluoride and fluoro phosgene which makes PFPE formulations somehow corrosive, especially on steel alloy compositions. Due to this fact, the high temperature corrosiveness should be carefully taken into account in the case of PFPE use.

11. Additives

Additives in lubricants enhance base oil functionalities. Additive technology is in broad scale based on organic chemistry syntheses. From their origin they are found by chance, less than by a real scientific approach. Nevertheless, literature about their reactions is innumerous from the very beginning. [2, 3, 9]

Modern additive technology commenced in the early 20th century and has progressed continuously due to advanced organic chemistry syntheses. Upcoming modern spectrometry has been used to clarify their structures and their reaction at different metal sites. [2, 3, 9]

Beyond the basic and industrial reaction mechanism studies, mixtures of additives have been studied extensively by industry and science over the years. Such studies reveal the mechanisms of compatibility and incompatibility of additives acting together at a given application. [9]

For example a functional mismatch is caused by diverse demanding addressed to additives, e.g.: additives acting against corrosion may interfere with additives that have to prevent metal surfaces against fretting or welding.

Modern additive technology is inevitable to reach the "for-life" goal of modern technologies. As "for-life" might be understood in a different manner by users, additive packages are developed during the decades adapted to a given customer demanding. For example, the demanding to get automatic transmission gear oil performances is achieved by additive packages that may not fit for wind turbine or paper mill applications. Hence, additives and their mixtures have to be selected carefully for each purpose. [2, 3, 8, 9]

In general there are no rules up to now to predict additive performances at a given technical application. As a consequence formulations have to be tested in forecast extensively to assure its functionality. Such testing is addressed by international and national regulations.

Additives may cover a distinct structure-property relationship. Since there are no scientific rules declaring on how a chemical structure of an additive causes a function all variations in additives have to be validated by tests.

Additive technologies have been revised many times during their history, either due to a change in demanding or due to their toxicity. Toxicity is a severe problem in additive technology, since no one knows their real long term biological and ecological effects. [9]

Since the validation of those different chemical additive structures causes tremendous costs, it is a fact, that additive free technologies or additive technologies with marginal content level are favored as future solutions.

The following chapter addresses additive technologies concerning extreme pressure, anti-wear functions and also corrosion-protecting and antioxidants.

11.1. Extreme Pressure (EP) and Anti-Wear (AW) additives

11.1.1. General

Extreme Pressure (EP) and Anti-Wear(AW) Additives are functional chemicals in lubricants with the task to separate metal surfaces in the case of heavy loading and to improve their resistance toward wear in the case of oil film break in the contact [9].

Machinery elements that start to run or stop due to emergency show pronounced loading due to a lack of lubrication, e.g. the oil does not separate the metal surfaces and the protection of the oil film drops down. At that point EP and AW additives are supposed to jump into the arena by causing reaction layers preventing the metal from direct rupture or welding.

Their chemical structures are found by chance. For example observations during drilling and maching show that tools perform better if lubrication is carried out by use of sulfurized oils derived from vegetables, mixed and heated with sulfur.

Later on intense research the nature and reaction started including modern surface spectrometry techniques. The transformation of EP/AW additives as a function of the nature of the surfaces, their loading, contact geometry, temperature and their structure shows a clear picture of structure-activity relationship. Also additives perform as a function of their chemical structure, but also as a function of their solubility in base oil and as a function of other additives being present. In that sense, it is shown that additives either may prolong service life but are also capable to shrink life.

11.1.2. Sulfur additives

Sulfur acts as a powerful extreme pressure additive. The high reactivity, especially toward copper makes it unlike to use sulfur as element in tribology.

Sulfur embedded in organic framework acts as a powerful Extreme Pressure additive. Choosing appropriate organic structures the activity toward copper drops down. However, using sulfurized additives copper deactivation should be present anyhow.

Sulfur is added either by reaction of reactive organic precursors like alkenes and their derivatives by heating up with the element, or by polymerization sequences with activated sulfur precursors such as di-sulfur dichloride. Doing so, all kinds of unsaturated specie gives reaction products leading to sulfurized specie. Prominent representatives are reaction products of Isobutene with Disulfur Dichloride, or reaction products with terpenes (Figure 51) but also unsaturated carboxylic acid esters, like rapseed oil:

Sulfurized Additives (S-Additives) are often used together with phosphoric acid esters, since the synergistic between those additives are known from the past. Doing so, gear oils may contain S-Additives with amine phosphate esters. Also extreme pressure additives containing

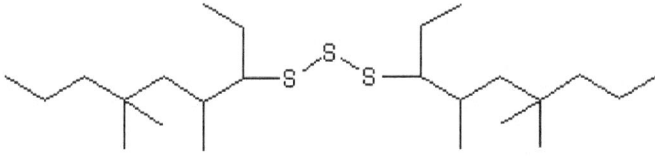

Figure 51. Didodecyltrisulfide as polysulfide representative

Dialkyl-Thiophosphoricacidesters are prominent representatives in sulfur additive chemistry. (Figure 52).

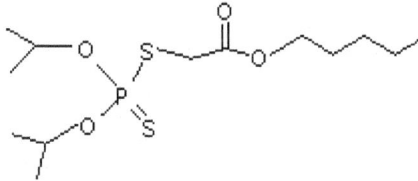

Figure 52. Thiophosphoric Acid Ester

11.1.3. Dithiophosphates

Zinc- and Molybdenum dithio phosphates (ZndtP- ModtP)

Zincdithiophosphate (ZndtP) represent a prominent group of EP/AW additives. They derive from the neutralization of Thiophosphoric Acids, obtained by ring opening of Phosphorous pentasulfide with alcohols, with Zn-Carbonate or Hydroxides. As a fact, the ZndtP differ strongly by their carbon-chain length. A couple of variants are achieved by choosing different alcohols in the ring opening sequence of Phosphorous (V) sulfide. From the structural perspective, ZndtP may be regarded as chelate complexes rather than a salt (Figure 53).

Molybdenumdithiophosphates contains a Molybdenum [μ-oxo] Core, distinct compared to ZndtP (Figure 54).

11.1.4. Dithiocarbamates

Similar to Dithiophosphates, Chelat Complexes from Zinc, Molybdenum but also Bismuth and others may be formed by reaction of Thiocarbamic Acid with the metal precursors. Dithiocarbamic Acid is synthesized via addition of amines to Carbondisulfide. By varying the chain length of the amine different dithiocarbamates are achieved (Figure 55 and Figure 56):

Various alcohols

HO

Phosphorous (V) Sulfide

+

Thiophosphoric Acid

Zn Carbonate

ZndtP

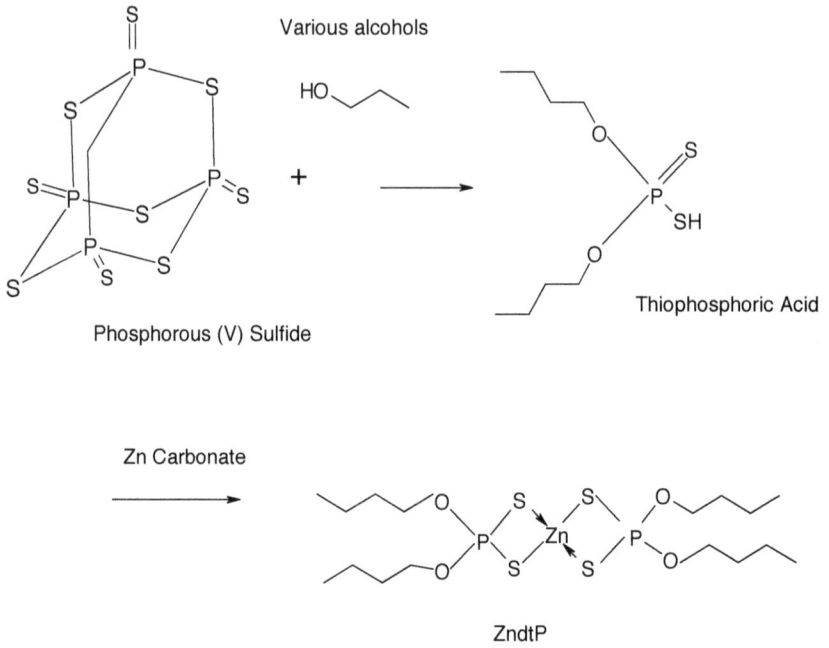

Figure 53. ZndtP from neutralization of Thiophosphoric Acid with Zn Carbonate

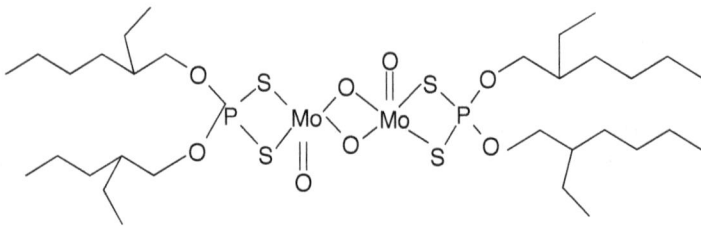

Figure 54. Molybdenumdithiophosphate

11.2. Corrosion protection

11.2.1. General

Within this chapter only iron as a chief element in technical application is considered.

Generally metal surfaces tend to corrosion if water, oxygen and probably salts, like sodium chloride are present. Corrosion may take place either by cathodic reduction of oxygen or by

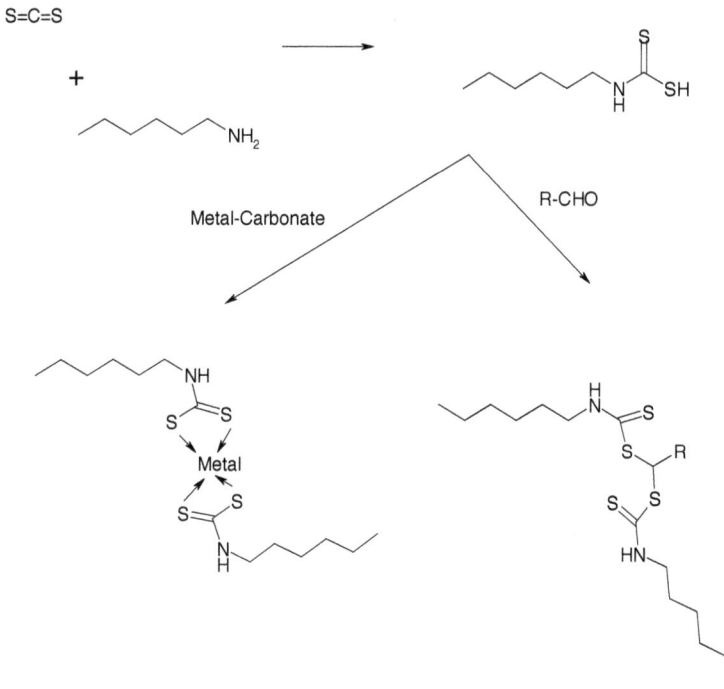

Figure 55. Syntheses of dithiocarbamates

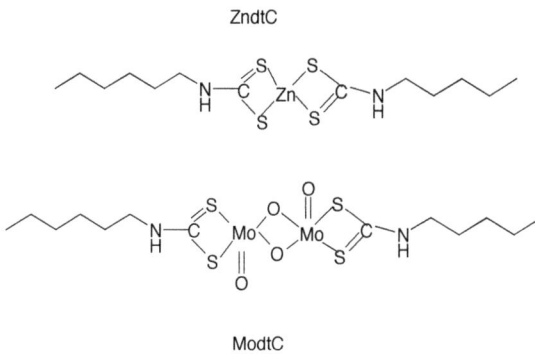

Figure 56. Zincdithiocarbamate (ZndtC) and Molybdenumdithiocarbamate (ModtC)

anodic oxidation of the metal. Charges, either positive (anode) or negative (cathode) pass the surface layer. [2, 3, 9]

Charge transport from the metal toward the outer region is hindered by the surface potential (over potential). Thus, corrosion processes have to overcome this potential and start after a certain induction period. Once, if this potential has been overcome the corrosion starts without hindrances by successive material transport. Materials transport ends up in a drastic change of the surface, mainly accompanied by a loss.

For iron as metal, the transport of the metal ends up in a flaky layer (Rust) that permits water and oxygen to penetrate. Due to this effect the rust process ends up in a total damage of the metal, especially in an environment that boosts corrosive processes.

Counteracting corrosion, the initial processes of charge transportation have to be blocked. Doing so, the over potential, e.g. the natural barrier of charges passing the surface has to be increased by creating additional layers on the metal surface (Passivation) or by creation of stable, insoluble complexes, formed by interaction of the surface atoms with a complex builder.

Passivation of iron surfaces and enhancing the over potential is achieved by deposition of chromium layers that cause a thin and gas-dense closed layer on the metal. Thus, chromium is a powerful inhibitor toward corrosion processes. As the charge transport phenomena occurring on iron surfaces are cathodic or anodic and vice versa, this process could also be stopped by offering an anodic victim like a zinc coating.

Additives that create a corrosion protection are in general dissolved in a carrier base-oil that spreads over the surface. Due to their adapted functional groups a physical binding toward the surface starts to create a layer. In order to create an appropriate corrosion protection this layer has to be packed dense to avoid the penetration of water and oxygen. This is realized by strong dipolar groups and oil soluble tails with a marginal demand in lateral spacing, e.g. long, - unbranched alkyl chains.

Else, passivation also is achieved by placing insoluble complex builders onto the iron surface, like phosphates are. Iron phosphate builds up a close dense insoluble layer on the surface.

Restriction of iron phosphate is indicated by the fact that, under certain conditions, phosphates start to get reduced forming posphanes. Phosph4anes strongly affect metals due to segregation of phosphorous at grain boundaries and releasing hydrogen into the metal. Hydrogen is detrimental to the microstructure by inducing, e.g. hydrogen enhanced local plasticit (HELP) or hydrogen induced cracking (HIC). Presence of phosphanes by reduction of phosphates takes place in acidic and reducing environment, e.g. presence of hydrogen sulfide, chlorides and others.

The following chapter will show some of the most prominent representatives of corrosion protectors.

11.2.2. Sulfonate-chemistry

General

Sulfonates derive from sulfonic acids by neutralization with alkali, earth alkali –metals but also with metals from the transition group, for example zinc. Principally each sulfonic acid may be neutralized. In technical applications mainly alkyl benze sulfonic acids and dodecyl-sulfonic acid are neutralized. Production starts from alkenes out of petrol chemistry by addition sulfuric acid or SO_3.

Neutralization with either sodiumhydroxide, Calciumcarbonate, Magnesiumcarbonate or Bariumcarbonate leads to sulfonates: A = Sodiumsulfonate, B = Calciumsulfonate and with excess Carbonate to over based Calciumsulfonate (B'), Magnesiumsulfonate (C) and Bariumsulfonate (D) (Figure 57).

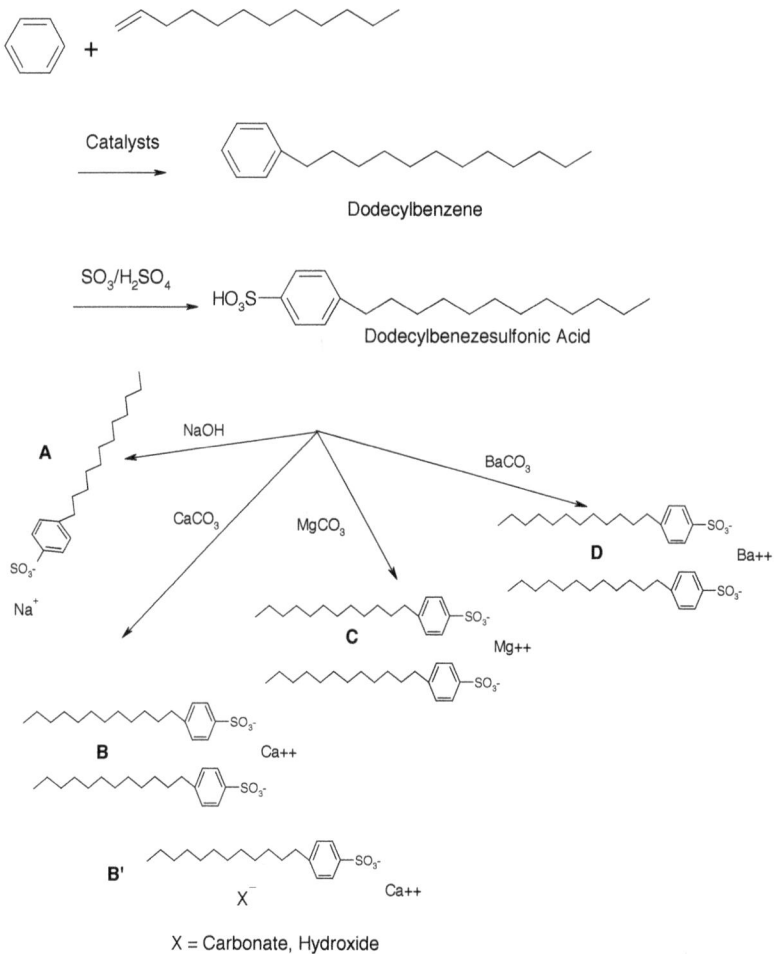

Figure 57. Sulfonic Acids and their Salts

11.2.3. Carboxylic acids and derivatives

Carboxylic Acid and their derivatives, e.g. esters may act as metal corrosion protectors. While carboxylic acids are supposed to cause corrosion, some of them prevent. Rust preventing carboxylic acids are derivatives from α-Aminoacids, like N-Oleylglycine. (Figure 58)

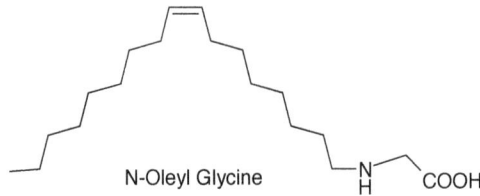

N-Oleyl Glycine $\quad$ N-H COOH

Figure 58. N-Oleylglycine

N-Oleylgylcine acts as powerful emulsifier, even at low dosage. Rust protecting is due to the spread of water in the formulation over a big volume. N-Oleylglycine, even at low percentages also counteracts with EP/AW additives, driving their activity down.

Carboxylic Acids, derived from Phenoles such as Nonyl-phenoxiaceticacid is a non emulsifying corrosion protector but under prohibition, due to its irritating effects (Figure 59).

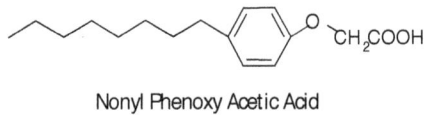

CH$_2$COOH

Nonyl Phenoxy Acetic Acid

Figure 59. Nonylphenoxyaceticacid

Succinic Acid Derivatives, such as Succinic Half Ester of Octanole are powerful metal protectors, but also strong counteracting with EP/AW additives. Synthesis is carried out by reacting succinic acid anhydride with alcoholes (Figure 60).

Carboxylates, derived from neutralizing carboxylic acids with transition metals like Zinc, Lead, Bismuth lead to corrosion protection. Common acids are Napthenic acids or medium chain carboxylic acids like octanoic acid (Figure 61).

11.2.4. Amine phosphate esters

Amine Phosphate Esters may act as anti-corrosion additives in addition to their anti-wear properties. Due to their synergistic properties and due to the fact, that certain amine phosphates are allowed as additives for incidental food contact, they are often found in all kind of lubricants (Figure 62).

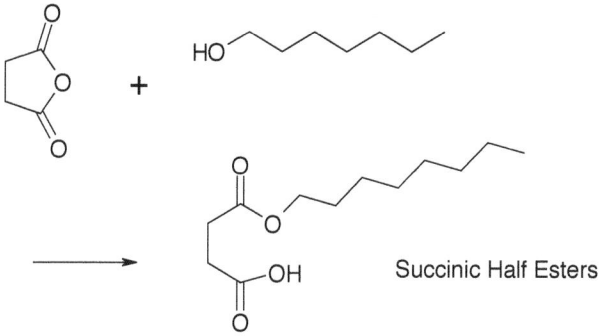

Figure 60. Succinic Half Ester

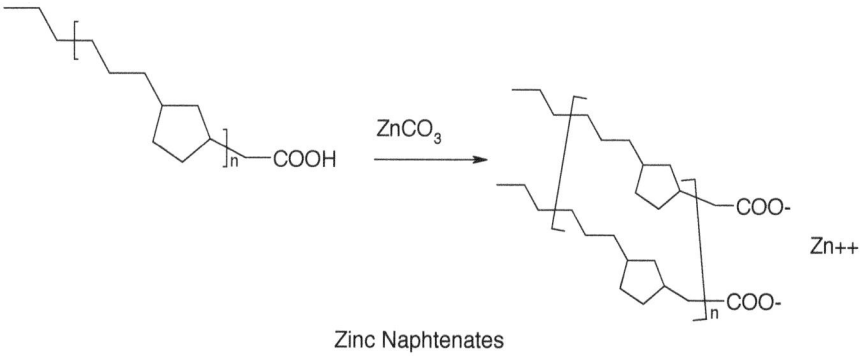

Figure 61. Zn (Bi) Carboxylates (Napthenate and Octoate)

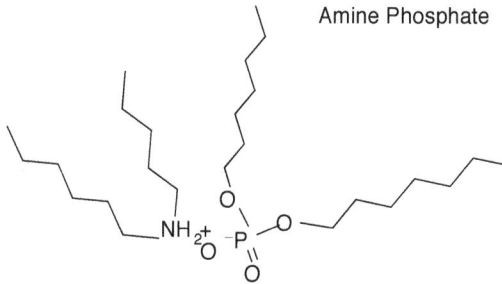

Figure 62. Amine Phosphate Structure

Amine Phosphates are powerful activators of copper and zinc and cause leaching of those metals from brass cages in bearings. Adding Amine Phosphates copper deactivators like benzotriazoles have to be present (Figure 63).

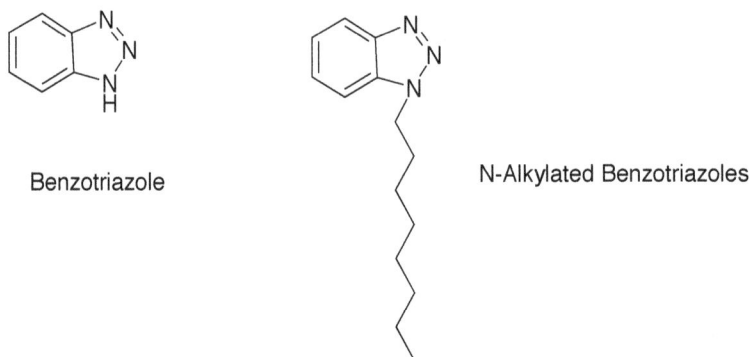

Benzotriazole   N-Alkylated Benzotriazoles

Figure 63. Benzotriazole and N-Alkylbenzotriazoles as Cooper Passivators

11.3. Antioxidants (AO)

AO prevent lubricants from oxygen attack. Oxygen is, by nature, a diradical that undergoes several transitions. Electron uptake from metal surfaces by a cathodic transfer, leads to varieties of activated oxygen specie, powerful attacking hydrocarbon sites by abstraction of hydrogen, leading to peroxides, and carbon radicals. The carbon radical itself starts to stabilize by abstraction of hydrogen leaving an alkene as new product [10]. (Figure 64)

Due to radical stabilization the new formed alkene starts to continue the oxidation by sequential abstraction of hydrogen, forming di-, tri- and polyalkenes, but also benzene rings. Apart from the hydrogen abstraction, also oxidation takes place by attacking carbon radicals by oxygen. At least the products created by such this procedures are carbonyl compounds, e.g. alcohols, ketones, aldehydes, carboxylic acids and sometimes esters. PAO oxidation at metal surfaces, e.g. iron beyond 120°C results in the formation of lactones (esters that come up by internal reaction between an alcohol group and terminal carboxylic group) (Figure 65).

Hence oxidation sequences dramatically change the original hydrocarbon chain. If once started it is self accelerating till new, different and stable products are reached. Oxidation is unselective and takes place everywhere in the chain. Hence, plenty of products are formed by radical oxygen assisted processes.

Antioxidants in general prevent the base oil, quenching the oxygen attack by formation of stable radicals. Stabilization of the radicals is realized by a delocalization of the persistent AO radical, created by oxygen attack due to the presence of aromats in the structure (Figure 66).

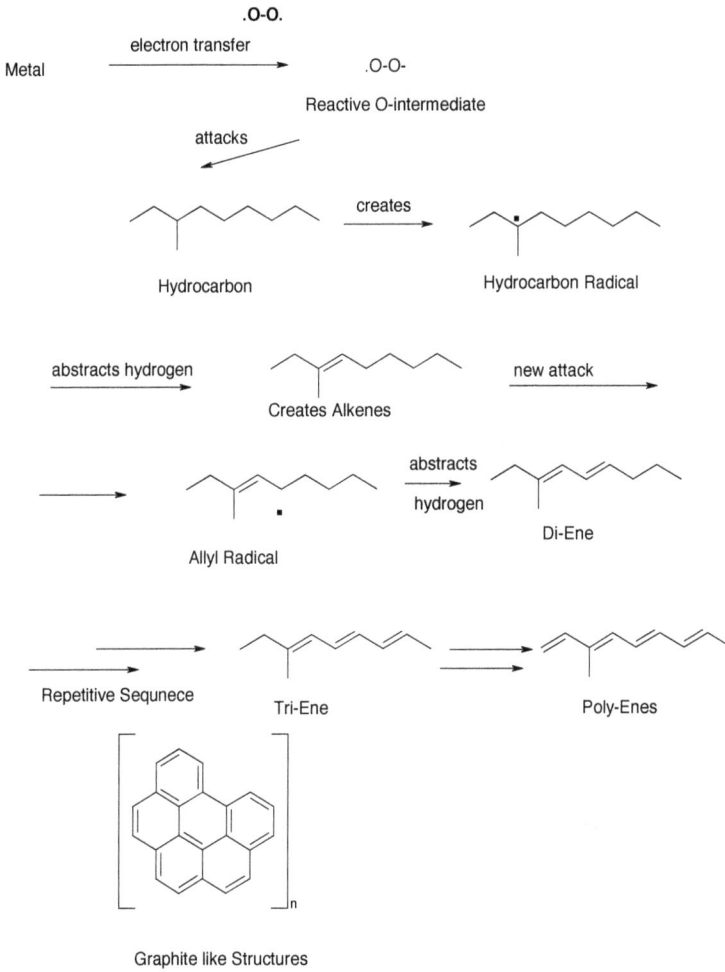

Figure 64. Oxygen – Hydrocarbon Attack sequence

The AO radical subsequently stabilizes to form new products like quinones. The quinone structure may form a dark colored charge transfer complex with the original antioxidant. Very often this causes strong discoloration of AO stabilized lubricants since the charge transfer complexes are very intense in color. Sometimes, for example in the case of poly-urea greases, such charge transfer complexes may interfere with the grease structure in terms of solidification.

Persistent radicals formed by AO are dangerous in some cases. In the case of their accumulation in the system they are able to boost oxidation rather than to prevent. Dosage of AO hence

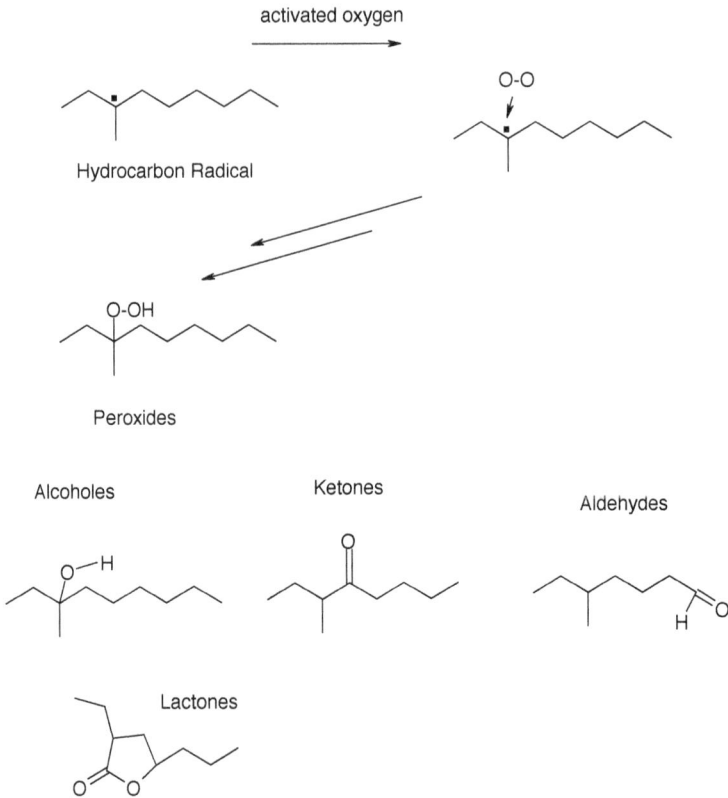

activated oxygen

O-O

Hydrocarbon Radical

O-OH

Peroxides

Alcoholes   Ketones

Aldehydes

Lactones

Figure 65. Oxidation sequence of Hydrocarbons toward carbonyl compounds

should be carefully tested. Formation of either charge transfer complexes or oxidation products by the presence of AO may cause increased formation of sludge in the lubricant if the dosage balance is not appropriate.

Nearly all AO contain aromats as a base principle. Prominent AO candidates are Butylhy-droxitoluene (BHT) (A), Alkyldiphenylamine), Phenyl-α-Naphtylamine (PAN) (C) and various others (Figure 67).

12. Greases

12.1. General remarks

Greases are defined apart from their chemical composition by the manufacturing processes. Thickener and oil, getting heated by stirring, start to dissolve. Getting cold, the process of

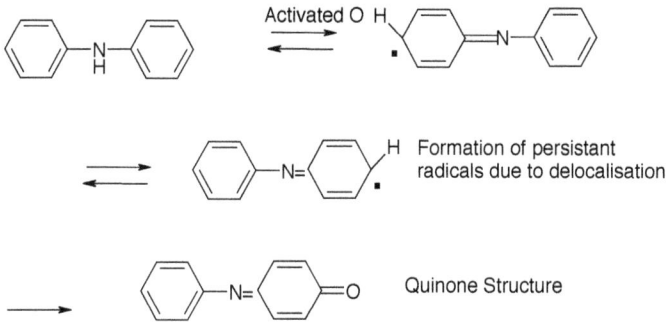

Figure 66. Principal delocalization of radicals created by oxygen attack.

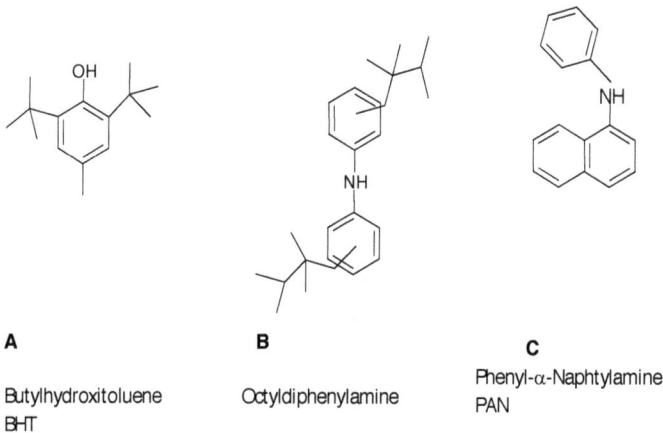

A	**B**	**C**
Butylhydroxitoluene BHT	Octyldiphenylamine	Phenyl-α-Naphtylamine PAN

Figure 67. Structures of AO: (A): BHT, (B) Alkyldiphenylamine, (C) PAN

stirring leads to a raw material where amorphous and crystalline structures are merged. The amount of crystals and amorphous materials depends on the nature of the raw materials on the one side and on the rate of heating and cooling on the other side. Rapid cooling causes homogeneous and amorphous structure, as particles are not able to grow to a large size. The raw grease, as effect of the mixture of solid structures has to be homogenized carefully. Homogenization leads to a smoothened appearance of the grease with a scale distribution of thickener particles as effect of the cooling process. Slow cooling generally leads to material with large sized particles as an effect of nucleation and crystal growth. Oil embedding in such structures is different due to the solid structure of the thickener. Stiffness and flowing capability may change as an effect of the merged structure. Greases, even in the case of identical chemical composition may differ significantly by their manufacturing process. Stiffness of greases is defined by the NLGI grade declaration, measured by penetration of standard cone into the grease. The deeper it's penetration the more liquid the grease will be. To get a constant value, the grease is worked by 60 strokes, then tempered to 25 °C and measured by cone penetration. NLGI grades are presented in table 2: [3]

NLGI Grade	Cone Penetration in 1/10 mm
OOO	445 - 475
OO	400 - 430
O	355 - 385
1	310 - 340
2	265 - 295
3	220 - 250
4	175 - 205
5	130 - 160
6	85 - 115

Table 2. NLGI Grades of Greases

12.1.1. Oil bleeding

Within grease the base oil is bound in different states. Some oil is weakly bound to the thickener nuclei and gets easy released. Oil, bound in micelles and large structures with van-der-Waals and dipolar bonding releases less. Oil release takes place due to centrifugal effects in speeding machinery elements, e.g. bearings, creeping across walls e.g. sealings enhanced by temperature. Successive loss of oil in grease may lead to its change in performance, accompanied by a malfunction. Oil bleeding is measured with different techniques. Within the most popular one the grease is sat on a sieve and pressed by a static load through it a given temperature. Bleeding is measured as a function of time. For bearings the long term bleeding rate should be less than 5 % per weight in 7 days. [3]

12.1.2. Dropping point

Greases - if heated - start to get liquid at a certain point. Molten grease will leak out at sealing edges and may cause a malfunction of the grease. For bearings the thumb rule is given by dropping point minus 50 °C as the upper point of applicability. [3]

12.2. Soap based greases

Greases are soft solids, created by a thickener that gelates in suitable base oils. Gelling takes place by intense mixing of thickeners with the base oil, often accompanied by heating till the gelation is reached [3]. (Figure 68):

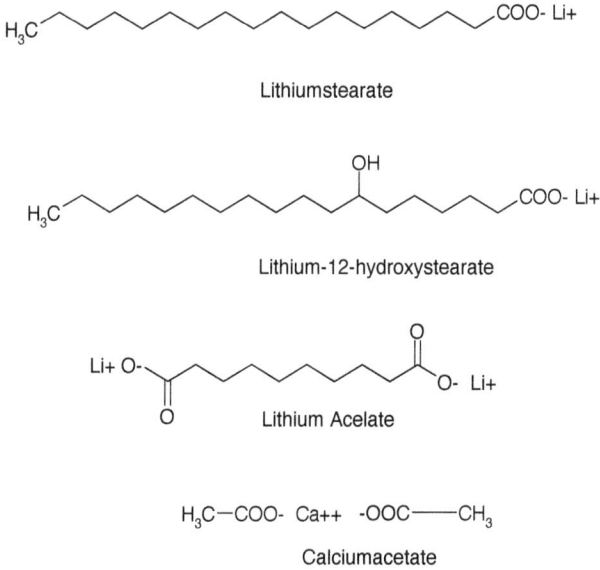

Lithiumstearate

Lithium-12-hydroxystearate

Lithium Acelate

$$H_3C-COO-\ Ca++\ \ -OOC-CH_3$$

Calciumacetate

Figure 68. Prominent representatives of thickeners for grease production

Thickeners are all substances where gelling in the base oil is achievable. Prominent representatives are lithium and calcium salts of carboxylic acids, for example Lithium Stearate, Lithium-12-hydroxistearate, Calciumstearate, Calcium-12-hydroxistearate but also Calciumacetate. Lithium Complex Greases are created by the co-existence of lithium-12-hydroxistearate with dicarboxylic acids like Acelaic or sebacaic acid.

Calcium Complex Greases are composed by calcium acetate, Calcium Stearate and calcium-12-hydoxistearate as thickeners.

Salts of magnesium, barium and alumina are used for grease production but to minor extent.

12.3. Di and Polyurea greases (PU-greases)

Urea Greases are often called PU-Greases in technical language.

Urea structures are realized by adding amines to isocyanates (Figure 69):

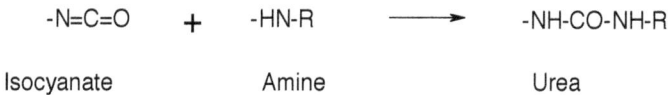

-N=C=O   +   -HN-R   ⟶   -NH-CO-NH-R

Isocyanate   Amine   Urea

Figure 69. Urea Formation

Di-Urea grease production take aromatic Isocyanates, like Diphenylmethane Isocyanate (Methylenbisdi-isocyanate, MDI) reacted with various aliphatic amines, like Cyclohexyla-mine, Alkylamines from C8 to C18 chain length.

Synthesis of the thickeners and grease formation is carried out simultaneously. Ester Oils, like trimellitic acid esters facilitate the synthesis by solving the precursors before the reaction takes place (Figure 70):

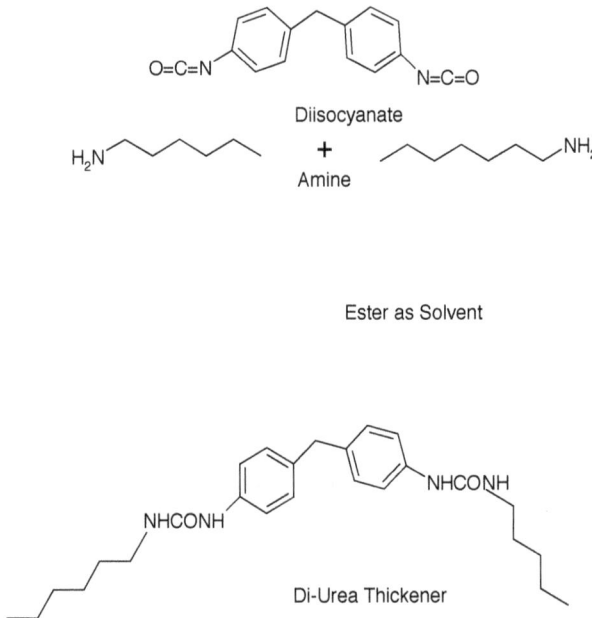

Figure 70. Formation of Di-Urea Grease

Tetra- and polyurea Greases are created by mixing Di-Isocyanates like MDI or Toluenediiso-cyanates (TDI) with diamines, like ethylene diamine and monoamines, like Octadecylamine in suitable base oils (Figure 71):

Urea Greases offer plenty nitrogen-hydrogen bridges within their structures. Concordant with the presence of temperature resistant aromatic nuclei and in junction with high temperature resistant base oils, they represent the group of high temperature grease "per se". As to the high variability of taking precursor amines, PU greases offer the possibility to adapt the grease to a given application, much more than soap greases do.

Polyurea Greases that start from tallow amine, tolyenediisocyanate and ethylene di-amine are in accordance with the US FDA regulations H1 (incidental food contact) if H1 base oil (like

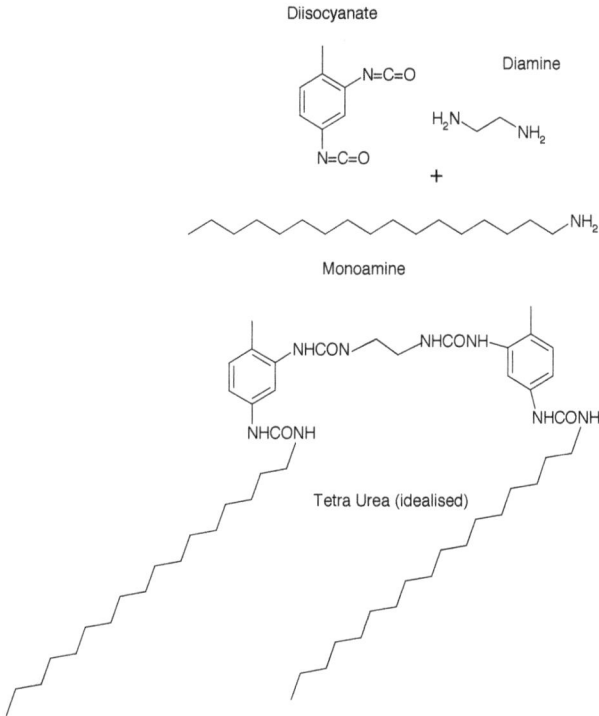

Figure 71. Formation of Tetra-and Polyurea Greases

white mineral oil or PAO) is used. Also the modern EU REACH regulations are valid for polymeric structure.

As the polymeric degree increase, the thickeners may get insoluble and crystalline. Greases are no longer available due to this because a lack of gelling. Due to this fact, variances of PU Greases are restricted.

MDI and especially TDI are ought to be highly toxic by inhalation. Production of PU greases have to take care, than none of the precursors are free in air, nor present in the grease.

Some isocyanates tend to polymerize during production, rather than to react with the amine, especially at the end of the syntheses. Polymeric Isocyanates may remain in the grease and cause severe toxicitiy, especially if the greases are up -heated.

PU Greases are very sensitive toward ingress of OH – groups (e.g. alkalines, water, polygly-coles) as the nitrogen-hydrogen bridging is disturbed. Ingress of such pollutants may cause a change in consistency. Polyglycoles, if heated emit aldehydes that interfere with the NH groups in PU greases. This reaction may end up in making the solid PU liquid! PU greases thus should be monitored to those facts (Figure 72):

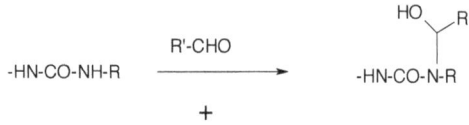

$$
\text{-HN-CO-NH-R} \quad \xrightarrow{\text{R'-CHO}} \quad
\begin{array}{c}
\text{HO} \diagdown \quad \diagup \text{R'} \\
| \\
\text{-HN-CO-N-R}
\end{array}
$$

+

Figure 72. Reaction of PU Grease and Aldehydes

Other incompatibilities of PU Greases arise from mixtures with clay thickeners due to the presence of either OH (Si-OH) or NH functional groups if the clay is modified by organic amines.

12.4. Other thickeners

12.4.1. Clay greases-structure and use

Clay Thickeners derive from Alumina-Silicates. Due to their high surface and modification they are suitable for gelling base oils, e.g. Esters, Napthenic Base Oils, sometimes Silicones and Phosphoric Acid Esters. Clay Structure is generated by tetrahedral arrangement of Silica with insertion of alumina (see figure) in layers of approximately 1-2 nm distance. Water and other cations may be inserted in the space in between the two layers. Other cations, e.g. magnesium, may also be inserted in between. [3] (Figure 73).

Gelling takes place by adhesion and insertion of organic molecules in the structure, assisted by polar additives like propylene carbonate. Clay grease is produced by multiple milling the clay with appropriate base oil by addition of water suppliers like glycerol or Propylene carbonate at temperatures below 100°C. If water is lost the structure may break down during the manufacturing. Doing so, the grease produced is a buttery solid with no dropping point.

12.4.2. Use of clay greases

Clay greases are used for applications where the grease should not move out and for special high temperature applications, e.g. cement industry in slow motion bearings. Due to the inertness of the inorganic structure toward alkaline and acids, clay greases are preferred in applications where water, alkaline and acids are present, e.g. chain or bearing lubrication with such ingress. Clay is declared as safe for incidental food contact and allowed for lubricants in food industry (USDA H1 regulated) in junction with base oils like white mineral oil, PAO or esters that are allowed for this purpose.

12.4.3. Restrictions in the use of clay greases

Restrictions for the use of clay greases are the presence of Lithium, - Calcium or Polyurea Greases that may interfere with the hydrogen bonding of the clay structure. Mixtures of clay

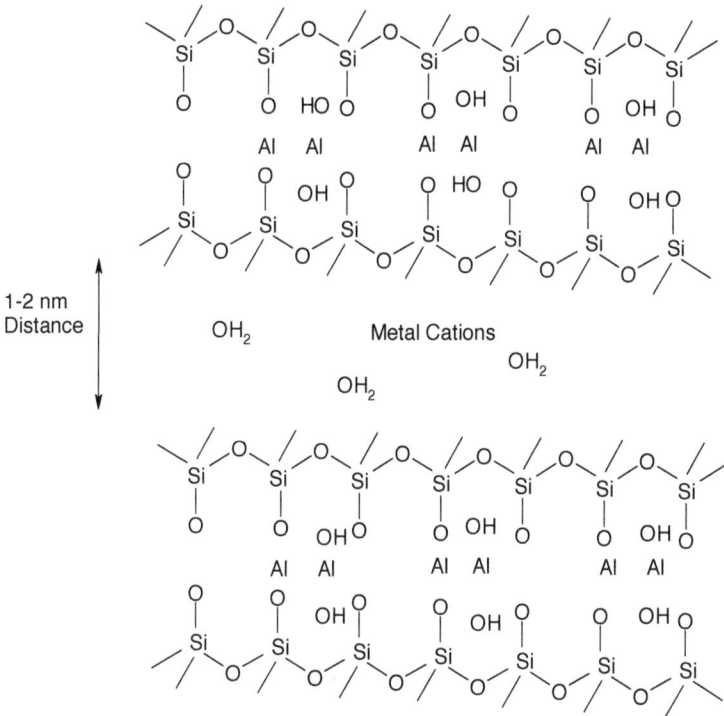

Figure 73. Estimated basic structure of clay

and conventional greases should be evaluated very carefully. Clay greases are restricted in bearing lubrication strictly due to over rolling speed. In general the speed factor is limited to ndm (Average of outer and inner diameter of the bearing times the speed (revolution per minute)) of 100.000. Only slow moving bearings could bear clay lubrication.

13. Silica

Silica is in use for thickeners as amorphous material, obtained by flame decomposition of Silica Tetrachloride (Figure 74):

Silica, due to its powerful surface activity may be used as powerful thickener in low percentage for each kind of base oil. Greases obtained by mixing silica with base oils are transparent. The inorganic structure causes no dropping point for such greases. Silica Thickened greases cause steep and irreversible thickening by heating up due to the increase of internal hydrogen bonding. They never should be in use for high speed and high temperature rotating bearings,

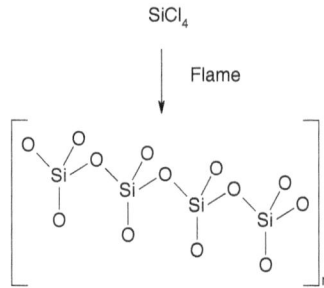

Figure 74. Principal formation of amorphous SiO2 by flame combustion

since they block their motion. The ndm (Average of Bearing Size times revolution per minute) is restricted to 100.000, hence slow motion. Due to the possible entrance of water, silica thickened grease is poorly water stable and should not be in use in applications where water (especially hot water) and alkalines are present. Alkalines react with silica to silicates, starting its degradation.

14. Polytetrafluoroethylene (PTFE)

PTFE is a convenient thickener in base oils for the purpose of incidental food contact, low friction properties and high temperature. The fluorine entity causes low activity toward oxygen. PTFE Grease is used in oxygen application (valves under oxygen impact), especially with PFPE.

15. Conclusion

Tribology is highly guided by physics and chemistry of the lubricants. Functionality of lubricants is given by their physics and their chemical structure. Modern understanding of lubrication hence allows the construction of lubricants appropriate to a given application to a certain extent. Under the conditions of full lubrication their physical properties, e.g. viscosity, viscosity-temperature and viscosity –pressure properties dominate over the chemical structure. Under such circumstances, the lubricant takes away heat (cooling function) from the mating contacts, but also wear and debris (cleaning function). Within a running – in period some reaction layers of lubricant constituents (additives) may be created. Basically those layers stay constant over time and do not change. On the other hand, if lubrication undermines the given roughness's of the mating partners, or overtakes the natural temperature limit given by the restrictions of organic chemistry (e.g. temperatures beyond 150°C), chemistry starts to perform reaction scenario highly related to the nature of the chemical structure of the ingredients in the lubricant. The basic reactions found here are radical reactions, as a fact of the presence of oxygen and iron. Within such radical reaction sequences hydrogen is abstracted, alkenes and alkynes are formed and their oxidation products (aldehydes, ketones, carboxylic acids and their derivatives). Additives, in general improve the lubricants by expanding their limits.

In general, lubrication fundamentals in tribology have overcome the alchemy of the past by numerous efforts taken by the scientific community.

Author details

Walter Holweger*

Schaeffler Technologies AG & Co.KG, R&D Central Materials, Germany

References

[1] Rudnik L.R., editor. Synthetics, Mineral Oils, and Bio-Based Lubricants. Boca Raton: CRC Press; 2005.

[2] Dresel W., Mang T., editors: Lubricants and Lubrication. 2nd Edition. Weinheim: Wiley-VCH; 2007.

[3] Klamann D. Schmierstoff und verwandte Produkte. Weinheim: VCH-Verlag; 1982.

[4] Mortier R.M., Fox M.F., Orszullik T.M., editors. Chemistry and Technology of Lubricants   Dordrecht: Springer;   2010.   http://link.springer.com/book/ 10.1007/978-1-4020-8662-5/page/1 (accessed 27 December 2012).

[5] Dowson D., Taylor C., Childs T., Dalmaz G. editors. Lubricants and Lubrication. In: Tribology Series 30 : Proceedings of the 21st Leeds-Lyon Symposium on Tribology. Amsterdam : Elsevier; 1995.

[6] Bloch, H.P., Practical Lubrication for Industrial Facilities. Lilburn: Fairmont Press; 2000.

[7] Stepina V., Vesely V. Lubricants and Special Fluids. Amsterdam: Elsevier; 1992.

[8] Lansdown A.R., Lubrication and lubricant selection: a practical guide. 3rd Edition. John Wiley & Sons; 2004.

[9] Rudnick L. R., editor. Lubricant Additives: Chemistry and Applications, 1st Edition. New York:Marcel Dekker, 2003.

[10] March, J., Advanced Organic Chemistry: Reactions, mechanisms, Structure. New York: Wiley-VCH; 1992.

Chapter 3

Some Aspects of Grease Flow in Lubrication Systems and Friction Nodes

Maciej Paszkowski

Additional information is available at the end of the chapter

1. Introduction

An optimally designed lubrication system should reliably distribute a lubricant to particular reception points. The distribution must be precise and preferably fully automated. It is particularly difficult to design such a system if it is to supply a lubricant to heavily loaded machines, featuring a considerable (even up to a few hundred) number of different kinds of friction nodes distributed in a non-linear way, with long distances between each other. An additional problem might be hard conditions of the environment in which the system is expected to work (high or low temperature, high air humidity, etc.). All the conditions make the task of the designing of a reliable central lubrication system a real challenge, even for an experienced designer specialising in that particular field. While building a lubrication system it is important to define fundamental parameters which will determine its reliable operation. One of the crucial things is to select an appropriate grease that ought to reduce friction resistance and wear of the friction nodes, to protect them against the influence of the environment, as well as to guarantee the lowest possible flow resistance during its distribution (mainly through the lubricating conduits). An inappropriately selected grease, in terms of its rheological and tribological properties, and also the dynamically changing working conditions, may lead to a considerable increase in the opearting costs of the machine.

The chapter contains the most important information on the structure of greases along with the discussion of the influence of the thickener's microstructure on the behaviour of the lubrication formula mainly in lubrication systems, but also in reception points, namely in the friction nodes (including roller bearings). It also refers to the problems of the influence of the mechanical stability of a thickener's microstructure on the quality of lubricating the roller bearings and their service life. Additionally, the chapter presents works on generating the lubrication film on the friction nodes working surface. Also, the fundamental problems connected with the grease rheology (the grease performance in the fixed flow rate conditions and the notion of the linear and the non-linear viscoelasticity) have been discussed in the text. The state of the art knowledge on the fundamental phenomena observed in the lubrication systems, including the mechanisms of the thixotropic changes in the grease microstructure at shearing and relaxation, as well as the forming of the boundary layer at the grease flowing through the conduits in the main line of the lubrication system have been presented.

2. Microstructure of the lubricating greases

Lubricating greases are rheologically complex two-phase non-Newtonian fluids. They are chemically and physically heterogeneous. The dispersive phase is normally a mineral oil, a synthetic oil or a vegetable oil, whereas the dispersed phase is a thickener and, depending on the needs, solid additives. The particles of the thickeners can vary in their dimensions. Soaps, for instance, do not normally exceed 100 μm in length, and their diameter is not shorter than 0.1 μm and not longer than 0.5 μm [1-3]. The lithium and calcium soaps particles are definitely bigger than the sodium soaps particles. The isometric aggregates of bentonite clay and mica are approximately 0.5 μm in width and 0.1 μm in thickness [1]. The solid additives have similar dimensions. Due to the size of the thickener particles, the greases acquire the characteristics of the mechanically dispersed (suspension) or the colloidal system. The thickener particles size depends primarily on the process of the grease production [4,5], as well as on the conditions in the friction node. During the shearing of a grease in the friction node, the size of the particles can change considerably. According to Boner [5], in order to achieve the optimum tribological and rheological properties of a lubricating compound, greases containing variously dispersed thickener particles ought to be mixed. It concerns particularly the lubricants thickened with metal soaps. Too high particles dispersion of the thickener in the grease can negatively influence its lubricating properties. Such particles do not present enough capability of making spatial, three-dimensional structures resulting from the physicochemical interactions. Too long particles of the thickener cause too high an increase of the consistency of the lubricating grease and lead to its easy breaking both in the lubrication systems and the tribological pairs.

Microstructure of the lubricating greases with soap thickeners can be compared to a sponge with a lubricating oil (Figure 1). It makes a three-dimensional, coherent network of interconnected particles (flow units), in the literature referred to as *floccules*. The oil is locked in the free spaces of the microstructure through the mechanical occlusion, the capillary phenomena as well as the molecular attraction between the thickener and the polar components of oil [6]. It is estimated that the amount of the oil locked in the microstructure of the soap greases can amount even to 75%. The soap particles making the microstructure, from the chemical viewpoint, are associated molecules [7], namely groups of identical molecules generated as a result of the dipol-dipol type of interaction or the hydrogen bonds [8]. The final skeleton of the microstructure – shaped in the dispersion center – is made in situ in the process of crystallization of the soap particles and/or through nucleation, namely making crystallite nucleuses and their further growth [5]. The shape of the thickener particles and their surface topography can be different, depending on the kind of soap used and the particles' size. If they exceed the size of 1 μm, their structure is rough and resembles twisted ropes. The crystallites of the colloidal size are definitely smoother and less twisted [3]. The microscope photographs also show a clear difference in the surface structure of the lithium, sodium and calcium thickener floccules [7,9]. The calcium soap particles are rougher than the particles of the lithium and sodium soaps, independently of their length [5]. In the case of the non-organic thickeners, the grease microstructure is made of numerous individual aggregates popularly called *open card-house* [10], for instance bentonite (Figure 2), mica or vermiculite. Such particles have a skeleton structure which resembles heterogeneous, fuzzy-edged, curled flakes or straight plates piled one on the top of the other [11].

The shape of the soap particles and other organic or non-organic thickeners, their anisometry (in reference to their length and lateral dimensions), as well as their dispersivity and the percentage in the full volume of the grease, greatly influence the physical properties of the ultimate lubricating compound [7,12-14]. The last two factors are critical for the easiness of the making of different energy connection types between the elements of the microstructure. Apart from providing the appropriate consistency, thickeners influence the way the lubricating grease flows, changes its shape or the type of flow resistance (pumpability) it presents. It is particularly important in the central lubrication systems where the lubricant is often transported in long conduits to particular receiving points.

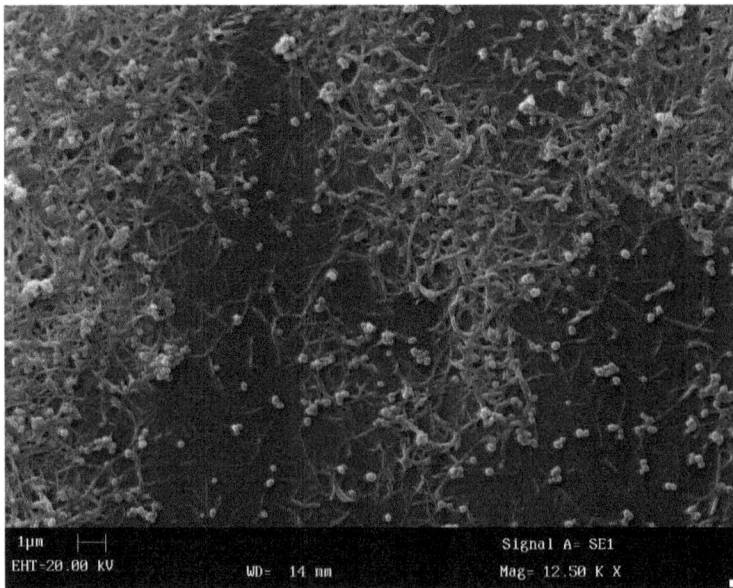

Figure 1. Floccules of the lithium thickener with particles of polytetrafluoroethylene (PTFE) improving the tribological properties of the lubricant. The microphotograph was taken by means of a scanning electron microscope (SEM) in secondary electron imaging (SEI) mode at the magnification of 12 500 times.

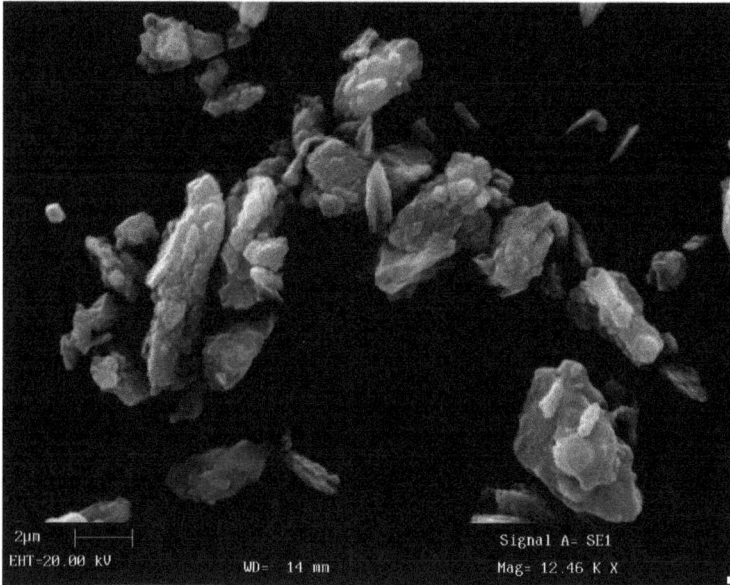

Figure 2. Bentonite clay aggregates making the *open card-house* structure.The microphotograph was taken by the scanning electron microscope (SEM) in secondary electron imaging (SEI) mode at the magnification of 12 000 times.

The mechanical durability and the stability of the microstructure of the thickener in a lubricating grease determines the friction reduction, the protection of the lubricated surfaces as well as the grease performance during the mechanical loading in the friction node and work at ultimately high temperatures. Cann [15] analyzed thin layers of the lithium lubricants without the solid additives on treadmill roller bearings with the use of attenuated total reflectance – Fourier transform infrared (ATR-FTIR) spectroscopy. The research showed that the oil gets out of the lubricated roller bearing as a result of damage in the thickener's microstructure. In the case of the untouched microstructure, the oil is connected with the thickener by means of the capillary forces. When the microstructure is damaged, it gets out of the bearing, washing out individual soap crystallites. Such an oil undergoes oxidation and evaporation easier, which lowers its lubrication quality. Additionally, its rheological properties are much worse than the properties of the base oil, and its kinematic viscosity is significantly reduced [16]. Similar conclusions were drawn by Farcas and Gafitanu [17]. Based on experimental research, Farcas and Gafitanu presented a mathematical model correlating the degree of the thickener microstructure destruction with the service life of the roller bearing. Their SEM photographs of the lubricating greases subjected to shearing in the roller bearing showed a damaged microstructure of the grease with the base oil bled out of it. According to the researchers, a crucial role in the process of the microstructure destruction is played by temperature. Above 60-70 °C, along with its increase by each 10-15 °C, the service life of the bearing falls by half. Above 70 °C, the bearing's failures are most frequently caused not by the contact fatigue, but predominantly by the degradation of the thickener's microstructure and worsening of the grease's properties. Vinogradov et al. proved that interactions between the thickener particles and the microstructure state may influence the increase of the friction moment in the roller bearing [18].

The thickener in the lubricating grease is responsible also for the thickness of the lubrication film on the working surfaces of the friction joints. Spikes, Cann, Wiliamson and Kendal in their works [19,20] presented results of the research on the interaction of the thickener's particles with the treadmill sprinkled with titanium oxide. They observed an increase in the thickness of the lubricating film in the elastohydrodynamic conditions of the collaboration of a steel ball with the treadmill. The measurement was done by means of Alström's method. In the case of the lubricants thickened with metal soaps, the biggest capability of making the lubrication film was recorded in the lithium grease. The thickness of the generated grease layer, apart from the kind of thickener, was also determined by the flow speed of the lubricant between the treadmill of the roller bearing and the rolling element. The thickness of the observed thickener's layer was 100 up to approximately 250 nm, depending on the kind of the lubricating grease. Long-lasting tests showed that the lubrication film's thickness generated on the bearing's treadmill, at a constant load, falls with time up to the stabilization at the level from 20 to 50 nm.

Apart from the thickener, the lubricating grease microstructure can also contain solid additives. Their main function is to improve the tribological properties of the lubricant. The solid additives are normally graphite, molybdenum disulfide (MoS_2), polytetrafluoroethylene (PTFE) as well as metal powders. The percentage of the solid additives in the grease usually does not exceed 5%. Research results indicate the benefits of the increase of the percentage of the solid additives in the lubricating greases. Of course, there are also works which show the advantages of the synergism of the powdered PTFE as well as tin and copper in the lubricating compounds [21,22].

Solid additives significantly influence the rheological properties of the lubrication compound [23,24]. The lubricants thickened with the polar lithium soap without the solid additives show quite high values of the structural viscosity and of the shear stress. After the enriching the lubricants with the graphite thickeners and MoS_2 the lowering of the parameters' values occurs. The additives reduce the shear stress both in the bulk of the grease as well as in the boundary layer. Research show that PTFE does not influence the change of the values of the parameters in the lithium soap. In the case of the lubricants thickened with the bentonite clay, solid additives cause the increase of the yield stress both close to the wall and in the bulk of the grease. The most influential determinant of the values increase is the graphite thickener. Adding the PTFE in this case does not significantly determine the structural changes thus generated lubrication compound.

3. The basic rheological properties of the lubricating greases

3.1. Behaviour of the lubricating greases in the steady flow conditions

Greases, depending on their characteristic structure, are shear-thinning (pseudoplastic) and rheounstable fluids having a yield point. They show very complex rheological properties.

Lubricating greases, at very low shear rates behave like a Newtonian fluid of a constant viscosity η_0, being a slope of a tangent to the flow curve, equal to the shear rate going to zero ($\eta_0 = \lim_{\dot\gamma \to 0} \eta(\dot\gamma)$). At the shear rate going to infinity, lubricating greases also behave like Newtonian fluids, but of viscosity η_∞ close to the dispersion phase, namely of the base oil ($\eta_\infty = \lim_{\dot\gamma \to \infty} \eta(\dot\gamma)$). The viscosity η_∞ is a slope of a straight line being a flow curve asymptote. In the bracket of the indirect shear rates, the curve of the lubricating grease flow curve is described with the following equation:

$$\eta = \frac{\tau}{\dot\gamma}, \tag{1}$$

where: η - dynamic viscosity, τ - shear stress, $\dot\gamma$ - shear rate.

Figure 3 illustrates results of the rheological measurements made on the lithium greases of a various thickener percentage, showing the shear stress changes and of the structural viscosity as a function of the shear rate. Additionally, the theoretical curves generated from the Herschel-Bulkley's [25] and Carreau-Yasuda's [26] dependences have been presented. Values of the parameters defining the theoretical curves traced according to the abovementioned dependences for the greases of a various thickener percentage have been presented in Table 1. The greases were thickened with lithium 12-hydroxystearate, and the oil base was ORLEN OIL SN-400 (ORLEN OIL, Cracow, Poland) mineral oil. The flow curves were defined at temperature of 25 °C. The tests were repeated five times, and their results were statistically elaborated at the confidence level $p = 0.95$, using t-Student test. Additionally, correlation coefficient values were calculated. The experiment was carried out by means of Physica Anton-Paar MCR 101 rotational rheometer. The rheometer was working in the cone-and-plate system (CP-25-1, 25 mm, 1°) at a constant measuring gap height of 49 μm.

Thickener content (%)	Herschel-Bulkley model				Carreau-Yasuda model					
	τ (Pa)	k_c (Pas)	m	R_{xy}	$\eta_0 - \eta_\infty$ (Pas)	η_∞ (Pas)	λ (s)	a	n	R_{xy}
4.0	120.27	5.826	0.66964	0.99983	1852.4	0.38843	47.456	2	0.20073	0.99600
5.0	190.52	83.155	0.37458	0.99958	4189.5	0.20014	47.457	2	0.23465	0.99668
6.0	216.00	83.867	0.38633	0.99973	4742.5	0.26093	47.939	2	0.22985	0.99811
6.5	241.47	212.430	0.29112	0.99982	9090.5	0.18576	47.940	2	0.20100	0.99919
7.0	672.60	148.930	0.35189	0.99750	15225.0	0.26328	47.455	2	0.17468	0.99696
8.0	712.54	285.550	0.28687	0.99592	21453.0	0.25487	47.423	2	0.16048	0.99944
9.0	785.29	230.630	0.37532	0.99956	16023.0	0.54032	47.458	2	0.21285	0.99743

Table 1. Values of the parameters determining the theoretical curves traces, defined according to dependences (1) and (2) for the greases of the thickener content 4-9%.

a)

b)

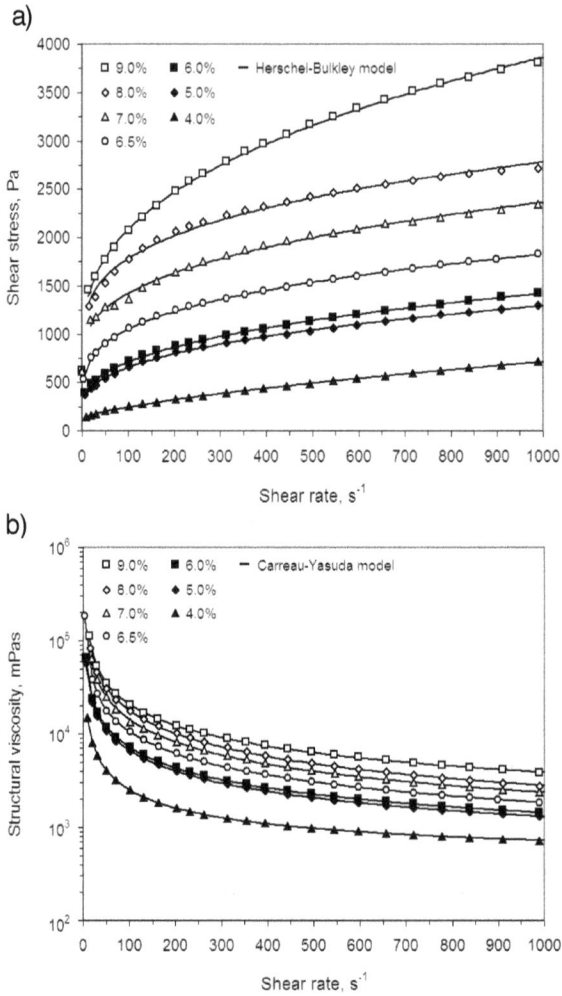

Figure 3. Dependences of the shear stress (a) and the structural viscosity (b) in a function of the shear rate for the lithium greases based on the mineral oil, with a various thickener percentage. The diagrams, in order to achieve a greater clarity, do not contain the confidence intervals.

The Herschel-Bulkley (2) dependence takes into account the existence of a yield stress τ_0 value, namely such a value of the shear stress above which the lubricating grease starts flowing:

$$\tau = \tau_0 + (k_C \cdot \dot{\gamma})^{\frac{1}{m}}, \qquad (2)$$

where: τ_0 –yield stress, k_C – consistency factor of the thickener in the grease, m – nondimensional index exponent. Carreau-Yasuda (3) dependence enables approximation of the course of changes in the structural viscosity as a function of the shear rate with viscosity η_∞ and η_0 included:

$$\eta(\dot{\gamma}) = \eta_\infty + (\eta_0 - \eta_\infty) \cdot [1 + (\lambda \cdot \dot{\gamma})^a]^{\frac{n-1}{a}}, \tag{3}$$

where: $\eta_\infty = \lim_{\dot{\gamma} \to \infty} \eta(\dot{\gamma})$, $\eta_0 = \lim_{\dot{\gamma} \to 0} \eta(\dot{\gamma})$, λ – time-constant, a and n – nondimensionalparameters. For the shear-thinning fluids, a parameter normally equals 2 [27].

The flow curves clearly illustrate the fact that the greases have the pseudoplastic properties. It results from the degradation of the lubricating greases' microstructure as well as from the further orientation of the dispersed particles of the thickener. During relaxation of the stress in the lubricating grease, a considerable entanglement of the thickener particles is observed. However, during shearing of the grease, one can record straightening and untangling of the particles which are directed along the current line. With the increase of the shear rate, the effect is more and more visible. The reduction of the internal friction resulting from a smaller size of the particles and a limited activity between them is observed. At very high shear rates, the total orientation of the particles in the grease is achieved. The internal friction remains constant at a low level. The structural viscosity of the grease then gets close to the viscosity of the base oil.

Figure 3 shows that at a large enough percentage of the dispersion, the interaction between the thickener particles may cause formation of the spatial microstructure resistant to the shear stress which does not exceed certain value. Below the value, the lubricating grease behaves like an elastic solid. When yield stress τ_0 (yield point) is reached, the microstructure gets damaged and the grease starts behaving as a viscous liquid. When the shear stress in lubricating grease becomes lower than the limit value, the immediate reconstruction of the microstructure takes place. The determination of the yield stress value in the lubricating grease is very important because it enables the determining of its usefulness for the application in the central lubrication systems of machines [9,28] or in the friction nodes [7].

3.2. Linear and non-linear viscoelasticity

When a lubricating grease undergoes the periodic stress oscillation, its response in a form of a relative deformation will be shifted in the phase by δ angle. The angle, called the phase angle shift, falls in the range from $0°$ to $90°$. The response of the lubricating grease determined by the extortion is influenced primarily by: complex modulus $|G^*|$ and complex viscosity $|\eta^*|$. The complex modulus is defined as a sum of the real and the imaginary part:

$$|G^*| = G' + iG'', \tag{4}$$

where: G' - storage modulus (elastic component in the phase with the stress), G'' - loss modulus (viscous component shifted in the phase in relation to the stress). Components G', G'' are equal:

$$G' = \frac{\sigma_0}{\gamma_0}\cos\delta, \tag{5}$$

$$G'' = \frac{\sigma_0}{\gamma_0}\sin\delta. \tag{6}$$

Storage modulus G' is a measure of the energy stored and recovered per each deformation cycle, whereas loss modulus G'' is a measure of the energy lost per a sinusoidal deformation cycle. What results from equations (5) and (6) is that phase angle shift δ can be defined as:

$$\tan\delta = \frac{G''}{G'}. \tag{7}$$

The dispersion effects can also be described by means of complex viscosity η^*. The complex viscosity, similarly to the complex modulus can be formulated as:

$$\left|\eta^*\right| = \eta' - i\eta'', \tag{8}$$

where:

$$\eta' = \frac{G_0}{\gamma_0\omega}\sin\delta, \tag{9}$$

$$\eta'' = \frac{G_0}{\gamma_0\omega}\cos\delta. \tag{10}$$

The real part of complex viscosity η' is called the viscous component or, less frequently, the dynamic viscosity, whereas imaginary part η'' is called the elastic component. The measurement of quantities G', G'' and η', η'' enables the rheological characteristics of the greases in terms of their viscoelastic properties.

Figure 4 illustrates example results of a strain sweep test of Nanolubricant 2010 (Orapi, Saint-Vulbas, France) commercialgrease, thickened with the lithium 12-hydroxystearate, based on a synthetic oil with the addition of tungsten disulfide nanoparticles. The deformation range of the grease was 0 to 100% at 1 Hz oscillation frequency. The measurements were done at 25 °C. The oscillation measurements made it possible to determine the limit of the linear viscoelasticity (marked in the diagram with the vertical line). The reasearch showed that plateau modulus G^0_N for Nanolubricant 2010 was 75 743 Pa at deformation γ_C equal 0.336%. The review of the methods used for the determining of the linear viscoelasticity limit (the critical point) on the basis of the dynamic-oscillation tests is widely discussed in [29].

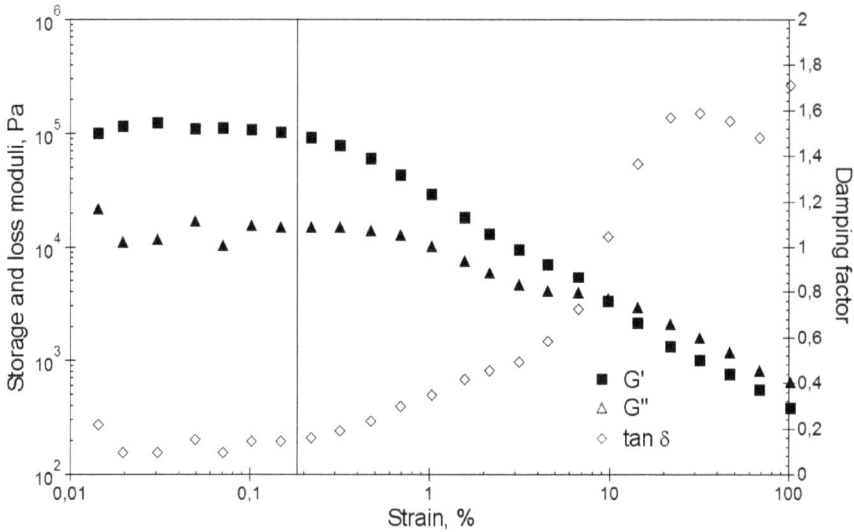

Figure 4. The storage and loss moduli and the phase angle shiftas a function ofNanolubricant 2010 grease strain.

Figure 5 illustrates the influence of the soap thickener (lithium 12-hydroxystearate) percentage in the lubricating grease based on the mineral oil on (a) value of critical strain γ_C and (b) of shear stress τ in the critical point, as a function of the percentage of the thickener. The diagrams show that the range of the linear viscoelasticity depends on the percentage of the thickener in the grease – the smaller percentage of the thickener, the wider range of the linear viscoelasticity. In the case of the shear stress in the critical point, there is an adverse situation. The research was carried out by means of a stress/strain controlled rotational rheometerPhysica Anton-Paar MCR 101. The rheometer was working in the cone-and-plate system (CP-25-1, 25 mm, 1°) at a constant measuring gap of 49 μm.

The rheological dynamic-oscillation research (which is a combination of the creep and relaxation experiments) are currently the basic tool at evaluating the structure of dispersion and its mechanical stability as well as the behaviour of the lubricating grease in the start phase of the flow. At high frequencies and amplitudes, the dynamic-oscillation tests can be used for the determining of the speed at which the microstructure is created and damaged under the

a)

b)

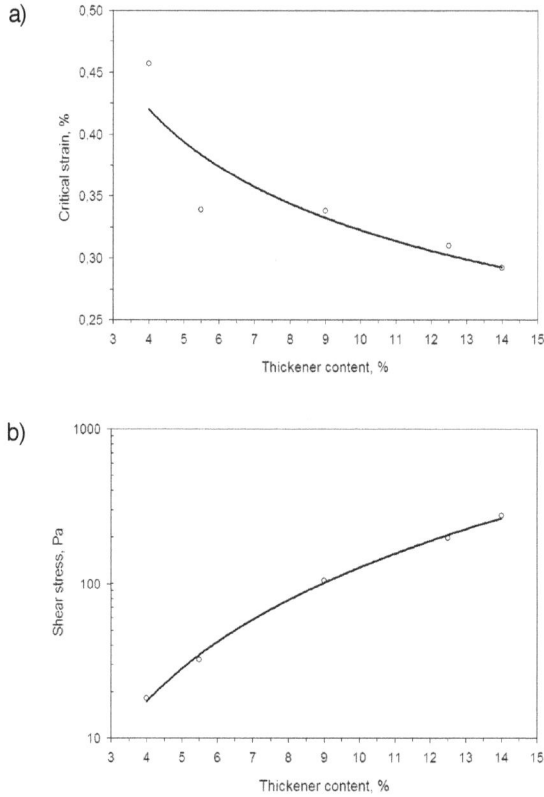

Figure 5. Critical strain γ_c (a) and shear stress τ in the critical point (b),as a function of the percentage of the lithium thickener in the lubricating grease

influence of oscillation. Carrying out such research is important for the evaluation of the lubricating greases' behaviour at the start-up phase of the friction nodes and of the flow in the lubrication systems conduits. The dynamic-oscillation research on the lubricating greases has been, among others, conducted by Yeong et al. [14] as well as Delgado [12], Martín-Alfonso et al. [30,31].

4. Mechanism of thixotropic changes in the microstructure of the lubricating grease during its shearing and relaxation

The term 'thixotropy' comes from the Greek words *thixis* (mix, shake) and *trepo* (revolve, change), and it was first suggested by Péterfi [32] (the original name – *thixitropy)* [33]. The phenomenon is defined as an isothermal decrease in structural viscosity during shearing followed by an increase in the viscosity and the re-solidification of the substance once shearing ends [34]. The phenomenon of thixotropy concerns suspensions of the internal microstructure made of particles which undergo aggregation or flocculation. In such systems, there are physical interactions between particles, which, during relaxation of a substance, determine the creation of a spatial, cross-linked microstructure. They are primarily the Van der Waals-London attractive forces, the electrostatic repulsive forces (particularly important when the dispersed phase particles are anisometric in shape) as well as the steric forces. Also, the Brownian motion may have an important part to play in the process, but their appearance is limited only to colloids. A wide description of the intermolecular interactions which are common in thixotropic systems can be found in the publications of Efremov et al. [35,36] as well as of Sonntag's [37] and Scheludko's [38]. During shearing, the microstructure made of the dispersed phase particles, as a consequence of their mutual interaction, gets disintegrated forming bigger aggregates and individual particles, floccules, suspended freely in the dispersion medium. The review of the references concerning the thixotropy of suspensions can be found, for instance, in the publications of Mewis and Barnes [10,39].

The lubricating greases also show thixotropy. The structural viscosity of the lubricating grease decreases systematically during its shearing in the tribological node or during its flow through lubrication system conduits. The standstill of the grease leads to its re-solidification and a little increase in its structural viscosity. The decrease of viscosity during flowing of the grease is reduced to a certain minimum value, primarily depending on the viscosity of the base oil, a kind and amount of the thickener as well as the shear rate. At high shear rates, the structural viscosity of the grease reaches the value close to the base oil viscosity. According to Czarny [1] "Both in the slide bearings and in the roller bearings, the high-speed ones in particular, the rheological properties of the lubricant, so its hydrodynamic lift force, are determined mainly by the base oil. The thickener influences the force at the start-up and at the stopping of the bearing.". Due to the complexity of the processes of degradation and reconstruction of the thickener's microstructure in the lubricating greases, scientific works which would thoroughly explain the processes have not been published so far.

The first publications on thixotropy of the lubricating greases appeared at the beginning of the 1950s. In 1951, Moore and Cravathpusblished their work [13] on the mechanical process of the soap fibre disruption in the lubricating greases caused by the shearing force and on the influence of the disruption on the change of the consistency of the greases. The research concerned the lubricants thickened with the barium, sodium, lithium and calcium soaps. The degradation of the soap fibre consisted in the working of the samples in a penetration apparatus, as well as on their long-lasting disintegration in a rolling apparatus. During shearing, at different intervals, the measuring of the greases micropenetration level was performed. The tests showed that the biggest differences in the consistency of the greases were observed in the commercial grease thickened with the barium soap, whereas the smallest differences were recorded for the lithium grease. Additionally, a series of photographs of single soap fibres by means of an electron microscope was taken. The photographs showed a considerable difference in the fibre length before and after shearing for most of the tested samples. Moore and

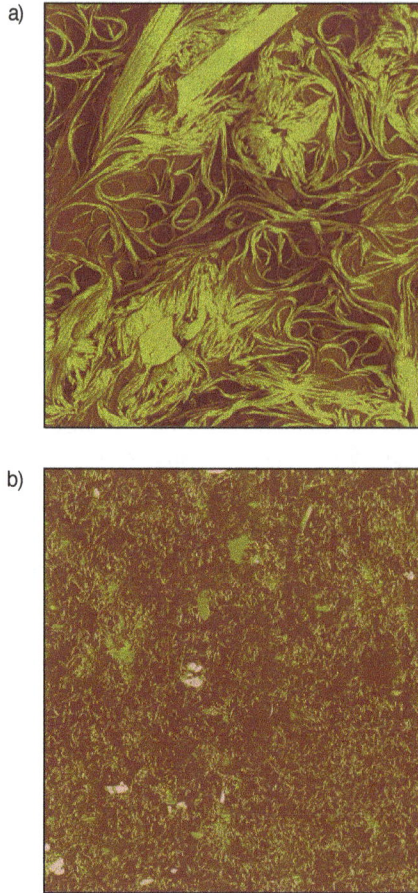

Figure 6. Microstructure of the lithium thickener in the paraffin oil [40]: (a) fresh grease, (b) grease sheared at 120 °C for 10 hours at $\dot{\gamma}$ = 0.01 s^{-1} shear rate. The microphotograph taken by means of the atomic force microscope (AFM). Scanning area 20 μm x 20 μm.

Cravath proposed a soap fibre disruption intensity model at different times of the external shearing force activity.

A similar analysis to the one conducted by Moore and Cravath was carried out by Renshaw [41]. He was particularly interested in the lithium thickened greases based on the ester oil, which showed high shear resistance. The measuring of the micropenetration was done directly before and after the 9-hour long shearing of the grease in a bearing at the speed of 10 000 rev/min and after 100 000 double strokes of the piston in the penetration apparatus. In the first case, Renshaw observed in the electron microscope, a considerable shortening of the lithium thickener's fibre and slight changes in the micropenetration values. After the working of the grease in the penetration apparatus, the situation was exactly reverse, namely the changes in the micropenetration were remarkable but the particles were shortened only to a minor extent. The research showed that the intensity of degradation in the thickener's microstructure depends not only on the mechanical strength of individual fibres, but mainly on the shearing way and on the direction of the fibres at the flow of the grease.

Sirianni et al. [42] investigated the question of the influence of size of the calcium and barium soaps and their concentration on the change of the lubricating greases *thixotropy coefficient*. For the experiment, they used a theory proposed earlier by Goodeveand Whitfield [43] which was aimed at the determination of the thixotropy of the black suspension in the mineral oil [44]. According to the theory, the speed of the microstructure reconstruction (proportional to the thickener's concentration) equals the speed of its degradation. The thickener's concentration, however, is proportional to the difference between apparent viscosity η and grease equilibrium viscosity η_0 following the total degradation of the cross-linked microstructure of the thickener at the infinitely high shear rate.

$$\eta - \eta_0 = \frac{\theta}{\dot{\gamma}} = \frac{F_T}{\dot{\gamma}}, \tag{11}$$

where: η – apparent viscosity, η_0 – equilibrium viscosity, $\dot{\gamma}$ – shear rate, θ – thixotropy coefficient, F_T – momentum generated by the intermolecular forces transmitted by the unit of surface in the unit of time.

The theory assumes that at the increase of $1/\dot{\gamma}$ expression, apparent viscosity η rises until it reaches the plateau. According to Goodeve and Whitfield, the thixotropy coefficient is a source of the non-Newtonian effects, which is directly correlated with the intermolecular bonds disruption effects, the average life span of the bonds and the change of size of the interacting molecules in the thixotropic process. In order to calculate the thixotropy coefficient, Goodeve assumed that the bonds between the molecules of the dispersed phase are subject to Hooke's law.

Moses and Puddington [45], using the same theory, determined the influence of the thickener's percentage (10%, 14% and 19.7%), the shear rate (from 1 400 s^{-1} to 250 000 s^{-1}), temperature (from 35 °C up to the value at which a visible sedimentation occurs) as well as the size-reduction of the thickener particles, on the change of the *thixotropy coefficient* of the lithium, sodium and aluminium greases. Different sizes of the soap fibre were generated by means of the colloid mill. For determining equilibrium viscosity η_0 a metal capillary was used.

Hahn, Ree and Eyring [46], while observing behaviour of the macromolecular polymers suspensions, formulated a theory (later called *sinh law*) which ultimately was widely used for the quantity description of the phenomena of the lubricating greases microstructure degradation and reconstruction, where the soap is a thickener. They took into account the fact that the thickener's structure contains the so called entangled particles (aggregates which create a three-dimensional network and make the solution acquire the non-Newtonian fluid properties) and the not entangled particles (single floccules which make the solution acquire the Newtonian fluid properties). The number of the former and the latter depends on the temporary shear stress. The flow curve equation used by Hahn et al. was:

$$\tau = \frac{\chi_1 \beta_1}{\alpha_1}\dot{\gamma} + \frac{\chi_2}{\alpha_2}\sinh^{-1}(\beta_2\dot{\gamma}), \tag{12}$$

where: χ_1 – share of the not entangled particles, χ_2– share of the entangled particles, $1/\alpha_1$ – the not entangled particles shear stress, $1/\alpha_2$ – the entangled particles shear stress, $1/\beta_1$ – shear intensity for the not entangled particles, $1/\beta_2$ – shear intensity for the entangled particles, $\dot{\gamma}$– shear rate.

Hahn and the collaborators assumed that at the increase, and next at the decrease of the shear rate, the free energy of the soap particles changes, and it is directly dependent on their strain energy. The cross-linked particles aggregates featuring the non-Newtonian characteristics get untangled making individual floccules, freely suspended in the dispersion medium, featuring the Newtonian characteristics. The process is reversible. Two years later, the scientists presented the equation of the changes kinetics [47]:

$$-\frac{d\chi_2}{dt} = \chi_2 k_f e^{\frac{\eta_0 w^2}{kT}} - \chi_1 k_b e^{\frac{-(1-\eta_0)(w')^2}{kT}}, \tag{13}$$

where: k_f – parameter determining the time of the entangled particles $\rightarrow$ not entangled particles reaction; k_b – parameter determining the time of the not entangled $\rightarrow$ entangled particles; w – elastic strain energy of the entangled particles, w' – elastic strain energy of the not entangled particles, η_0 – viscosity of the entire thixotropy system at zero shear rate, k – Boltzman constant, T – temperature.

Thus, Hahn et al. elaborated theoretical flow curves for the lubricating greases during the disruption and reinforcenment of the thickener's structure. The models have been widely applied for the determination of the thixotropic properties of, for instance, motor oils [48]. Validity of the assumptions made at the making of the mathematical dependences was proven by the experimentally developed curves. The mathematical model proposed by Hahn was also used by Utsugi et al. [49]. They analyzed the thixotropy of the calcium greases and the silica gel thickened greases. At the determining of the thixotropic hysteresis loop, Utsugi used the rotational rheometer. Bair [50] proved that the models proposed in the 1960s by Hahn and Eyring were still true. During the verification of the models, Bair used more technologically advanced rotational rheometers. He drew particular attention to the possibility of the dependences application for describing the thixotropy phenomenon of the lubricants working at the friction contact, in the elastohydrodynamic conditions.

Vinogradov, Deinega and Verbitsky [18] investigated the correlation between the changes of the friction momentum in the ball bearings and the thixotropic properties of the calcium and the lithium greases. At the research, they observed that the shape of the dispersed phase particles considerably influences the behaviour of the lubricating greases during their further deformation and shear-thinning. On the turn of the 1960s and 1970s, Vinogradov also investigated the lubricating greases using the polarimetry and other optical methods. It enabled the quantity evaluation of the level of the thickener microstructure reconstruction. He also conducted research on the thixotropic hysteresis loop for the calcium and lithium greases with the use of the rotational coaxial cylinder rheometers. His works are widely discussed by Froischteter in [7].

Sacchettini, Magnin, Piau and Pierrard [51] published in the Journal of Theoretical and Applied Mechanics a paper on the thixotropic properties of commercial greases: Shell Alvania, ELF Multi and ELF EL series (8305-8308 and 9309) of NLGI (National Lubricating Grease Institute) 2 consistency class. The analyzed lubricating greases showed a different fibre length of the soap thickener (lithium 12-hydroxystearate). The oil base in the greases was the mineral oil. Sacchettini et al. focused mainly on the research of change in the shear stress as a function of shearing time. They analyzed changes in the lithium thickener microstructure at the cyclic shearing lasting a few seconds at different shear rates (from 0.018 s^{-1} to 18 s^{-1}) and different grease relaxation time (from 2 minutes to 38 hours). They did not analyzed the structural viscosity increase of the greases at their relaxation, however. Apart from the thixotropy, Sacchettini et al. investigated the wall effects.

Czarny, in his paper [52], presented results of the research on the process of degradation and reconstruction of the thickener microstructure in the commercial calcium greases (STP) and the lithium greases (ŁT-4S2). He used the method of cyclic shearing of the lubricating grease at specified time intervals, at a constant shear rate. The research on the grease was carried out without removing it from the rheometer head. He thus simulated the conditions observed in the conduits of lubrication system. In 1990, Czarny investigated the influence of temperature on the thixotropy of the lubricating greases [53]. He proposed a mathematical description of the energetic changes which take place in the microstructure of the thickener during its shearing at various temperatures. Czarny observed that along with the temperature increase, the structural viscosity of the lubricating grease is reduced, and the activity of the thickener particles increases. According to Czarny, the motion of the particles along with the increase in the lubricant temperature, a certain energetic barrier, which Czarny called *the activation energy*, must be overcome. Czarny also carried out the research on the degradation of the thickener microstructure by means of the Couette rotational rheometer, and on the influence of temperature on the process. He investigated the bentonite greases, the calcium greases (STP, Kalton EP1), the lithium greases (Shell Alvania EP2, ŁT4-S2) as well as the lithium-calcium greases. In accordance with the experimental research and the theory proposed by Czarny, when temperature increases, the thickener intermolecular bonds get weakened, making a cross-linked, compact microstructure, and the activity of single crystallites freely suspended in the base oil increases. As a result, the lubricating greases, in the process of long-lasting shearing, take on the Newtonian fluid characteristics.

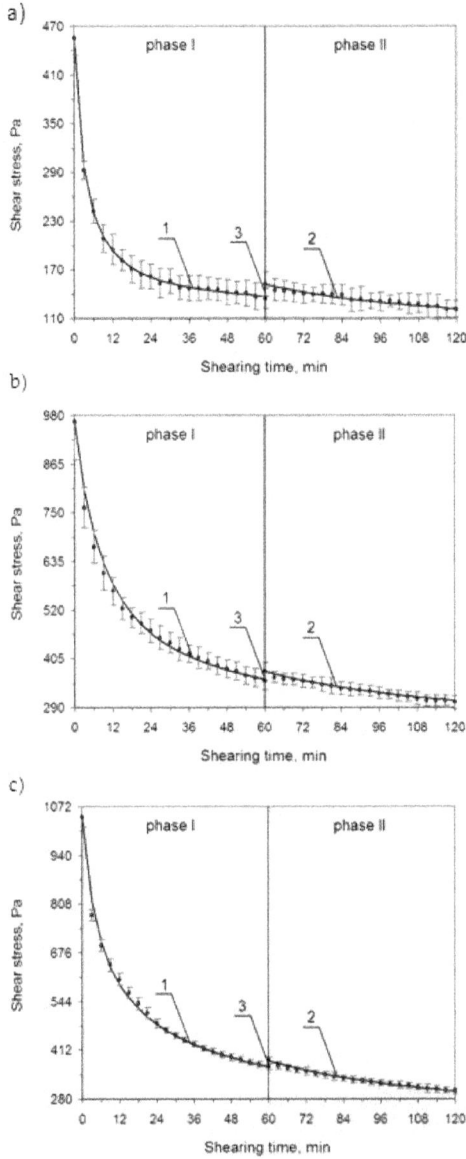

Figure 7. Dependence of the shear stress as a function of shearing time at constant shear rate $\dot{\gamma} = 8.1 \ s^{-1}$ ($p = 0.95$) for the greases with the thickener percentage: (a) 6%, (b) 7%, (c) 8%. 1 – the first stage of shearing, 2 – the second stage of shearing, 3 – resolidification process of the grease in its relaxation phase.

Figure 7 illustrates the thixotropy curves for the lubricating greases thickened with the lithium 12-hydroxystearate based on the mineral oil, received by means of the measuring method proposed by Czarny. The grease shearing time in two cycles was 120 minutes in total. The grease relaxation time between the shearing cycles was 24 hours. The research was repeaed five times. The experiment was carried out with the use of Rheotest 2.1 rotational co-axial cylinder rheometer working at 2 mm measuring gap. For the measurements, a right handed helix shaped grooved polytetrafluoroethylene internal cylinder, with 95 0.5 mm-deep grooves was used. Such a surface structure enabled a visible reduction of the Weissenberg effect and of the slip effect at the cylinder wall, which could have negatively influenced the measurements.

The energetic interpretation of the lubricating greases thixotropy was presented by Kuhn [54]. He focused primarily on those cases of thixotropy of the lubricating greases which occur during the hydrodynamic lubrication of the slide bearing. According to Kuhn, the energy loss during the friction shearing of the grease is a function of the lubricant's qualities and the conditions it is sheared in. If the lubricant's qualities change in the process of the fluid friction, *the energy density* (energy/volume) also changes:

$$e_{rh_e} = \eta \dot{\gamma} \left(\frac{\bar{d}_e}{h_0^*} \right),$$ (14)

where: e_{rh_e} – rheological density of energy in the microcontact region (J/mm³), η –dynamic viscosity (Pas), $\dot{\gamma}$ – shear rate (s⁻¹), d_e – average diameter of the microcontact (mm), h_0^* – average lubrication film thickness (mm).

Using the dependence presented above, Kuhn estimated friction coefficient μ which was:

$$\mu = \frac{e_{rh}}{p_r} i_{rh},$$ (15)

where: e_{rh} – sum of the rheological energy densities in the region of all microcontacts (J/mm³), i_{rh} – friction intensity (non-dimensional value), p_r – pressure (MPa).

Kuhn also observed a similarity between the curve of changes in the rheological energy density as a function of time at different shear rates and the curves received by Czarny [55] on the experimental way. Later in his work, Kuhn developed the energetic model of the lubricating greases thixotropy [56,57], and together with Balan suggested the experimental ways of estimating the rheological density of energy with the use of the rotational cone-and-plate rheometer [58]. In [59], he defined the tribological properties of the lithium and calcium greases based on the mineral oil by means of the energetic parameters, and also presented the relations between the viscoelastic properties of the lubricating grease, the friction in the frictional contact and the elastic strain energy accumulation and the shear stress at the shearing of the grease.

Kuhn's research proved that in the quantification of the tribological processes, the energy loss resulting from the structural changes which take place in the frictional contact region must be taken into account.

5. The rheological properties of the boundary layer in the lubricating greases

Lubricating greases are, on the one hand, required the best possible slide and anti-corrosion properties, and on the other hand, the lowest possible and independent of conditions flow resistance at the supplying of the friction nodes. The problem of the flow resistance of the lubricating greases is currently particularly important in the context of lubrication systems. It results primarily from the fact of a higher and higher automation of the systems. The flow resistance have negative influence on the proper functioning of the central lubrication systems. An important factor reducing the resistance in the lubrication systems is the appearance in the region of the lubricating grease contact with the conduit wall, of the so called *the boundary layer* (or *the wall effect*) which has different rheological properties than the rest of the grease's volume. The phenomenon of the boundary layer formation while the suspension flows through the axisymmetric conduits is frequently referred to as *Segré-Silberberg's effect* or the *sigma effect*. For the first time, the phenomenon was thoroughly described in 1962 [60]. The influence of the boundary layer formation on the fall of the suspension flow resistance is particularly visible in the small diameter axisymmetric conduits where the suspension flow is laminar. It leads to the decrease in the pressure gradient along the conduit. Simultaneously with the increase in the conduit diameter, the boundary layer influence considerably decreases [61].

The first mentions of the boundary layer in the lubricating greases appeared in 1960. Bramhal and Hutton [62] carried out the research on the influence of the wall types on the boundary layer formation. The research was conducted with the use of a plunger viscometer. While doing the research, they recorded a slip in the grease, which occurred in the area of the walls. Bramhal and Hutton observed inconsiderable influence of the wall's material and the wall surface topography on the grease slip. They explained the observed effect as a consequence of repulsion of the soap particles suspended in the base oil from the walls.

An attempt of explaining the phenomenon of the boundary layer formation in the lubricating greases within the area of the lubricating grease contact with the lubrication system conduit wall and the working surface of the sliding pair was made by Czarny. In [52,63] he proves that the boundary layer formation in the lubricating greases depends mainly on the surface material which the lubricating grease reacts with. He observed the changes of the shear stress in the commercial greases in the area of the wall made, among others, of methyl polymethacrylate, polytetrafluoroethylene, polyamide, brass, bronze, duralumin and cast iron. The phenomenon of the boundary layer formation was particularly visible at the low shear rates. According to Czarny, the boundary layer formation is an outcome of the thickener's particles adsorption on the surface of the material. Thus, during the flow of the grease, a toroidal space depleted of the thickener is created. The number of the adsorbed thickener particles depends on the kind of grease and on the collaborating materials, and is inversely proportional to the thickener percentage.

Czarny and Moes [28] proposed a dependence describing the change of the shear stress in the lubricating grease as a function of the distance from the wall:

$$\tau = \tau_0 \left\{ 1 - e^{-\left(\frac{z+d}{s}\right)} \right\}, \qquad (16)$$

where: τ_0 – yield stress, z – normal wall coordinate, d – surface layer thickness, s – thickness of the boundary layer with the exponentially decreasing yield stress, down to its minimum value on the wall $\tau = \tau_w$ (Figure 8).

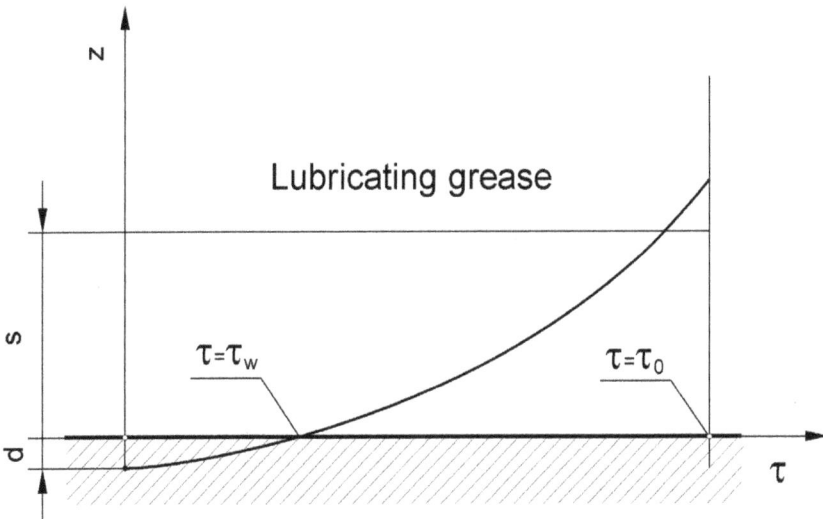

Figure 8. Stress distribution τ in the grease boundary layer as a function of the distance to wall z, according to Czarny and Moes [28].

According to Czarny, the model can be used for practical purposes at specific design solutions of lubrication systems. Parameter s and d values, needed for the determination of the stress distribution, which depend on the kind of grease, temperature of the tribological pair, the kind of the pair's wall material and its surface topography can be experimentally defined.

Czarny and Moes, during their research on the lithium soap thickened grease – Alvania EP2, found that the value of the yield stress in the grease mass τ_0 and at the wall τ_w rise along with the increase of the grease consistency level. Simultaneously, the increase determines the reduction in the thickness of boundary layer s and surface layer d. The observed boundary layer thickness s and surface layer d equaled, in the case of Alvania EP2, respectively 3.0-9.5 and 0.6-1.5 μm. According to Czarny, the boundary layer formation in the lubricating greases is catalyzed by different thickener particles size as well as the lubricating grease temperature. Along with the temperature increase, the thickness of the absorbed layer d increases. Czarny explains the increase by an easier adsorption of the active soap particles on the material surface at a higher temperature. The thickest, 25 μm boundary layer of Alvania EP2 was observed at temperature of 373 K. It might be interesting to know that, for instance at temperature of 273 K, the boundary layer was approximately 2 μm thick.

Vinogradov et al. [64,65] investigated the shear stress decrease in the lubricating grease at the wall. The research was carried out on the lithium and calcium soap as well as ceresine thickened greases. All the researched lubricating greases formed the boundary layer. The experiments were conducted with the use of the rotational cone-and-plate rheometer and of the capillary rheometer. They found that the thickness of the boundary layer in the grease is determined by the stress which is present there. Based on the Herschel-Bulkley model, they defined dependences which enable the determination of the grease flow parameters in the gap and in the axisymmetric conduit, including the wall effect.

The research of the lubricating grease flow in the conduits was conducted also by Swartz and Hardy [66]. They recorded a decrease in resistance of the lubricating grease flow through the steel axisymmetric conduits. According to their hypothesis, it was caused by bleeding of the base oil on the conduit wall and by repulsing the thickener particles from the wall. The hypothesis, however, was not experimentally verified.

Biernacki in [67-69] investigated the problem of influence of the wall material and of the grease temperature on the boundary layer formation. He studied the phenomenon of the boundary layer formation in the commercial greases, in the area of the walls made of plastics (polyamide, polyvinyl chloride, methyl polymethacrylate, polytetrafluoroethylene) as well as the walls made of metal (steel, copper, mazak, duralumin, aluminium) of different surface topography. The research was carried out mainly with the use of the rotational coaxial cylinder rheometer. In the lithium and calcium greases he recorded lower shear stress values at the walls of the cylinders made of metal than at those made of plastics. The differences in the shear stress values were particularly visible in the low shear rates range and at the low temperature. With the increase of temperature and the shear stress, the differences in the values of the shear stress in the greases depending on the wall material disappear. In the polymer and aluminium greases Biernacki did not observed considerable differences in the shear stress values recorded at the surface of the walls made of different materials.

Biernacki also investigated the problem of the boundary layer thickness in the complex lithium and bentonite greases in the area of two copper and polypropylene plates which approach parallel to each other. The experiment showed that the lithium grease at the copper plate forms a 65 μm thick boundary layer. The depleted part of the lithium thickener was formed 27 μm from the plate surface. Biernacki did not observe the boundary layer formation in the lithium grease at the polypropylene plate. In the bentonite grease, the boundary layer was not formed in any of the cases.

Delgado et al. in [70-72] did research on the influence of air, roughness of the steel conduit wall surface, as well as the conduit diameter on the pressure drop at the horizontal flow of the grease. They observed that the roughness of the conduit wall influences the formation of the boundary layer in the lubricating grease. The reduction of the surface roughness considerably reduced the pressure gradient. The differences grew with the conduit diameter reduction or with the grease flow rise. Delgado et al. also found out that the air present in the lubricating grease causes the reduction of the wall slip. According to their hypothesis, it is caused by the fact that the air in the lubricating grease generates considerable differences in the moistening of the internal walls of the conduit. The research was carried out mainly on the complex soap lubricants.

6. Conclusions

A conclusion of the presented research is that the flow of the plastic greases in the lubrication systems depends not only on the design of the systems but also on the kind of grease. Greases are non-Newtonian fluids of a highly complex structure, and therefore the to carry out the evaluation process of their behaviour in particular conditions is multi-stage and very difficult. The flow resistance of the grease in particular elements of the lubrication system depend, among others, on the structural viscosity of the grease. All greases get thinned, to a larger or smaller extent, in the shearing process. At the distribution of the grease to the friction nodes in a particular period of time, the flow resistance gets reduced as a result of the grease thickener's microstructure degradation. The decrease in the grease flow resistance is strictly correlated with the decrease of the shear stress in the grease. In the first seconds of supplying a new portion of the grease into the main conduit of the lubrication system, the pump must in the first place overcome the flow resistance generated by the grease, which is connected with the occurrence of the yield stress in the grease. The value of the stress, as well as the characteristics of the shear thinning for a particular grease, among other factors, depend on the flow rate generated by the lubrication pump, but also on the physical and chemical activity of the environment. An important determinant on the generating the flow resistance in the lubrication systems are the physical-chemical phenomena connected with the interaction of the grease thickener particles suspended in the base oil. One of them is a phenomenon of the grease thixotropy, consisting in its gel→sol→gel transition, namely the reduction in the grease structural viscosity at its flow and a partial thickener microstructure reconstruction in the phase of the grease relaxation, between the consecutive cycles of its distribution to the friction nodes. The phenomenon is crucial in the case of the greases which stay in the conduits and dividers of the lubrication system. It can be dangerous for the complex systems, working cyclically (for instance, the central lubrication systems of big mining machines). A short-term, uncontrolled increase in the flow resistance in such systems may lead to the excessive energy consumption by the pump, and in the extreme cases, to damage of the charging unit, and consequently, to the system failure.

Critical for the question of the flow resistance is also the material used for the making of the lubrication conduits as well as their surface topography. Some materials (copper, steel and cast iron in particular) show the increased adsorption of the particles of the thickener on their surface. It results in the generation of certain thickener depleted zones, in a shape of a ring, presenting a lower structural viscosity. The flow resistance at the start of the pump dosing the grease depending on the material of which the lubrication conduit is made, can be reduced even by half. In the case when the roughness of the internal surface finish of the conduit is higher or equal to the thickness of the thickener depleted layer, then the grease flow resistance is considerably higher than the one which would be observed at the smooth surface finish. While designing the lubrication systems, it is critical to pay attention to whether the roughness of the lubrication conduit surface are as low as possible. It is important to mention here that the pressure drops connected with the resistance, in the majority of the lubrication system elements are local in their character, whereas in the lubrication conduits, there are constant pressure drops. In the case of the particularly long conduits, the flow resistance for the entire lubrication system is predominant there. Other elements which can increase the lubricant flow resistance as well as the pressure in the system are various types of gaps in dividers. Also in this case, their shape, section area and the material of which they are made are important. Similarly complex is a problem of determining the resistance generated by the very friction nodes and defining the pressure necessary to deliver the required grease portion to the bearing. The most important design characteristics of the delivered grease flow resistance reduction bearings are primarily their shape, dimensions of the bearings' working side surfaces, the distribution and way the lubrication gaps are cut, as well as the circumferential backlash.

Nomenclature

m – index exponent

α_1 – the not entangled particles shear stress, Pa

α_2 – the entangled particles shear stress, Pa

β_1 – shear intensity for the not entangled particles, s^{-1}

β_2 – shear intensity for the entangled particles, s^{-1}

γ_C – critical strain, %

$\dot{\gamma}$ – shear rate, s^{-1}

η – apparent viscosity (dynamic viscosity), Pas

η_0 – viscosity at zero shear rate, Pas

η_∞ – viscosity at infinite shear rate, Pas

η' – viscous component, Pas

η'' – elastic component, Pas

$|\eta*|$ – complex viscosity, Pas

θ – thixotropy coefficient

λ – time-constant, s

μ – friction coefficient

τ – shear stress, Pa

τ_0 – yield stress, Pa

χ_1 – share of the not entangled particles

χ_2 – share of the entangled particles

a – rheological parameter

d – surface layer thickness, µm

d_e – average diameter of the microcontact, mm

e_{rh} – sum of the rheological energy densities in the region of all microcontacts, J/mm³

e_{rh_e} – rheological density of energy in the microcontact region, J/mm³

G' – storagemodulus, Pa

G'' – lossmodulus, Pa

G^0_N – plateau modulus, Pa

$|G*|$ – complex modulus, Pa

h_o* – average lubrication film thickness, mm

i_{rh} – friction intensity

k – Boltzmann constant, J/K

k_b – parameter determining the time of the not entangled → entangled particles

k_c – consistency factor of the thickener in the grease, Pas

k_f – parameter determining the time of the entangled particles → not entangled particles reaction m – index exponent

n – index exponent

p_r – pressure, MPa

s – thickness of the boundary layer with the exponentially decreasing yield stress, µm

T – temperature, K

w – elastic strain energy of the entangled particles

w' – elastic strain energy of the not entangled particles

z – normal wall coordinate

Author details

Maciej Paszkowski

Address all correspondence to: maciej.paszkowski@pwr.wroc.pl

Wroclaw University of Technology, Institute of Machines Design and Operation, Wroclaw, Poland

References

[1] Czarny, R. Lubricating Greases. Warsaw: WNT Publishers, (2004) (in Polish).

[2] Vold JM. Colloidal Structure in Lithium Stearate Greases. Physical Chemistry 1956;60(4) 439-442.

[3] Farrington BB, Davis WN. Structure of Lubricating Greases. Industrial and Engineering Chemistry 1936;28(4) 414-416.

[4] IshchukYuL.Lubricating Grease Manufacturing Technology.New Delhi: New Age International Limited Publishers, 2005.

[5] Boner CJ. Modern Lubricating Greases, Broseley: Scientific Publication; 1976.

[6] Czarny R. Effect of Changes in Grease Structure on Sliding Friction, Industrial Lubrication and Tribology 1995;47 3-7.

[7] Froischteter GB, Trilisky KK, IshchukYuL, Stupak PM. Rheological and Thermophysical Properties of Greases. London: Gordon & Breach Science Publishers, 1989.

[8] Basiński A et al. Polish Dictionary of Chemical Terminology. Warsaw: WNT Publishers; 1975 (in Polish).

[9] Czarny R. Investigation into Phenomena Accompanying the Flow of Greases in Lubrication Systems. Wroclaw: Publishing House of the Wroclaw University of Technology; 1992 (in Polish).

[10] Mewis J. Thixotropy – A General Review, Journal of Non-Newtonian Fluid Mechanics 1979;6 1-20.

[11] Luckham PF, Rossi S. The Colloidal and Rheological Properties of Bentonite Suspensions. Advances in Colloid and Interface Science 1999;82 43-92.

[12] Delgado MA, Valencia C, Sanchez MC, Franco JM, Gallegos C. Influence of Soap Concentration and Oil Viscosity on the Rheology and Microstructure of Lubricating Greases. Industrial and Engineering Chemistry Research 2006;45(6) 1902-1910.

[13] Moore RJ, Cravath AM. Mechanical Breakdown of Soap-base Greases, Industrial and Engineering Chemistry 1951;43(12) 2892-2897.

[14] Yeong SK, Luckham PF, TadrosThF. Steady Flow and Viscoelastic Properties of Lubricating Grease Containing Various Thickener Concentrations. Journal of Colloid and Interface Science 2004;274 285-293.

[15] Cann PM. Grease Degradation in Bearing Simulation Device. Tribology International 2006;39 1698-1706.

[16] Cousseau T, Graca BM, Campos AV, Seabra JHO. Influence of Grease Rheology on Thrust Ball Bearings Friction Torque, Tribology International 2012;46 106-113.

[17] Farcas F, Gafitanu MD. Some Influence Parameters on Grease Lubricated Rolling Contacts Service Life. Wear 1999;225-229 1004-1010.

[18] Vinogradov GV, DeinegaYuF, VerbitskyYaA. Thixotropy of Greases and Low-speed Operation of Rotational Plastoviscometers and Rolling Bearings.RheologicaActa 1967; 6(3) 252-259.

[19] CannPM, Spikes HA. Film Thickness Measurements of Lubricating Grease Under Normally Starved Conditions. NLGI Spokesman 1992;56 (2) 21-26.

[20] Wiliamson BP, Kendal DR, Cann PM. The Influence of Grease Composition on Film Thickness in EHD Contacts. NLGI Spokesmen 1993;57(8) 13-18.

[21] Krawiec S. On the Mechanism of the Synergistic Effect of PTFE and Copper in a Lithium Grease Lubricant. Industrial Lubrication and Tribology 2011;63(3) 171-177.

[22] Krawiec S. The Synergistic Effect of Copper Powder with PTFE in a Grease Lubricant Under Mixed Friction Conditions. Archives of Civil and Mechanical Engineering 2011;11(2) 379-390.

[23] Czarny R, Paszkowski M. Einfluss von Zusatzen auf die rheologischen Eigenschaften der Schmierfette, 15th International Colloquium Tribology. Automotive and Industrial Lubrication, 2006, Stuttgart-Ostfildern, Germany.

[24] Czarny R, Paszkowski M. The Influence of Graphite Solid Additives, MoS2 and PTFE on Changes in Shear Stress Values in Lubricating Greases. Journal of Synthetic Lubrication 2007;24(1) 19-29.

[25] Herschel WH, Bulkley R. Measurement of Consistency as Applied to Rubber – Benzene Solutions. Proceedings of American Society of Testing Materials 1926;26, 621-633.

[26] Yasuda K, Armstrong R, Cohen RE. Shear Flow Properties of Concentrated Solutions of Linear and Star Branched Polystyrenes. RheologicaActa 1981;20 163-178.

[27] Bird R, Armstrong R, Hassager O. Dynamics of Polymer Liquids: Volume 1. Fluid Mechanics. New York: John Wiley & Sons; 1987.

[28] Czarny R, Moes H. Some Aspects of Lubricating Grease Flow, 3rd International Tribology Congress – Eurotrib '81, 1981, Warsaw, Poland.

[29] Liu Ch, He J, Ruymbeke E, Keunings R, Bailly C. Evaluation of Different Methods for the Determination of the Plateau Modulus and the Entanglement Molecular Weight. Polymer 2006;47 4461–4479.

[30] Martín-Alfonso JE, Valencia C, Sánchez MC, Franco JM, Gallegos C. Recycled and Virgin LDPE as Rheology Modifiers of Lithium Lubricating Greases: A Comparative Study. Polymer Engineering and Science 2008;48 1112–1119.

[31] Martín-Alfonso JE, Valencia C, Sánchez MC, Franco JM, Gallegos C. Rheological Modification of Lubricating Greases with Recycled Polymers from Different Plastics Waste. Industrial & Engineering Chemistry Research 2009;48 4136–4144.

[32] Péterfi T. Die Abhebung der Befruchtungsmembran bei Seeigeleiern. Archiv für Entwicklungsmechanik der Organismen 1927;112 660-695.

[33] Lagaly G, Beneke K. Eighty years of colloid science in Hungary and Germany. University Kiel, 2002.

[34] Freundlich H. Thixotropie. Paris: Hermann, 1935.

[35] Efremov IF, Us'yarov OG. The Long-range Interaction Between Colloid and Other Particles and the Formation of Periodic Colloid Structures. Russian Chemical Reviews 1976;45(5) 435-453.

[36] Efremov IF. The Dilatancy of Colloidal Structures and Polymer Solutions, Russian Chemical Reviews 1982;51(2) 160-177.

[37] Sonntag H. Lehrbuch der Kolloidwissenschaft. Berlin: VEB Deutscher Verlag der Wissenschaften, 1977.

[38] Scheludko A. Colloid Chemistry. Warsaw: WNT Publishers, 1969. Translated from Russian by Siedlecka Z (in Polish).

[39] Barnes HA. Thixotropy – a Review. Journal of Non-Newtonian Fluid Mechanics 1997;70 1-33.

[40] Franco JM, Delgado MA, Valencia C. Combined Oxidative-shear Resistance of Castrol Oil-based Lubricating Greases, 3rd Arnold Tross Colloquium, 2007, Hamburg, Germany.

[41] Renshaw TA. Effect of Shear on Lithium Greases and Their Soap Phase. Industrial and Engineering Chemistry 1955;47(4) 834-838.

[42] Sirianni AF, Moses GB, Puddington IE. Particle Shape in Thixotropic Suspensions. Canadian Journal of Chemistry 1951;29 166-172.

[43] Goodeve CF, Whitfield GW, The Measurement of Thixotropy in Absolute Units. Transaction of Faraday Society 1938;34 511-520.

[44] [44] Arnold JE, Goodeve CF. The Coefficient of Thixotropy of Suspensions of Carbon Black in Mineral Oil. Journal of Physical Chemistry 1940;44(5) 652-670.

[45] Moses GB, Puddington IE. Rheological Properties of Some Soap-Oil Systems, Canadian Journal of Chemistry 1951;29 996-1009.

[46] Hahn SJ, Ree T, Eyring H. A Theory of Thixotropy and its Application to Grease.NLGI Spokesman 1957;21(3) 12-20.

[47] Houser EA, Reed CE. Studies in Thixotropy II – The Thixotropic Behavior and Structure of Bentonite. Journal of Physical Chemistry 1937;41(7) 911-934.

[48] Hahn SJ, Eyring H, Higuchi I, Ree T. Flow Properties of Lubricating Oils Under Pressure. NLGI Spokesman 1958;21(3) 123-128.

[49] Utsugi H, Kim K, Ree T, Eyring H. Thixotropy Property of Lubricating Grease, NLGI Spokesman, NLGI 28th Annual Meeting, 1960, Chicago, USA.

[50] Bair S. Actual Eyring Models for Thixotropy and Shear-thinning: Experimental Validation and Application to EHD. ASME Journal of Tribology 2004;126 728-732.

[51] Sacchettini M, Magnin A, Piau JM, Pierrard JM. Charactérisation d'une graisse lubrifiante en écoulements viscosimétriques transitoires. Journal of Theoretical and Applied Mechanics, Numérospécial 1985; 165-199.

[52] Czarny R. Influence of the Lubricating Grease Structure on its Rheological Properties.Scientific Works of the Institute of Machine Design and Operation of Wroclaw University of Technology.Problems of Friction, Lubrication and Machine System Computerization. Wroclaw: Publishing House of Wroclaw University of Technology; 1989 (in Polish).

[53] Czarny R. Effect of Temperature on the Thixotropy Phenomenon in Lubricating Greases, Proceedings of the Japan International Tribology Conference, 1990, Nagoya, Japan.

[54] Kuhn E. An Energy Interpretation of Thixotropic Effect. Wear 1991;142(1991) 203-205.

[55] Czarny R. Einfluss der Thixotropie auf die rheologischen Eigenschaften der Schmierfette. Tribologie und Schmierungstechnik 1989;3 (36) 134-140.

[56] Kuhn E. Energy Investigation of the Rheological Wear of Lubricant Greases. Applied Rheology 1992;2(4) 252-257.

[57] Kuhn E. Description of the Energy Level of Tribologically Stressed Greases. Wear 1995;188(1995) 138-141.

[58] Kuhn E, Balan C. Experimental Procedure for the Evaluation of the Friction Energy of Lubricating Greases. Wear 1997;209(1997) 237-240.

[59] Kuhn E. Experimental Grease Investigation From an Energy Point of View. Industrial Lubricating and Tribology 1999;51(5) 246-251.

[60] Segré G, Silberberg A. Behavior of Macroscopic Rigid Spheres in Poiseuille Flow. Part II. Experimental Results and Interpretation.Journal of fluid mechanics 1962;14 136-157.

[61] Heywood NI, Cheng DC-H. Flow in Pipes. Part II. Multiphase flow.Physics in Technology 1984;15.

[62] Bramhall AD, Hutton JF. Wall Effect in the Flow of Lubricating Greases in Plunger Viscometers. British Journal of Applied Physics 1960;11 363-369.

[63] Czarny R. Influence of the Wall Type on Formation of the Grease Wall Effect, Conference materials, XXI Autumn Tribological School – Machine and appliance friction pairs lubrication. Contemporary tendencies in the research theory development, 1996, Lodz – Arturowek, Poland (in Polish).

[64] Vinogradov GV, Froishteter GB, Trilsky KK. The Generalized Theory of Flow of Plastic Disperse Systems with Account of the Wall Effect.RheologicaActa 1978;17(2) 156-165.

[65] Vinogradov GV, Froishteter GB, Trilsky KK, Smorodinsky EL. The Flow of Plastic Disperse Systems in the Presence of the Wall Effect.RheologicaActa 1975;14(9) 765-775.

[66] Swartz CJ, Hardy B. Mathematical Model of Grease Flow in Pipes. NLGI Spokesman 1991;55(3) 14 –17.

[67] Biernacki K. Influence of Surface Roughness and Temperature on the Lubricating Grease Flow in the Boundary Layer.Tribologia 2000;2 187-198 (in Polish).

[68] Biernacki K, Czarny R. Influence of the Wall Material and Temperature on the Rheological Properties of the Boundary Layer in the Lithium and Calcium Greases. Hydraulika i Pneumatyka 2002;4 24-25 (in Polish).

[69] Biernacki K. Analysis of the Wall Effect in Lubricating Greases.Tribologia 2006;5 111-129 (in Polish).

[70] Delgado MA, Franco JM, Partal P, Gallegos C. Experimental Study of Grease Flow in Pipelines: Wall Slip and Air Entrainment Effects. Chemical Engineering and Processing 2005;44 805–817.

[71] Ruiz-Viera MJ, Delgado MA, Franco JM, Gallegos C. Evaluation of Wall Slip Effects in the Lubricating Grease/Air Two-phase Flow Along Pipelines. Journal of Non-Newtonian Fluid Mechanics 2005;139 190–196.

[72] Ruiz-Viera MJ, Delgado MA, Franco JM, Sanchez MC, Gallegos C. On the Drag Reduction for the Two-phase Horizontal Pipe Flow of Highly Viscous Non-Newtonian Liquid/Air Mixtures: Case of Lubricating Grease. International Journal of Multiphase Flow 2006;32 232–247.

Boundary Lubrication Applications

Chapter 4

Titanium and Titanium Alloys as Biomaterials

Virginia Sáenz de Viteri and Elena Fuentes

Additional information is available at the end of the chapter

1. Introduction

Bone and its several associated elements – cartilage, connective tissue, vascular elements and nervous components – act as a functional organ. They provide support and protection for soft tissues and act together with skeletal muscles to make body movements possible. Bones are relatively rigid structures and their shapes are closely related to their functions. Bone metabolism is mainly controlled by the endocrine, immune and neurovascular systems, and its metabolism and response to internal and external stimulations are still under assessment [1].

Long bones of the skeletal system are prone to injury, and internal or external fixation is a part of their treatment. Joint replacement is another major intervention where the bone is expected to host biomaterials. Response of the bone to biomaterial intervenes with the regeneration process. Materials implanted into the bone will, nevertheless, cause local and systemic biological responses even if they are known to be inert. Host responses with joint replacement and fixation materials will initiate an adaptive and reactive process [2].

The field of biomaterials is on a continuous increase due to the high demand of an aging population as well as the increasing average weight of people. Biomaterials are artificial or natural materials that are used to restore or replace the loss or failure of a biological structure to recover its form and function in order to improve the quality and longevity of human life. Biomaterials are used in different parts of the human body as artificial valves in the heart, stents in blood vessels, replacement implants in shoulders, knees, hips, elbows, ears and dental structures [3] [4] [5]. They are also employed as cardiac simulators and for urinary and digestive tract reconstructions. Among all of them, the highest number of implants is for spinal, hip and knee replacements. It is estimated that by the end of 2030, the number of total hip replacements will rise by 174% (572,000 procedures) and total knee arthroplasties are projected to grow by 673% from the present rate (3.48 million procedures) [6]. This is due to the fact that human joints suffer from degenerative diseases such as osteoarthritis (inflammation in the bone joints), osteoporosis (weakening of the bones) and trauma leading to pain or loss in function. The degenerative diseases lead to degradation of the mechanical properties of the bone due to excessive loading or absence of normal biological self-healing process. Artificial biomaterials are the solutions to these problems and the surgical implantation of these artificial biomaterials of suitable shapes help restore the function of the otherwise functionally compromised structures. However, not only the replacement surgeries have increased, simultaneously the revision surgery of hip and knee implants have also increased. These revision surgeries which cause pain for the patient are very expensive and also their success rate is rather small. The target of present researches is developing implants that can serve for much longer period or until lifetime without failure or revision surgery [7]. Thus, development of appropriate material with high longevity, superior corrosion resistance in body environment, excellent combination of high strength and low Young's modulus, high fatigue and wear resistance, high ductility, excellent biocompatibility and be without citotoxicity is highly essential [8] [9].

In general, metallic biomaterials are used for load bearing applications and must have sufficient fatigue strength to endure the rigors of daily activity. Ceramic biomaterials are generally used for their hardness and wear resistance for applications such as articulating surfaces in joints and in teeth as well as bone bonding surfaces in implants. Polymeric materials are usually used for their flexibility and stability, but have also been used for low friction articulating surfaces. Titanium is becoming one of the most promising engineering materials and the interest in the application of titanium alloys to mechanical and tribological components is growing rapidly in the biomedical field [10], due to their excellent properties.

This chapter is focused on the use of titanium and its alloys as biomaterials from a tribological point of view. The main limitation of these materials is their poor tribological behavior characterized by high friction coefficient and severe adhesive wear. A number of different surface modification techniques have been recently applied to titanium alloys in order to improve their tribological performance as well as osseointegration. This chapter includes the most recent developments carried out in the field of surface treatments on titanium with very promising results.

2. Biomaterial properties

The main property required of a biomaterial is that it does not illicit an adverse reaction when placed into services, that means to be a biocompatible material. As well, good mechanical properties, osseointegration, high corrosion resistance and excellent wear resistance are required.

2.1. Biocompatibility

The materials used as implants are expected to be highly non toxic and should not cause any inflammatory or allergic reactions in the human body. The success of the biomaterials is mainly dependent on the reaction of the human body to the implant, and this measures the biocompatibility of a material [11]. The two main factors that influence the biocompatibility of a material are the host response induced by the material and the materials degradation in the body environment (Figure 1). According to the tissue reaction phenomena, the biocompatibility of orthopedic implant materials was classified into three categories by Heimke [12], such as "biotolerant", showing distant osteogenesis (bone formation with indirect contact to the material); "bioinert", showing contact osteogenesis (bone formation with direct contact to the material), and "bioactive", showing bonding osteogenesis (bone formation with chemical or biological bonding to the material).

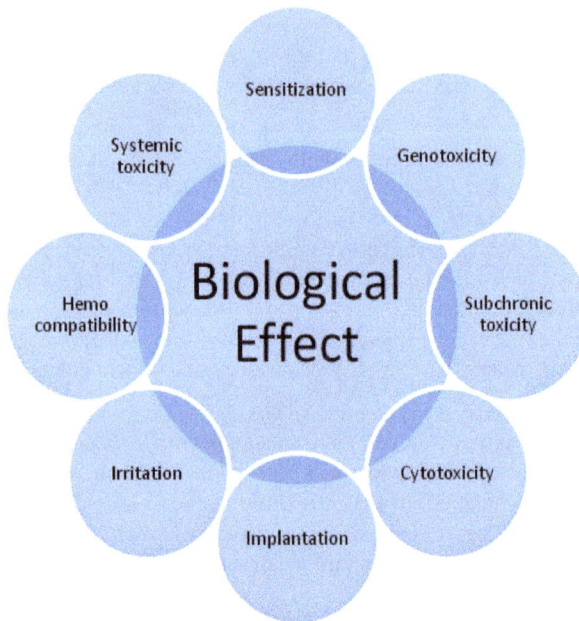

Figure 1. Biological effects of a biomaterial

When implants are exposed to human tissues and fluids, several reactions take place between the host and the implant material and these reactions dictate the acceptability of these materials by our system. The issues with regard to biocompatibility are (1) thrombosis, which involves blood coagulation and adhesion of blood platelets to biomaterial surface, and (2) the fibrous tissue encapsulation of biomaterials that are implanted in soft tissues.

2.2. Mechanical properties

The most important mechanical properties that help to decide the type of material are hardness, tensile strength, Young's modulus and elongation. An implant fracture due to a mechanical failure is related to a biomechanical incompatibility. For this reason, it is expected that the material employed to replace the bone has similar mechanical properties to that of bone. The bone Young´s modulus varies in a range of 4 to 30 GPa depending on the type of the bone and the direction of measurement [13] [14].

2.3. Osseointegration

The inability of an implant surface to integrate with the adjacent bone and other tissues due to micromotions, results in implant loosening [15]. Osseointegration (capacity for joining with bone and other tissue) is another important aspect of the use of metallic alloys in bone applications (Figure 2). A good integration of implant with the bone is essential to ensure the safety and efficacy of the implant over its useful life. It has been shown in previous studies [16], that enhancement of the bone response to implant surfaces can be achieved by increasing the roughness or by other surface treatments [17]. Although the precise molecular mechanisms are not well understood, it is clear that the chemical and physical properties of the surface play a major role in the implant – surface interactions through modulation of cell behavior, growth factor production and osteogenic gene expression [18] [19] [20].

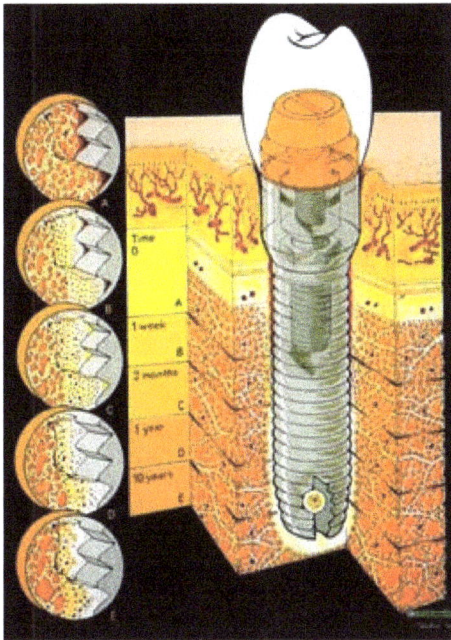

Figure 2. Schematic drawing of the principles of osseointegration [21]

Furthermore, it is known that even if initial implant stability is achieved, the bone may retreat from or be isolated from the implant because of different reasons or situations listed below:

1. Reaction of the implant with a foreign body as debris from implant component degradation or wear, or to toxic emissions from the implant [22]

2. Damage or lesion to the bone through mechanical trauma surgery

3. Imposition of abnormal or unphysiological conditions on the bone, such as fluid pressures or motion against implant components

4. Alteration to the mechanical signals encouraging bone densification; strain reductions or stress-shielding of replaced or adjacent bone.

2.4. High corrosion resistance

All metallic implants electrochemically corrode to some extent. This is disadvantageous for two main reasons: (1) the process of degradation reduces the structural integrity and (2) degradation products may react unfavorably with the host. Metallic implant degradation results from both electrochemical dissolution and wear, but most frequently occurs through a synergistic combination of the two [23] [24]. Electrochemical corrosion process includes both generalized dissolution uniformly affecting the entire surface and localized areas of a component.

Metal implant corrosion is controlled by (1) the extent of the thermodynamic driving forces which cause corrosion (oxidation/reduction reactions) and (2) physical barriers which limit the kinetics of corrosion. In practice these two parameters that mediate the corrosion of orthopedic biomaterials can be broken down into a number of variables: geometric variables (e.g., taper geometry in modular component hip prostheses), metallurgical variables (e.g., surface microstructure, oxide structure and composition), mechanical variables (e.g., stress and/or relative motion) and solution variables (e.g., pH, solution proteins and enzymes) [25].

The corrosion resistance of a surgically implanted alloy is an essential characteristic since the metal alloys are in contact with a very aggressive media such as the body fluid due to the presence of chloride ions and proteins. In the corrosion process, the metallic components of the alloy are oxidized to their ionic forms and dissolved oxygen is reduced to hydroxide ions.

The corrosion characteristics of an alloy are greatly influenced by the passive film formed on the surface of the alloy and the presence of the alloying elements.

2.5. Wear resistance

Wear always occurs in the articulation of artificial joints as a result of the mixed lubrication regime. The movement of an artificial hip joint produces billions of microscopic particles that are rubbed off cutting motions. These particles are trapped inside the tissues of the joint capsule and may lead to unwanted foreign body reactions. Histocytes and giant cells phagocytose and "digest" the released particles and form granulomas or granuloma-like tissues. At the boundary layer between the implant and bone, these interfere with the transformation process of the bone leading to osteolysis. Hence, the materials used to make the femoral head and cup play a significant role in the device performance. Since the advent of endoprosthetics, attempts have been made to reduce wear by using a variety of different combinations of materials and surface treatments.

Nowadays, the materials used for biomedical applications are mainly metallic materials such as 316L stainless steel, cobalt chromium alloys (CoCrMo), titanium-based alloys (Ti-6Al-4V) and miscellaneous others (including tantalum, gold, dental amalgams and other "specialty" metals). Titanium alloys are fast emerging as the first choice for majority of applications due to the combination of their outstanding characteristics such as high strength, low density, high immunity to corrosion, complete inertness to body environment, enhanced compatibility, low Young´s modulus and high capacity to join with bone or other tissues. Their lower Young´s modulus, superior biocompatibility and better corrosion resistance in comparison with conventional stainless steels and cobalt-based alloys, make them an ideal choice for bio-applications [26]. Because of the mentioned desirable properties, titanium and titanium alloys are widely used as hard tissue replacements in artificial bones, joints and dental implants.

3. Titanium and titanium alloys

The elemental metal titanium was first discovered in England by William Gregor in 1790, but in 1795 Klaproth gave it the name of titanium. Combination of low density, high strength to weight ratio, good biocompatibility and improved corrosion resistance with good plasticity and mechanical properties determines the application of titanium and its alloys in such industries as aviation, automotive, power and shipbuilding industries or architecture as well as medicine and sports equipment.

Increased use of titanium and its alloys as biomaterials comes from their superior biocompat-ibility and excellent corrosion resistance because of the thin surface oxide layer, and good mechanical properties, as a certain elastic modulus and low density that make that these metals present a mechanical behaviour close to those of bones. Light, strong and totally biocompatible, titanium is one of the few materials that naturally match the requirements for implantation in the human body. Among all titanium and its alloys, the mainly used materials in biomedical field are the commercially pure titanium (cp Ti, grade 2) and Ti-6Al-4V (grade 5) alloy. They are widely used as hard tissue replacements in artificial bones, joints and dental implants. As a hard tissue replacement, the low elastic modulus of titanium and its alloys is generally viewed as a biomechanical advantage because the smaller elastic modulus can result in smaller stress shielding.

Other property that makes titanium and its alloys the most promising biomaterials for implants is that titanium-based materials in general rely on the formation of an extremely thin, adherent, protective titanium oxide film. The presence of this oxide film that forms spontaneously in the passivation or repassivation process is a major criterion for the excellent biocompatibility and corrosion resistance of titanium and its alloys.

Concerning the medical applications of these materials, the use of cp (commercially pure) Titanium is more limited to the dental implants because of its limited mechanical properties.

In cases where good mechanical characteristics are required as in hip implants, knee implants, bone screws, and plates, Ti-6Al-4V alloy is being used [27] [28]. One of the most common applications of titanium alloys is artificial hip joints that consist of an articulating bearing (femoral head and cup) and stem [24], where metallic cup and hip stem components are made of titanium. As well, they are also often used in knee joint replacements, which consist of a femoral and tibial component made of titanium and a polyethylene articulating surface.

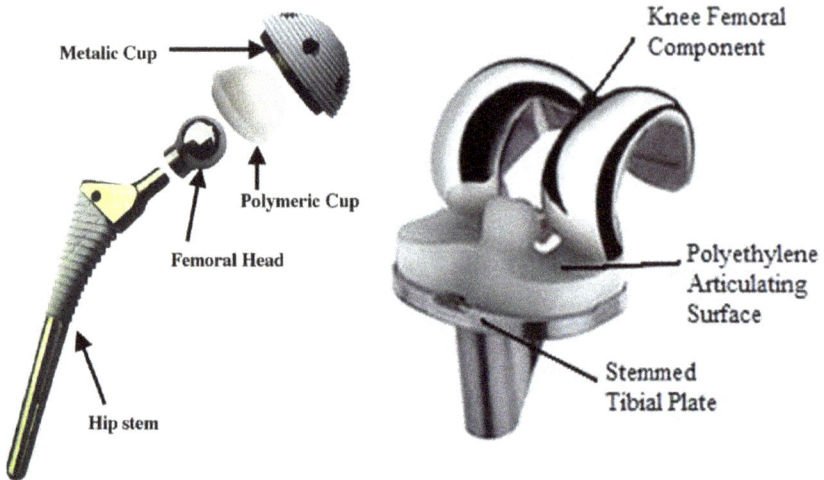

Figure 3. Schematic diagram of artificial hip joint (left) and knee implant [29] (right)

3.1. Wear problems in titanium and titanium alloys

The fundamental drawback of titanium and its alloys which limits wider use of these materials include their poor fretting fatigue resistance and poor tribological properties [30] [31], because of its low hardness [32]. Their poor tribological behavior is characterized by high coefficient of friction, severe adhesive wear with a strong tendency to seizing and low abrasion resistance [33]. Titanium tends to undergo severe wear when it is rubbed between itself or between other materials. Titanium has tendency for moving or sliding parts to gall and eventually seize. This causes a more intensive wear as a result of creation of adhesion couplings and mechanical instability of passive layer of oxides, particularly in presence of third bodies (Figure 4). Owing to this effect, in cases of total joint replacements made of titanium head and polymer cup, the 10%-20% of joints needs to be replaced within 15-20 years and the aseptic loosening accounts for approximately 80% of the revisions [34]. The reason for the failure of the implants is due to the high friction coefficient of these materials that can lead to the release of wear debris from the implant into the bloodstream that results in an inflammation of the surrounding tissue and gives rise to the bone resorption (osteolysis) [35] [36], which ultimately leads to loosening of the implant and hence the implant has to be replaced by a new one.

Contact
load force

Direction
of sliding

Wear
debris

Transfer film
Depends on environment
(chemistry, humidity,
temperature, etc.)

Tribo-modified
wear track and
possible subsurface
damage

Substrate

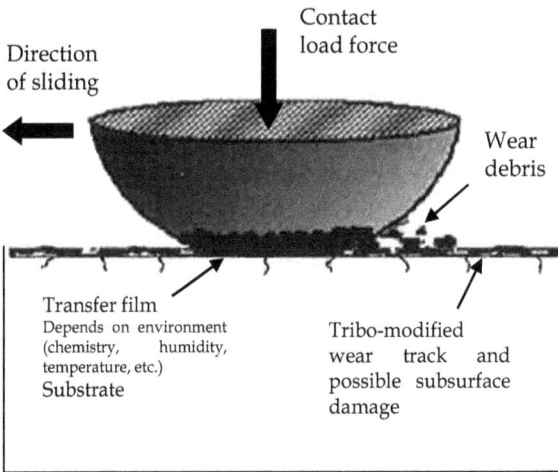

Figure 4. Schematic representation of a sliding tribological coating with the presence of third bodies [37]

3.2. Corrosion behaviour of titanium and titanium alloys

All metals and alloys are subjected to corrosion when in contact with body fluid as the body environment is very aggressive owing to the presence of chloride ions and proteins. A variety of chemical reactions occur on the surface of a surgically implanted alloy. The metallic components of the alloy are oxidized to their ionic forms and dissolved oxygen is reduced to hydroxide ions.

Most metals and alloys that resist well against corrosion are in the passive state. Metals in the passive state (passive metals) have a thin oxide layer (TiO_2 in case of titanium) on their surface, the passive film, which separates the metal from its environment [38]. Typically, the thickness of passive films formed on these metals is about 3-10 nm [39] and they consist of metal oxides (ceramic films). The natural oxide is amorphous and stoichiometrically defective. It is known that the protective and stable oxides on titanium surfaces (TiO_2) are able to provide favorable osseointegration. The stability of the oxide depends strongly on the composition structure and thickness of the film [40].

Because of the presence of an oxide film, the dissolution rate of a passive metal at a given potential is much lower than that of an active metal. It depends mostly on the properties of the passive film and its solubility in the environment. These films which form spontaneously on the surface of the metal prevent further transport of metallic ions and/or electrons across the film. To be effective barriers, the films must be compact and fully cover the metal surface; they must have an atomic structure that limits the migration of ions and/or electrons across the metal oxide–solution interface; and they must be able to remain on the surface of these alloys even with mechanical stressing or abrasion, expected with orthopedic devices [25].

The relatively poor tribological properties and possible corrosion problems have led to the development of surface treatments to effectively increase near-surface strength, improving the hardness and abrasive wear resistance thereby reducing the coefficient of friction as well as avoiding or reducing the transference of ions from the surface or bulk material to the surrounding tissue.

3.3. Osseointegration of titanium and titanium alloys

When an implant is surgically placed within bone there are numerous biological, physical, chemical, thermal and other factors functioning that determine whether or not osseointegration will occur.

Titanium and its alloys have been widely used for dental and orthopedic implants under load-bearing conditions because of their good biocompatibility coupled with high strength and fracture toughness. Despite reports of direct bonding to bone, they do not form a chemical bond with bone tissue. For the last decade, various coatings have been attempted to provide titanium and its alloys with bond-bonding ability, which spontaneously bond to living bone. Hydroxyapatite plasma spray coatings are widely used in cementless hip replacement surgery, but the hydroxyapatite coating, although exhibiting a very good biocompatibility, presents some disadvantages including delamination of the coating layer from the substrate, difficulties in controlling the composition of the coating layer and degradation of the coating layer itself, which can release debris becoming a source of third body wear [41].

A strong and durable bone to implant connection can be achieved by the formation of a stable bone tissue at the bone-implant interface by proper implant surface treatments, as can be electrochemical deposition, dipping and physical vapor deposition techniques [42].

3.4. Surface treatments of titanium and titanium alloys

Surface engineering can play a significant role in extending the performance of orthopedic devices made of titanium several times beyond its natural capability.

The main objectives of surface treatments mainly consist of the improvement of the tribological behaviour, corrosion resistance and osseointegration of the implant. There are coatings for enhanced wear and corrosion resistance by improving the surface hardness of the material that can be applied by different surface modifications techniques such as surface oxidation, physical deposition methods like ion implantation and plasma spray coatings, as well as thermo-chemical surface treatments such as nitriding, carburizing and boriding [43] [44].

Great efforts have been devoted to thickening and stabilizing surface oxides on titanium to achieve desired biological responses. The biological response to titanium depends on the surface chemical composition, and the ability of titanium oxides to absorb molecules and incorporate elements. Surface topography plays a fundamental role in regulating cell behavior, e.g. the shape, orientation and adhesion of cells.

One possible alternative to solve tribological problems and which is going to explain more detail consists of protecting the alloy surface by means of biocompatible Diamond-Like Carbon

(DLC) coatings. "Diamond-Like Carbon" is a generic term referring to amorphous carbon films, deposited by either Physical Vapor Deposition (PVD) or Plasma-Enhanced Chemical Vapor Deposition (PECVD). DLC coatings basically consist of a mixture of diamond (sp^3) and graphite (sp^2). The relative amounts of these two phases will determine much of the coating properties. They are thus metastable and mostly amorphous, "crystalline" clusters being too small or too defective to reach graphite or diamond structures. Both the mechanical and the tribological properties of DLC coatings have been studied for about 30 years, and several different types of DLC coatings can currently be found. DLC films are attractive biomedical materials due to their relatively high hardness, low friction coefficient, owing to the solid lubricant because of its graphite and amorphous carbon contents [31], good chemical stability and excellent bio and hemocompatibility [45] [44] [46] [47]. Cells are seen to grow well on these films coated on titanium and other materials without any cytotoxicity and inflammation.

Oxidation remains the most popular technique for the surface modification of Ti alloys; these oxide layers on titanium are commonly produced by either heat treatment [48] [49] [50] or electrolytic anodizing [51]. Thermal oxidation results in the formation of a 15-30 μm thick titanium dioxide layer of the rutile phase. However, due to their long-term high temperature action, thermal diffusion processes can also lead to the formation of a diffusion sub-layer consisting of an oxygen solid solution in α-Ti, and development of phase segregation and coalescence which may cause substrate embrittlement and worsened mechanical and/or corrosion performance.

Conventional anodic oxidation, which is carried out in various solutions providing passivation of the titanium surface, generates thin films of amorphous hydrated oxide or crystalline TiO_2 in the anatase form [52]. These films exhibit poor corrosion resistance in some reducing acids and halide solutions, while rutile generally possesses much better protective properties. However, recent developments in high voltage anodizing allow the production of crystalline rutile/anatase films at near to ambient temperature [53]. By anodic oxidation, elements such as Ca and P can be imported into the surface oxide on titanium and the micro-topography can be varied through regulating electrolyte and electrochemical conditions. The presence of Ca-ions has been reported to be advantageous to cell growth, and in vivo data show implant surfaces containing both Ca and P enhance bone apposition on the implant surface.

Furthermore, there are alternative methods to improve the biocompatibility such as biocompatible chemicals [54] and materials such as ceramics for coating. In some studies, titanium surfaces were modified using phosphoric acid in an "in vitro" study to improve the biocompatibility of dental implants. Results indicated that pretreatment of the implant with phosphoric acid caused no citotoxicity to the osteoblasts [55]. Micro arc oxidation method in phosphoric acid on titanium implants provided chemical bonding sites for calcium ions during mineralization [56].

Moreover, there have been developed coatings for high osseointegration. Hydroxyapatite (HA) coating is a proven method to improve the implants' mechanical bonding [57] [58], biocompatibility and improve the osseointegration. The higher the degree of osseointegration, the higher is the mechanical stability and the probability of implant loosening becomes smaller. The process of osseointegration depends upon the surface properties such as surface chemistry, surface topography, surface roughness and mainly the surface energy. TiO_2, calcium phosphate, titania/hydroxiapatite composite and silica coating by the sol-gel method can be applied on the surface of the titanium and titanium alloys. Plasma Electrolytic Oxidation (PEO) or Micro-Arc Oxidation (MAO) technique is used for the synthesize TiO_2 layer. This technique is based on the modification of the growing anodic film by arc micro-discharges, which are initiated at potentials above the breakdown voltage of the growing oxide film and move rapidly across the anode surface. This technology provides a solution by transforming the surface into a dense layer of ceramic which not only prevents galling but also provides excellent dielectric insulation for contact metals, helping to protect them against aggressive galvanic corrosion. PEO process transforms the surface of titanium alloys into a complex ceramic matrix by passing a pulsed, bi-polar electrical current in a specific wave formation through a bath of low concentration aqueous solution. A plasma discharge is formed on the surface of the substrate, transforming it into a thin, protective layer of titanium oxide, without subjecting the substrate itself to damaging thermal exposure.

Among all the above mentioned surface treatments, Diamond-Like Carbon coating and Plasma Electrolytic Oxidation are the most promising ones applied on titanium surfaces. These two treatments are explained in more detail in the following sections.

3.4.1. Diamond-like carbon coatings

In some biomedical applications continuously sliding contact is required, subjecting the implant to aggressive situations. To achieve and maintain higher efficiency and durability under such increasingly more severe sliding conditions, protective and/or solid coatings are becoming prevalent.

These coatings can generally be divided in two broad categories [59] : "soft coatings", which are usually good for solid lubrication and exhibit low friction coefficients, and "hard coatings", which are usually good for protection against wear, and exhibit low wear rates and hence longer durability (Figure 5).

It would thus seem to be difficult to associate low friction and high wear resistance with all types of coating in most tribological contacts. Some trade-offs can be found in combining both hard and soft materials in composite or multilayer coatings, which require complex procedures and further optimization of the deposition process. Nevertheless, a diverse family of carbon-based materials seems to "naturally" combine the desired set of tribological properties, providing not only low friction but also high wear resistance. These materials are widely known as the diamond and Diamond-Like Carbon (DLC) coatings. They are usually harder than most metals and/or alloys, thus affording very high wear resistance and, at the same time, impressive friction coefficients generally in the range of 0.05-0.2 [60] [61] [62]. In some cases, friction values lower than 0.01 have been reported [63] [64], offering a sliding regime often referred to as "superlubricity". These exceptional tribological abilities explain the increasing success of Diamond-Like Carbon coatings over the years, both in industrial applications and in the laboratory. The exceptional tribological behavior of Diamond-Like Carbon films appears to be due to a unique combination of surface chemical, physical, and mechanical interactions at their sliding interfaces [65].

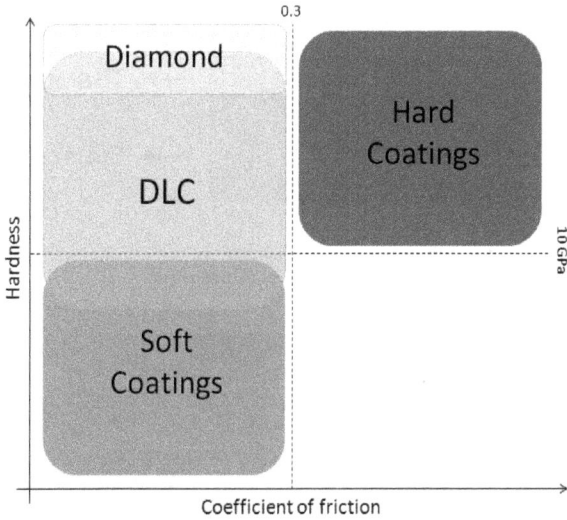

Figure 5. Classification of coatings with respect to hardness and coefficient of friction, highlighting the special case of carbon-based coatings

Since their initial discovery in the early 1950s, Diamond-Like Carbon coatings have attracted the most attention in recent years, mainly because they are cheap and easy to produce and offer exceptional properties for demanding engineering and medical applications. They can be used in invasive and implantable medical devices. These films are currently being evaluated for their durability and performance characteristics in certain biomedical implants including hip and knee joints and coronary stents.

Diamond-Like Carbon is the only coating that can provide both high hardness and low friction under dry sliding conditions. These films are metastable forms of carbon combining both sp2 and sp3 hybridizations, including hydrogen when a hydrocarbon precursor is used during deposition. The tribological behavior of Diamond-Like Carbon films requires a solid background on the chemical and structural nature of these films, which, in turn, depends on the deposition process and/or parameters. The chemical composition, such as the hydrogen and/ or nitrogen content or the presence of other alloying elements, controls the mechanical and tribological properties of a sliding pair consisting of DLC on one or both sliding surfaces [66]. For example, DLC samples containing different concentrations of titanium (Figure 6) have also been examined "in vitro" to obtain a biocompatible surface that is hard, preventing abrasion and scratching [67].

It is well known that Diamond-Like Carbon films usually present smooth surfaces, except maybe in the case of films formed by unfiltered cathodic vacuum arc deposition (Figure 7). Roughness of the films on industrial surfaces will then be mainly controlled by the substrate roughness and can therefore be minimized.

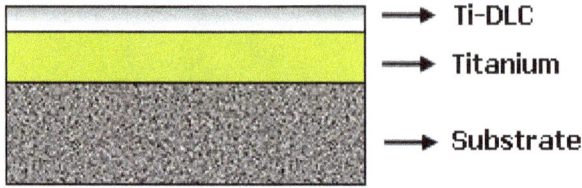

Figure 6. Scheme of titanium doped DLC coating. In this case, the first titanium layer was deposited in order to improve adhesion of DLC coating to the substrate and relax stress of the coating

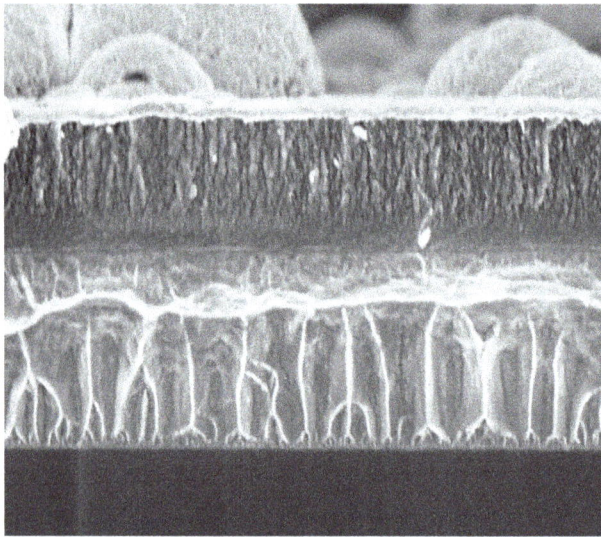

Figure 7. SEM (Scanning electron microscopy) micrograph of Ti-DLC coating deposited by physical vapour deposition technique using cathodic arc evaporation method

A frequently observed feature in tribological testing of Diamond-Like Carbon films is the formation of transfer layer. The formation of carbonous transfer layer on the sliding surface was observed to reduce the friction coefficient [68].

DLC coatings are usually applied by means of Cathodic Arc Evaporation Physical Vapor Deposition technology. An arc can be defined as a discharge of electricity between two electrodes. The arc evaporation process begins with the striking of a high current, low voltage arc on the surface of a cathode that gives rise to a small (usually a few microns wide) highly energetic emitting area known as a cathode spot. The localised temperature at the cathode spot is extremely high (around 15000 °C), which results in a high velocity (10 km/s) jet of vaporised cathode material, leaving a crater behind on the cathode surface.

The plasma jet intensity is greatest normal to the surface of the cathode and contains a high level of ionization (30%-100%) multiply charged ions, neutral particles, clusters and macro-particles (droplets). The metal is evaporated by the arc in a single step, and ionized and accelerated within an electric field. Theoretically the arc is a self-sustaining discharge capable of sustaining large currents through electron emission from the cathode surface and the re-bombardment of the surface by positive ions under high vacuum conditions.

If a reactive gas is introduced during the evaporation process dissociation, ionization and excitation can occur during interaction with the ion flux and a compound film will be deposited. Without the influence of an applied magnetic field the cathode spot moves around randomly evaporating microscopic asperities and creating craters. However if the cathode spot stays at one of these evaporative points for too long it can eject a large amount of macro-particles or droplets as seen above. These droplets are detrimental to the performance of the coating as they are poorly adhered and can extend through the coating.

A recent tribological study carried out about the effect of deposition of Diamond-Like Carbon coatings on a substrate of Ti-6Al-4V for knee implants has confirmed that these types of coating improve the tribological response of substrate decreasing the coefficient of friction (μ) (Table 1) and reducing the wear of the surface (Figure 8) [69]. For this study fretting tests were performed using alumina balls as counter body, bovine serum as lubricant and a continuous temperature of 37 °C, trying to simulate real environment.

Sample	$\mu \pm SD$ (standard deviation)	Disc Wear Scar, Maximum Depth (μm)
Ti-6Al-4V	0.86 ± 0.08	10 ± 3
Ti-DLC	0.24 ± 0.01	Polishing Effect

Table 1. Friction coefficients values and ball and disc wear scars measurements

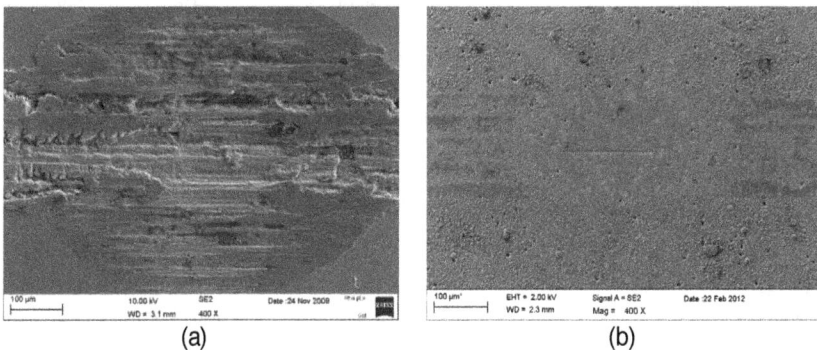

(a)   (b)

Figure 8. SEM micrographs of the fretting tests wear scars. Ti-6Al-4V (left), Ti-DLC (right)

3.4.2. Plasma electrolytic oxidation treatment

In biomedical application titanium is the most employed alloy due to its biocompatibility as an implant material, attributed to surface oxides spontaneously formed in air and/or physiological fluids [70]. Cellular behaviors, e.g. adhesion, morphologic change, functional alteration, proliferation and differentiation are greatly affected by surface properties, including composition, roughness, hydrophilicity, texture and morphology of the oxide on titanium [71] [72]. The natural oxide is thin (about 3–10nm in thickness [39]) amorphous and stoichiometrically defective. It is known that the protective and stable oxides on titanium surfaces are able to provide favorable osseointegration [73] [74]. The stability of the oxide depends strongly on the composition structure and thickness of the film [75].

On titanium and its alloys a thin oxide layer is formed naturally on the surface of titanium metal in exposure to air at room temperature [76] [77] [78]. Titania (TiO_2) exists in three polymorphic forms: rutile, anatase and brookite. Rutile, stable form of titania at ambient condition, possesses unique properties [79]. The metastable anatase and brookite phases convert to rutile upon heating. However, contact loads damage this thin native oxide film and cause galvanic and crevice corrosion as well as corrosion embrittlement. Moreover, the low wear resistance and high friction coefficient without applied protective coatings on the surface gravely limit its extensive applications. The most accepted technique for the surface modification of Ti alloys is oxidation. Anodizing produces anatase phase of titania that shows poor corrosion resistance in comparison with rutile phase. Recent developments in high voltage anodizing cause a crystalline rutile/anatase film at near to room temperature.

Attempts to improve surface properties of titanium and its alloys over the last few decades have led to development of Plasma Electrolytic Oxidation (PEO) technique by Kurze et al. [80] [81], which is a process to synthesize the ceramic-like oxide films at high voltages. This technique is based on the modification of the growing anodic film by spark/arc microdischarges in aqueous solutions (Figure 9), which are initiated at potentials above the breakdown voltage of the growing oxide film and move rapidly across the anode surface [53]. Since they rapidly develop and extinguish (within 10^{-4}-10^{-5} s), the discharges heat the metal substrate to less than 100-150 ℃. At the same time the local temperature and pressure inside the discharge channel can reach 10^{-3}-10^{-4} K and 10^{-2}-10^{-3} MPa, respectively, which is high enough to give rise to plasma thermo-chemical interactions between the substrate and the electrolyte. These interactions result in the formation of melt-quenched high-temperature oxides and complex compounds on the surface, composed of oxides of both the substrate material and electrolyte-borne modifying elements. The result is a porous oxide coating.

The PEO coating shows a significantly higher thickness (18 μm ± 4 μm) than PVD coatings and also a different morphology. The external part of the layer is porous (with pore diameter ranging from 3 to 8 μm) (Figure 10). The coating becomes increasingly compact on going towards the interface with the substrate. This kind of morphology leads to a relatively high surface roughness.

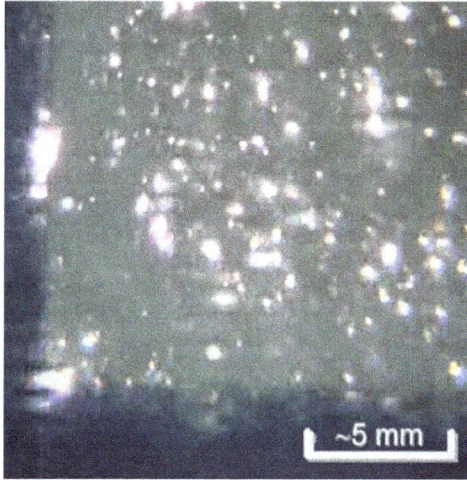

Figure 9. Photography of the arc micro-discharges in PEO process

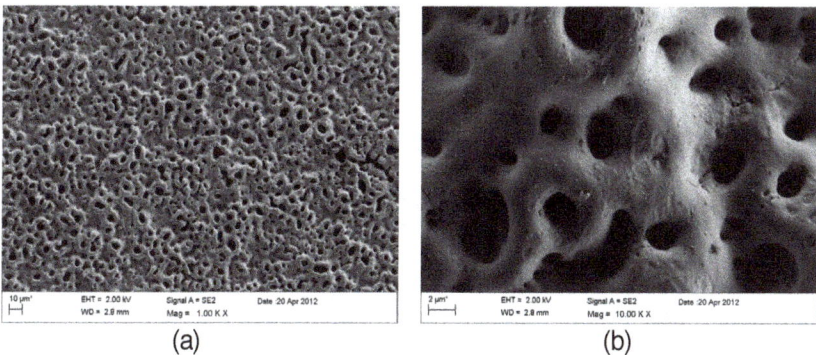

Figure 10. SEM micrographs of porosity of the external layer in PEO treatment. a) overview and b) detail

This method is characterized by the titanium surface, at near-to-ambient bulk temperature, into the high temperature titanium oxide (rutile) modified by other oxide constituents. Economic efficiency, ecological friendliness, corrosion resistance, high hardness, good wear resistance, and excellent bonding strength with the substrate are the other characteristics of this treatment [82] [83] [84].

The main conversion products formed by the PEO treatment are titanium oxides: rutile and anatase, typical anodic oxidation products of titanium. The structure and composition of anodic oxide films are known to be strongly dependent on film formation temperature and potential [85] [86]. In the case of PEO coatings, both the electrolyte composition and the current density regime have an influence on the phase composition and morphology of the anodic oxide layer [87]. A higher spark voltage causes a higher level of discharge energy, which provides a larger pore [88].

The influence of electrolyte characteristics on the phase composition of PEO films on titanium has previously been studied [89] [90]. It has been shown that surface layers composed of rutile, anatase, rutile/anatase, as well as oxides of electrolyte elements (e.g. Al_2O_3, MgO, WO_3), their hydroxides and complex oxides (e.g. Al_2TiO_5, $AlPO_4$, $CaWO_4$, $BaTiO_3$, $MnTiO_3$, etc.) can be produced.

Surfaces containing Ca and/or P induce osteoinduction of new bones and become bioactive. Ca and P ions can be incorporated into the layer, controlling the electrolyte employed during the electro oxidation process, and they further transform it into hydroxyapatite by a hydrothermal treatment [41].

One technique that could show the effect of the electrolyte in the chemical composition of the coating could be the EDS (Energy Dispersive Spectroscopy) technique. In the following graphs a comparative study can be observed. The results of different samples, uncoated cp Ti, a coating obtained with a commercial electrolyte and a coating prepared in an aqueous electrolyte containing calcium phosphate and β-glycerophosphate, are showed in the following spectrums. The Ca- and P-containing titania coatings produced by PEO improve the bioactivity of the titanium-constructed orthopedic implant [91]. In Figure 11, in spectrum b) and c) can be observed the difference in the calcium quantity presented into the coating.

The biological response to titanium depends on the surface chemical composition and the ability of titanium oxides to absorb molecules and incorporate elements [92]. Surface topography plays a fundamental role in regulating cell behaviour, e.g. the shape, orientation and adhesion of cells [93] [94]. As a surface begins to contact with biological tissues, water molecules first reach the surface. Hence, surface wettability, initially, may play a major role in adsorption of proteins onto the surface, as well as cell adhesion. Cell adhesion is generally better on hydrophilic surfaces. It is known that changes in the physicochemical properties, which influence the hydrophilicity of Ti dioxide, will modulate the protein adsorption and further cell attachment [39]. By anodic oxidation, elements such as Ca and P can be imported into the surface oxide on titanium and the micro-topography can be varied through regulating electrolyte and electrochemical conditions. The presence of Ca-ions has been reported to be advantageous to cell growth, and "in vivo" data show implant surfaces containing both Ca and P enhance bone apposition on the implant surface.

Some experiments carried out to study the tribological behaviour of the PEO-treated Ti-6Al-4V by means of dry sliding tests against PS (plasma sprayed) Al_2O_3–TiO_2 and compared with that of thin PVD coatings showed that the best tribological behavior, both in terms of low coefficient of friction and high wear resistance (i.e. low wear damage) was displayed by the PEO treated samples. The highest wear resistance was displayed by the PEO-treated samples, with negligible wear loss even under the highest applied load of 35 N. This good tribological behavior should be mainly related to the superior thickness of this coating that can better support the applied load.

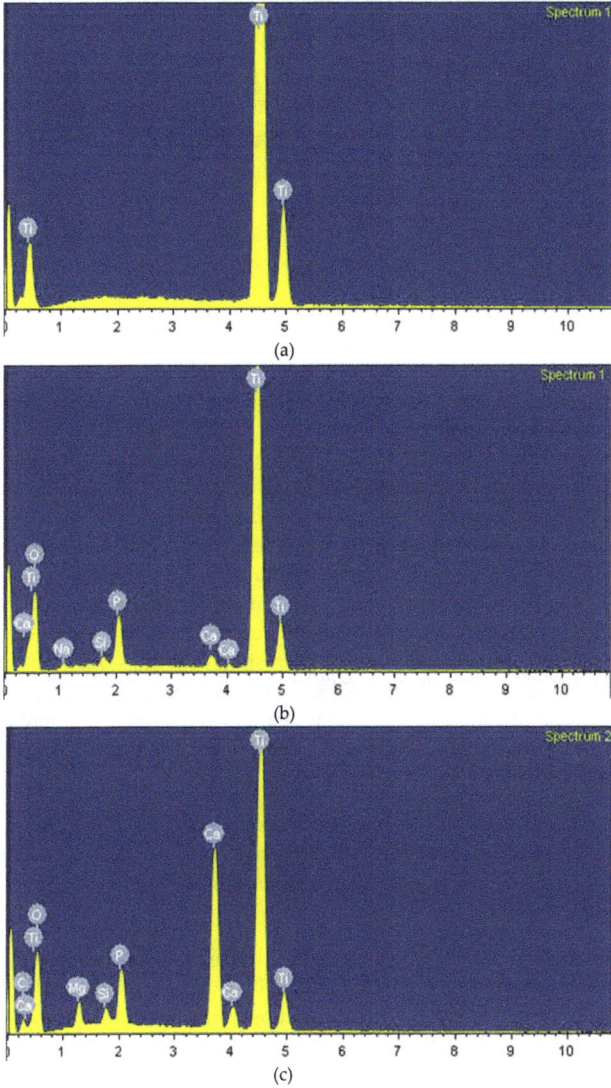

Figure 11. a) Microchemical analysis of cp Ti, b) microchemical analysis of coating prepared with commercial electrolyte, c) microchemical analysis of coating prepared with calcium phosphate and β-glycerophosphate electrolyte.

The PEO treatment leads to a very good tribological behavior, significantly reducing both wear and friction of the Ti-6Al-4V alloy, even under high applied loads (up to 35 N). This good tribological behaviour should be mainly related to the superior thickness of this coating, which can better support the applied load. The main wear mechanism is micro-polishing and the coating thickness dictates its tribological life [95].

Last studies carried out have concluded that the PEO surface treatments enhance the biological response "in vitro", promoting early osteoblast adhesion, and the osseointegrative properties "in vivo", accelerating the primary osteogenic response, as they confirmed by the more extensive bone-implant contact reached after 2 weeks of study [94].

4. Conclusions

Titanium and its alloys are considered to be among the most promising engineering materials across a range of application sectors. Due to a unique combination of high strength-to-weight ratio, melting temperature and corrosion resistance, interest in the application of titanium alloys to mechanical and tribological components is growing rapidly in a wide range of industries, especially in biomedical field, also due to their excellent biocompatibility and good osseointegration. In such application, components made from Ti-alloys are often in tribological contact with different materials (metals, polymers or ceramics) and media, under stationary or dynamic loading and at various temperatures. These contact loads can cause damage of the thin native oxide film which passivates the titanium surface; and the metal can undergo intensive interactions with the counterface material and/or the surrounding environment. These interactions can generate various adverse effects on titanium components, such as high friction or even seizure (galvanic and crevice corrosion) as well as corrosion embrittlement, which lead to the premature failure of the implanted systems. The development of new specialized surface modification techniques for titanium and its alloys is therefore an increasingly critical requirement in order to control or prevent these effects and improve osseointegration, hence extending the lifetime of the implant.

Physical Vapour Deposition (PVD) technique allows develop Diamond-Like Carbon coatings that can be doped with different elements as titanium, tantalum, silver... which are biocompatible and increase the corrosion and wear resistance of the substrate, diminishing friction coefficient.

Plasma Electrolytic Oxidation (PEO) technique provides a possibility for the variation of composition and structure of the surface oxide film and attracts special interest for the corrosion protection and the optimization of friction and wear of titanium alloys as well as enhance the osseointegration.

Acknowledgements

The authors acknowledge financial support from the Spanish Ministry of Science and Innovation obtained in the project: CSD2008-00023 FUNCOAT (in the frame of the CONSOLIDER INGENIO-2010 program) and from the Basque Government.

Author details

Virginia Sáenz de Viteri* and Elena Fuentes

*Address all correspondence to: virginia.saenzdeviteri@tekniker.es

IK4-Tekniker, Eibar, Spain

References

[1] Korkusuz, P. & Korkusuz, F. Hard Tissue – Biomaterial Interactions. In: Michael J. Yaszemski; Debra J. Trantolo; Kai-Uwe Lewandrowski; Vasif Hasirci, David E. Altobelli & Donald L. Wise. (ed.) Biomaterials in Orthopedics. United States of America: Marcel Dekker, Inc.; 2004. p1-40.

[2] Santavirta, S., Gristina, A. & Konttinen, YT. Cemented versus cementless hip arthroplasty: a review of prosthetic biocompatibility. Acta Orthopaedica Scandinavica 1992;63 225-232.

[3] Park, J.B. & Bronzino, L.D., (ed.) Biomaterials: principles and applications. Boca Raton, Florida: CC Press; 2003 p. 1-241.

[4] Ramakrishna, S., Mayer, J., Wintermantel, E. & Leong K. W. Biomedical applications of polymer-composite materials: A review. Composites Science and Technology 2001;61(9) 1189-1224.

[5] Wise, D.L. Biomaterials engineering and devices. Berlin: Human Press; 2000.

[6] Kurtz, S., Ong, K., Jau, E., Mowat, F. & Halpern, M. Projections of primary and revision hip and knee arthroplasty in the United States from 2005 to 2030. Journal of Bone and Joint Surgery – American Volume 2007;89 780-785.

[7] Geetha, M., Singh, A.K., Asokamani, R. & Gogia, A.K. Ti based biomaterials, the ultimate choice for orthopaedic implants- A review. Progress in Material Science 2009;54 397-425.

[8] Long, M. & Rack. H.J. Titanium alloys in total joint replacement – A materials science perspective. Biomaterials 1998;19 1621-1639.

[9] Wang, K. The use of titanium for medical applications in the USA. Materials Science and Engineering A - Structural Materials Properties Microstructure and Processing 1996;213 134-137.

[10] http://azom.com/article.aspx?ArticleID=108

[11] Williams, D.F. On the mechanisms of biocompatibility. Biomaterial 2008;29(20) 1941-2953.

[12] Heimke, G. & Stock, D. Clinical application of ceramic osseo – or soft tissue - integrated implant. Orthopedic Ceramic Implants 1984;4 1-19.

[13] Black, J. & Hastings G.W. Handbook of biomaterials properties. London UK: Chapman and Hall; 1998.

[14] Lawrence Katz, J. Anisotropy of Young's modulus of bone. Nature 1980;283 106-107.

[15] Viceconti, M., Muccini, R., Bernakiewicz, M., Baleani M. & Cristofolini, L. Large-sliding contact elements accurately predict levels of bone-implant micromotion relevant to osseointegration. Journal of Biomechanics 2000;33 1611-1618.

[16] Wennerberg, A. On surface roughness and implant incorporation. Göteborg. Sweden: Biomaterials/Handicap Research; Institute of Surgical Sciences, Göteborgs Universitet, 1996.

[17] Calrsson, L.V., Macdonald, W., Magnus Jacobsson, C. &. Albrektsson T. Osseointegration Principles in Orthopedics: Basic Research and Clinical Applications. In: Michael J. Yaszemski; Debra J. Trantolo; Kai-Uwe Lewandrowski; Vasif Hasirci, David E. Altobelli & Donald L. Wise. (ed.) Biomaterials in Orthopedics. United States of America: Marcel Dekker, Inc.; 2004 p223-240.

[18] Junker, R., Dimakis, A., Thoneick, M. &. Jansen, J.A. Effects of implant surface coatings and composition on bone integration: a systematic review. Clinical Oral Implants Research 2009;20 185-206.

[19] Kim, H.J., Kim, S.H., Kim, M.S., LeeJ, E.J., Oh, H.G., Oh, W.M., et al. Varying Ti-6Al-4V surface roughness induces different early morphologic and molecular responses in MG63 osteoblast-like cells. Journal of Biomedical Materials Research 2005,74A 366-373.

[20] Vlacic-Zischke, J., Hamlet, S.M., Friis, T., Tonetti, M.S. & Ivanovski, S. The influence of surface microroughness and hydrophilicity of titanium on the up-regulation of TGFb/BMP signalling in osteoblasts. Biomaterials 2011;32 665-671.

[21] http://360oandp.com/Technology-Osseointegration.aspx

[22] Montanaro, L., Campoccia, D. & Arciola, C.R. Nanostructured materials for inhibition of bacterial adhesion in orthopedic implants: a minireview. International Journal of Artificial Organs 2008;31 771-776.

[23] Black. J. Othopaedic Biomaterials in Research and Practice. New York: Churchill Livingstone; 1988.

[24] Jacobs, J.J., Gilbert, J.L. & Urban, R.M. Corrosion of metal orthopaedic implants. Journal of Bone and Joint Surgery – American Volume 1998;80 268-282.

[25] Hallab, N.J., Urban, R. M. & Jacobs, J.J. (2004). Corrosion and Biocompatibility of Orthopedic Implants, In: Michael J. Yaszemski; Debra J. Trantolo; Kai-Uwe Lewan-

drowski; Vasif Hasirci, David E. Altobelli & Donald L. Wise. (ed.) Biomaterials in Orthopedics. United States of America: Marcel Dekker, Inc.; 2004 p63-92.

[26] Liu, X., Chu, P.K. & Ding C. (2004). Surface modification of titanium, titanium alloys, and related materials for biomedical applications. Mater Sci Eng, Vo. R 47, (2004), pp. 49-121.

[27] Stadlinger, B., Ferguson, S.J., Eckelt, U., Mai, R., Lode, A.T., Loukota, R. & Sclotting F. Biomechanical evaluation of a titanium implant surface conditioned by a hydroxide ion solution. British Journal of Oral & Maxillofacial Surgery 2012;50 74-79.

[28] Subramani, K. & Mathew, R.T. Titanium Surface Modification. Techniques for Dental Implants – From Microscale to Nanoscale. Emerging Nanotechnologies in Dentistry. DOI: 10.1016/B978-1-4557-7862-1.00006-7.

[29] http://hss.edu/conditions_arthritis-of-the-knee-total-knee-replacement.asp

[30] Fraczek, T., Olejnik, M. & Tokarz, A. Evaluations of plasma nitriding efficiency of titanium alloys for medical applications. Metalurgija 2009;48(2) 83-86.

[31] Kustas, F.M. & Misra, M.S. Friction and Wear of Titanium Alloys, In: Scott D. Henry (ed.) Volume 18, Friction, Lubrication and Wear Technology. United States of America: ASM International; 1992. p. 1585-1598.

[32] Freese, H., Volas, M.G. & Wood, J.R. (2001). In: Brunette D.M., Tengvall P., Textor M., Thomsen P. (eds.) Titanium in Medicine. Springer: Berlin; 2001. p25-51.

[33] Yerokhin, A.L., Niea, X., Leyland, A. & Matthews, A. Characterization of oxide films produced by plasma electrolytic oxidation of a Ti–6Al–4V alloy. Surface & Coating Technology 2000;130 195–206.

[34] Malchau, H. & Herberts, P. Revision and re-revision rate in THR: a revision-risk study of 148,359 primary operations. Scientific exhibition, 65th annual meeting of the AAOS, New-Orleans, 1998.

[35] Chandra, A., Ryu, J.J., Karra, P., Shrotriya, P., Tvergaard,V., Gaisser, M. & Weik, T. Life expectancy of modular Ti6Al4V hip implants: Influence of stress and environment. Journal of the Mechanical Behavior of Biomedical Materials 2011;4 1990-2001.

[36] Wolford, L.M. Factors to consider in joint prosthesis systems, Proceedings (Baylor University Medical Center) 2006;19 232-238.

[37] Zabinski, J.S. & Voevodin, A.A. Ceramic and other hard coatings, In: Joze Vizintin, Mitjan Kalin, Kuniaki Dohda & Said Jahanmir (eds) Trybology of Mechanical Systems: A Guide to Present and Future Technologies. United States of America: ASME Press; 2004 p157-182.

[38] Landolt, D. Corrosion and Surface Chemistry of Metals. Lausanne Switzerland: EPFL Press; 2007.

[39] Neoh, K.G., Hu, X., Zheng, D. & Tang Kang, E. Balancing osteoblast functions and bacterial adhesion on functionalized titanium surfaces. Biomaterials 2012;33 2813-2822.

[40] Zhu, X., Chen, J., Scheideler, L., R. Reichl, R. & Geis-Gerstorfer, J. Effects of topography and composition of titanium surface oxides on osteoblast responses. Biomaterials 2004;25 4087-4103.

[41] Liu, F., Wang, F., Shimizu, T., Igarashi, K. & Zhao, L. (2005). Formation of hydroxyapatite on Ti-6Al-4V alloy by microarc oxidation and hydrothermal treatment. Surface & Coatings Technology 2005;199 220-224.

[42] Kokubo, T., Kim, H-M., Miyaji, F. & Nakamura, T. Preparation of bioactive Ti and its alloys via simple chemical surface treatment. Journal of Biomedical Materials Research 1996;32 409-417.

[43] Carapeto, A.P., Serro, A.P., Nunes, B.M.F., Martins, M.C.L., Todorovic, S., Duarte, M.T., André, V., Colaço, R. & Saramago, B. Characterization of two DLC coatings for joint prosthesis: The role of albumin on the tribological behavior. Surface & Coatings Technology 2010;204 3451-3458.

[44] Ma, G., Gong, S., Lin, G., Zhang, L. & Sun, G. A study of structure and properties of Ti-doped DLC film by reactive magnetron sputtering with ion implantation. Applied Surface Science 2012;258 3045-3050.

[45] Dowling, D.P. Evaluation of diamond-like carbon coated orthopedic implants. Diamond and Related Materials 1997;6 390-393.

[46] Zhang, L., Lv, P., Huang, Z.Y., Lin, S.P., Chen, D.H., Pan, S.R. & Chen, M. Blood compatibility of La_2O_3 doped diamond-like carbon films. Diamond and Related Materials 2008;17 1922-1926.

[47] Zheng, Y., Liu, D., Liu, X. & Li, L. Ti-TiC-TiC/DLC gradient nano-composite film on a biomedical NiTi alloy. Biomedical Materials 2008;3 044103-044109.

[48] Han, Y., Hong, S.H. & Xu, K.W. Porous nanocrystalline titania films by plasma electrolytic oxidation. Surface & Coatings Technology 2002;154 314–318.

[49] Huang, P., Wang, F., Xu, K. & Han, Y. Mechanical properties of titania prepared by plasma electrolytic oxidation at different voltages. Surface & Coatings Technology 2007; 201 5168–5171.

[50] Lange, R., Lüthen, F., Beck, U., Rychly, J., Baumann, A. & Nebe, B. Cell-extracellular matrix interaction and physic-chemical characteristics of titanium surfaces depend on the roughness of the material. Biomolecular Engineering 2002;19 255-261.

[51] Huang, P., Xu, K-W. & Han, Y. Preparation and apatite layer formation of plasma electrolytic oxidation film on titanium for biomedical application. Materials Letters 2005;59 185-189.

[52] Cigada, A., Cabrini, M. & Pedferri, P. Increase of the corrosion resistance of Ti6Al4V alloy by high thickness anodic oxidation. Journal of Materials Science – Materials in Medicine 1992;3 408-412.

[53] Yerokhin, A.L., Nie, X., Leyland, A., Matthews, A. & Dowey, S.J. Surface & Coatings Technology 1999;116.

[54] Nanci, A., Wuest, J.D., Peru, L., Brunet, P., Sharma, V., Zalzal, S. & McKee, M.D. Chemical modification of titanium surfaces for covalent attachment of biological molecules. J Biomed Mater, Vol. 40, (1998), pp. 237-242.

[55] Viorney, C., Guenther, H.L., Aronsson, B.O., Pechy, P., Descouts, P. & Gratzel, M. (2002). Osteoblast culture on polished titanium disks modified with phosphoric acids. Journal of Biomedical Materials Research 2002;62 149-155.

[56] Sul, Y.T., Johansson, C.B., Kang, Y., Jeon, D.G., Kang, Y., Jeong, D.G. & Albrektsson, T. Bone reaction to oxidized titanium implants with electrochemical anion sulphuric acid and phosphoric acid incorporation. Clinical Implant Dentistry and Related Research 2002;4 78-87.

[57] Cook, S.D., Thomas, K.A., Kay, J.F. & Jarcho, M. Hydroxyapatite-coated porous titanium for use as an orthopedic biologic attachment system. Clinical Orthopaedic and Related Research 1988;230 303-312.

[58] Rashmir- Raven, M.A., Richardson, D.C., Aberman, H.M. & DeYoung, D.J. The response of cancellous and cortical canine bone to hydroxyapatite-coated and uncoated titanium rods. Journal of Applied Biomaterials 1995;6 237-242.

[59] Holmberg, K. & Matthews, A. Coatings Tribology – Properties, Techniques and Applications in Surface Engineering. Amsterdam, The Netherlands: Elsevier; 1994.

[60] Grill, A. Wear 1993;168(1-2) 143.

[61] Grill, A. Surface & Coatings Technology 1997;94-95(1-3) 507.

[62] Erdemir, A. & Donnet, C. Trybology of diamond, diamond-like carbon, and related films. In: B. Bhushan (ed.) Handbook of Modern Tribology, Vol. 2. Materials Coatings. Boca Raton, Florida: CRC Press; 2001.

[63] Donnet, C., Belin, M., Augé, J.C., Martin, J.M., Grill, A. & Patel, V. Surface & Coatings Technology 1994;68-69 626.

[64] Erdemir, A., Erylmaz, O.L. & Fenske, G. (2000). Journal of Vacuum Science of Technology A – Vacuum Surfaces and Films 2000;18(4) 1987.

[65] Fontaine, J., Donnet, C. & Erdemir, A. Fundamentals of the Tribology of DLC Coatings. In: Christophe Donnet & Ali Erdemir (ed.) Tribology of Diamond-Like Carbon Films. USA: Springer; 2008. p139-154.

[66] Donnet, C. & Erdemir, A. Diamond-like Carbon Films: A Historical Overview. In: Christophe Donnet & Ali Erdemir (ed.) Tribology of Diamond-Like Carbon Films. USA: Springer; 2008. p1-12.

[67] Hauert, R., Knoblauch-Meyer, L., Francz, G., Schroeder, A. & Wintermantel, E. Surface & Coatings Technology 1999;120-121 291-296.

[68] Ronkainen, H. & Holmberg, K. Environmental and Thermal Effects on the Tribological Performance of DLC Coatings. In: Christophe Donnet & Ali Erdemir (ed.) Tribology of Diamond-Like Carbon Films. USA: Springer; 2008. p155-200.

[69] Sáenz de Viteri, V., Barandika, M.G., Ruiz de Gopegui, U., Bayón, R., Zubizarreta, C., Fernández, X., Igartua, A. & Agullo-Rueda, F. Characterization of Ti-C-N coatings deposited on Ti6Al4V for biomedical applications. Journal of Inorganic Biochemistry 2012;117 359-366.

[70] Williams, D.F. Titanium and titanium alloys. In: Williams DF (ed.) Biocompatibility of clinical implant materials, Vol. I. Boca Raton, Florida: CRC Press, Inc; 1981. p. 9-44

[71] Lampin, M., Warocquier-Clerout, R., Legris, C., Degrange, M. & Sigot-Luizard, M.F. Correlation between substratum roughness and wettability, cell adhesion, and cell migration. Journal of Biomedical Materials Research 1997;36 99-108.

[72] Lim, Y.J., Oshida, Y., Andres, C.J. & Barco, M.T. Surface characterization of variously treated titanium materials. International Journal of Oral & Maxillofacial Implants 2001;16 333–342.

[73] Keller, J.C., Stanford, C.M., Wightman, J.P., Draughn, R.A. & Zaharias, R. Characterization of titanium implant surfaces. III. Journal of Biomedical Materials Research 1994;28 939–946.

[74] Kieswetter, K., Schwartz, Z., Dean, D.D. & Boyan, B.D. The role of implant surface characteristic in the healing of bone. Critical Reviews in Oral Biology & Medicine 1996;7 329–345.

[75] Pouilleau, J., Devilliers, D., Garrido, F., Durand-Vidal, S. & Mahe, E. Structure and composition of passive titanium oxide films. Materials Science and Engineering 1997;B47 235–243.

[76] Fei, C., Hai, Z., Chen, C. & Yangjian, X. Study on the tribological performance of ceramic coatings on titanium alloy surfaces obtained through microarc oxidation. Progress in Organic Coatings 2009;64 264–267.

[77] Kuromoto, N.K., Simão, R.A. & Soares, G.A. Titanium oxide films produced on commercially pure titanium by anodic oxidation with different voltages. Materials Characterization 2007;58 114–121.

[78] Wanga, Y., Jiang, B., Lei, T. & Guo, L. Dependence of growth features of microarc oxidation coatings of titanium alloy on control modes of alternate pulse. Materials Letters 2004;58 1907–1911.

[79] Han, Y., Hong, S. & Xu, K. Synthesis of nanocrystalline titania films by micro-arc oxidation. Materials Letters 2002;56 744–747.

[80] Kurze, P., Krysman, W., Dittrich, K.H. & Schneider, H.G. Process characteristics and parameters of anodic oxidation by spark deposition (ANOF). Crystal Research and Technology 1984;19 973–979.

[81] Kurze, P., Dittrich, K.H., Krysman, W. & Schneider, H.G. Structure and properties of ANOF layers. Crystal Research and Technology 1984;19 93–99.

[82] Han, I., Choi, J.H., Zhao, B.H., Baik, H.K. & Lee, I. Micro-arc oxidation in various concentration of KOH and structural change by different cut off potential. Current Applied Physics 2007;7S1 23–27.

[83] Matykina, E., Berkani, A., Skeldon, P. & Thompson, G.E. Real-time imaging of coating growth during plasma electrolytic oxidation of titanium. Electrochimica Acta 2007;53 1987–1994.

[84] Wang, Y., Lei, T., Jiang, B. & Guo, L. Growth, microstructure and mechanical properties of microarc oxidation coatings on titanium alloy in phosphate-containing solution. Applied Surface Science 2004;233 258–267.

[85] Shibata, T. & Zhu, Y. C., Corrosion Science 1995;37(1) 133–144.

[86] Shibata, T. & Zhu, Y. C., Corrosion Science 1995;37(2) 253–270.

[87] Yerokhin, A.L., Nie, X., Leyland, A., Matthews, A. & Dowey, S.J. (1999). Surface & Coatings Technolog 1999;122 73–93.

[88] Shokouhfar, M., Dehghanian, C., Montazeri, M. & Baradaran, A. Preparation of ceramic coating on Ti substrate by plasma electrolytic oxidation in different electrolytes and evaluation of its corrosion resistance: Part II. Applied Surface Science 2012;258 2416-2423.

[89] Amin, M.S., Randeniya, L.K., Bendavid, A., Martin, P.J. & Preston, E.W. Amorphous carbonated apatite formation on diamond-like carbon containing titanium oxide. Diamond and Related Materials 2009;18 1139-1144.

[90] Yang, B., Uchida, M., Kim, H-M., Zhang, X. & Kokubo, T. Preparation of bioactive titanium metal via anodic oxidation treatment. Biomaterials 2004;25 1003-1010.

[91] Han, Y., Sun, J. & Huang, X. Formation mechanism of HA-based coatings by micro-arc oxidation. Electrochemistry Communications 2008;10 510-513.

[92] Letic-Gavrilovic, A., Scandurra, R. & Abe, K. Genetic potential of interfacial guided osteogenesis in implant devices. Dental Materials Journal 2000;19 99–132.

[93] Eriksson, C., Lausmaa, J. & Nygren, H. Interactions between human whole blood and modified TiO2-surfaces: influence of surface topography and oxide thickness on leukocyte adhesion and activation. Biomaterials 2001;22 1987–1996.

[94] Ravanetti, F., Borghetti, P., De Angelis, E., Chiesa, R., Martini, F.M., Gabbi, C. & Cacchioli, A. (2010). In vitro cellular response and in vivo primary osteointegration of electrochemically modified titanium. Acta Biomaterialia 2010;6 1014-1024.

[95] Ceschini, L., Lanzoni, E., Martini, C., Prandstraller, D. & Sambogna, G. Comparison of dry sliding friction and wear of Ti6Al4V alloys treated by plasma electrolytic oxidation and PVD coating. Wear 2008;26 86-95.

Chapter 5

Cryogenic Tribology in High-Speed Bearings and Shaft Seals of Rocket Turbopumps

Masataka Nosaka and Takahisa Kato

Additional information is available at the end of the chapter

1. Introduction

In recent years, as a rule, improvement of the reliability of liquid propellant rockets becomes an international technical problem for built-up of safe space transport systems. The high performance, liquid propellant rocket engines require high-pressured turbopumps to deliver extremely low temperature propellants of liquid oxygen (LO$_2$, boiling point 90 K) and liquid hydrogen (LH$_2$, boiling point 20 K) to a combustion chamber in engine [1]. In LO$_2$/LH$_2$ turbopumps, cryogenic high-speed bearings and rotating-shaft seals are very important parts to sustain high reliability of the high-rotating-shaft systems. The turbopump bearings are directly equipped in cryogenic propellants in pump side [2]. The shaft seal systems are also set up between the cryogenic pumps and the hot turbines to restrain the leakage of cryogenic propellants and hot turbine gas [3].

These bearing and shaft seals have to operate under poor lubricating conditions due to extremely small viscosity at cryogenic temperatures. Furthermore, the turbopump bearings and shaft seals have to overcome a severe high-speed operation that has several critical speeds demonstrating self-induced severe vibration of the rotating shaft. In order to develop turbo-pump bearings and shaft seals, many inexperienced technical and tribological problems must be solved for extremely low temperature and high speed of operational conditions. Such cryogenic tribological technology has been playing a key role in cryogenic turbopumps to achieve high reliability.

This chapter presents a topical review of cryogenic tribological studies (for about 30 years in Japan) on the research and development of the cryogenic high-speed bearings and shaft seals of rocket turbopumps [4, 5]. The high-speed bearings and shaft seals were continually studied for the LE-5 engine used in the Japanese H-I rocket (developed in 1986) and the LE-7 engine used in the H-II rocket (developed in 1994). The bearings and shaft seals used in LO$_2$/LH$_2$

turbopumps of the LE-5 and LE-7 had a rotational speed level of 50,000 rpm and had been studied and developed from the mid-1970 to the mid-1990. Specially, the all-steel bearings (made of AISI 440C) of the LH_2 turbopump of the LE-7 demonstrated high performance with high reliability at high-speed level at 2 million DN (40 mm x 50,000 rpm). The shaft seal systems in the LE-5/LE-7 turbopumps that used a mechanical seal, a floating ring seal (annular seal) and a segmented seal are also reviewed.

Furthermore, for future space transport systems to reduce launch cost and to increase efficiency, advanced rocket engines which are characterized by high durability (long life) and high performance (light weight) are required in recent years. Advanced bearing and shaft seal that have high durability, i.e., a long life of 7.5 hours for the turbopump bearings used in reusable space shuttle main engine (the SSME). Its required life is 15 times longer than that (30 minutes) of the turbopump bearings used in the LE-7. At the first time, the SSME turbopump bearings experienced a serious wear problem in LO_2 due to poor self-lubrication of the retainer [6]. In order to extend bearing life, the hybrid ceramic bearing with Si_3N_4 balls was used to reduce serious wear in the conventional all-steel bearing. A new type of the retainer having PTFE/ bronze-powder insert was also developed to obtain sufficient self-lubrication of the hybrid ceramic bearing. Consequently, the improvement of the SSME turbopump bearings needed a long time of about 20 years [7].

Today, ultra-high speed level above 100,000 rpm is required to make a small and light turbopump for advanced second-stage engine. These advanced research and development are actively underway. In Japan, a new type of hybrid ceramic bearing having Si_3N_4 balls with a single guided retainer demonstrated excellent performance at an ultra-high speed of 120,000 rpm (3 million DN) in LH_2 and recorded the world's top speed (in 2001) [8]. The result of this bearing was applied to the LH_2 turbopump (rotational speed, 90,000 rpm) of the RL60 demonstrator engine (in 2003). The RL60 demonstrator engine was developed in the USA with international collaboration (USA, Japan, Russia and Sweden) and the LH_2 turbopump was developed by a Japanese company [9]. In Europe, for the VINCI engine under development, high-DN hybrid ceramic bearing was tested in LH_2 at a speed of 70,000 rpm (2.8 million DN) and continuous studies on a high-DN bearing was conducted at DN up to 3.3 million (120,000 rpm) in LH_2 (in 2005) [10]. Furthermore, in Russia, for the developed RD0146 engine, its rotational speed of the main LH_2 turbopump was 123,000 rpm (3.08 million DN), but detail of its bearing was unknown (in 2003) [11].

This chapter also reviews advanced bearings and shaft seals which were studied from the mid-1990 to the mid-2000 after the development of turbopump bearings and shaft seals of the LE-7 [4,5]. It is typical that a long-life bearing with single-guided retainer demonstrated a long operation for 12 hours under 50,000 rpm. A hybrid ceramic bearing having single-guided retainer and Si_3N_4 balls was able to demonstrate ultra-high-speed performance at speeds up to 120,000 rpm and show excellent performance under 3 million DN. An annular seal made of an Ag plated steel ring also presented two-phase seal performance at speeds up to 120,000 rpm.

These historical reviews are intended to help the technical succession to next young generation who challenges research and development of the future space transportation system. These reviews are based on previous studies carried out by Japan Aerospace Exploration Agency (JAXA) at Kakuda Space Center. All materials used in this chapter have been published by papers.

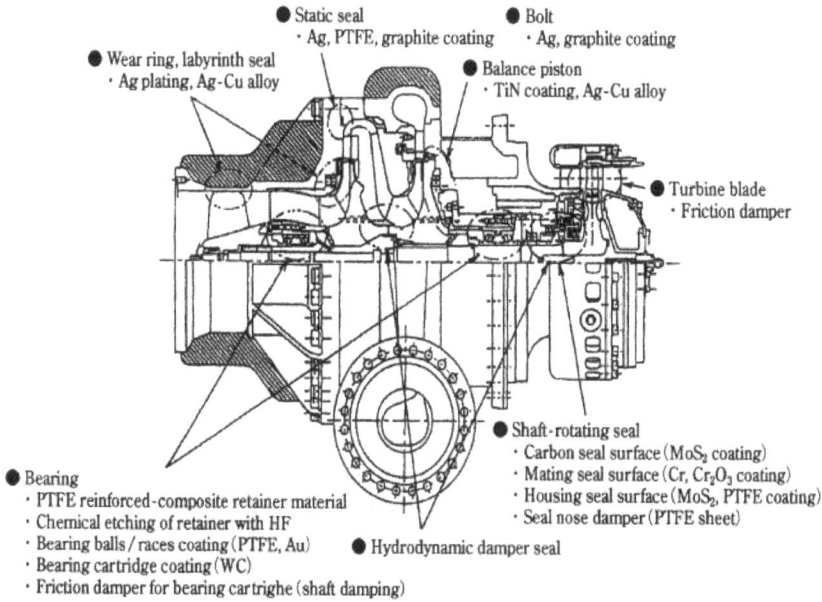

● Static seal
· Ag, PTFE, graphite coating

● Bolt
· Ag, graphite coating

● Wear ring, labyrinth seal
· Ag plating, Ag-Cu alloy

● Balance piston
· TiN coating, Ag-Cu alloy

● Turbine blade
· Friction damper

● Bearing
· PTFE reinforced-composite retainer material
· Chemical etching of retainer with HF
· Bearing balls / races coating (PTFE, Au)
· Bearing cartridge coating (WC)
· Friction damper for bearing cartrighe (shaft damping)

● Shaft-rotating seal
· Carbon seal surface (MoS₂ coating)
· Mating seal surface (Cr, Cr₂O₃ coating)
· Housing seal surface (MoS₂, PTFE coating)
· Seal nose damper (PTFE sheet)

● Hydrodynamic damper seal

Figure 1. Typical tribo-components and solid lubricants used in turbopumps

2. Bearings and shaft seals of turbopumps

2.1. Turbopumps and tribo–components

The LO_2/LH_2 turbopumps as well as the tribo-components, such as high-speed bearings and rotating shaft-seals, were studied and developed to use in the LE-5 and LE-7. In reference to the structure of the LH_2 turbopump of the LE-7, the tribo-components and solid lubricants used in the LE-5 and LE-7 turbopumps are typically indicated in Fig. 1 [4]. In addition, main design parameters of the turbopumps and DN values of bearings for the LE-5 and LE-7 are listed in Table 1 [5]. Here, the DN value that represents high-speed level of bearing is defined as the product of the inner-race bore diameter D (in mm) and the pump rotational speed N (in rpm). The rotor speed is typically restricted by the DN limits of the bearing.

Engine (thrust)	LE-5 (10 tons)		LE-7 (86 tons)	
Rocket	Second stage of H-1		First stage of H-2	
Engine cycle	Gas-generator cycle		Staged-combustion cycle	
Turbopump	LO$_2$	LH$_2$	LO$_2$	LH$_2$
Pump pressure [MPa]	5.2	5.5	17.4 (25.8)*	27.0
Pump flow rate [kg/s]	20	3.6	212 (46)*	36
Shaft rotational speed [rpm]	16,500	50,000	18,000	42,000
Bearing DN [mm x rpm]	49.5 x 10^4	125 x 10^4	81 x 10^4	168 x 10^4
Turbine pressure [MPa]	0.48	2.4	19.1	20.6
Turbine temperature [K]	690	840	810	830
Turbine gas flow rate [kg/s]	0.39	0.42	14.9	33.6
Shaft power [kW]	130	490	4,700	19,700
Weight [kg]	23	25	160	200

For pre-burner in bracket; ( )*

Table 1. Design parameters of turbopumps and DN values of bearings for LE-5 and LE-7

a.   *LE-5 turbopumps*

For the upper stage of the H-I rocket, the LE-5 had a gas-generator cycle with 10-ton thrust and its chamber pressure of 3.4 MPa was relatively low. Its engine cycle is not able to achieve a high engine performance due to an open cycle. For the LH$_2$ turbopump of the LE-5, the pump discharge pressure was relatively low at 5.5 MPa and the discharge flow rate was 51 liters/s. The turbine pressure was 2.4 MPa. The paired bearings of 25-mm bore operated at a speed of 50,000 rpm (1.25 million DN) and sustained the shaft power of 490 kW [12].

For the LO$_2$ turbopumps, the discharge pressure was 5.2 MPa and the discharge flow rate was 18 liters/s. The turbine pressure was 0.48 MPa. The paired bearings of 30-mm bore operated at a speed of 16,500 rpm and sustained the shaft power of 130 kW. Basic design and technology of the cryogenic tribo-components used in the small turbopumps was experimentally established under the development of the LE-5.

b.   *LE-7 turbopumps*

For next technical challenge in the first stage engine of the H-II rocket, the LE-7 had a staged-combustion cycle (similar to that of the SSME) with 100-ton thrust and a high chamber pressure of 13 MPa. Its engine cycle can obtain high performance due to a closed engine cycle. For the high-pressure, large LH$_2$ turbopump of the LE-7, the pump discharge pressure was increased to 27 MPa, and the discharge LH$_2$ flow rate was 510 liters/s. The turbine pressure was relatively high at 20.6 MPa. The paired bearings of 35-mm bore were at the inducer side, and the paired bearings of 40-mm bore were at the turbine side. These bearings operated at a speed of 42,000 rpm (1.68 million DN) and sustained the shaft power of 19,700 kW [13,14].

For the LO$_2$ turbopumps, the discharge pressure was 18 MPa for the main pump and 26 MPa for the preburner pump, respectively. The total discharge LO$_2$ flow rate was 240 liters/s. The turbine pressure was 19.1 MPa. The paired bearings of 32-mm bore were located at the inducer side and the paired bearings of 45-mm bore were at the turbine side. These bearings operated at a speed of 18,000 rpm and sustained the shaft power of 4,700 kW [14,15].

c. *Tribo-components in turbopumps*

As shown in Fig. 1, it is important to prohibit severe friction and wear in cryogenic environment that various solid lubricants are applied to the frictional parts in static and dynamic tribo-components. Since the turbopums are operated under large power conditions connecting with high fluid and mechanical vibration, it must pay attention that many components in contact are sure to generate relative motion and resulted in severe adhesive conditions. It needs proper lubrication to avoid severe frictional adhesion of assembled parts used in cryogenic environment.

The rotor of turbopump is directly supported by two sets of self-lubricated ball bearings in cryogenic pump fluid. The shaft seal of turbopump is installed between the cryogenic pump and the hot gas turbine. The shaft seal system must seal the cryogenic propellants and the combustion gases (steam with rich hydrogen gas) safely and securely. High-speed components, such as bearings, shaft seals, Labyrinth seals, wear rings and balance pistons, used the proper solid lubricants to protect them from severe friction and wear in the reduction (LH_2) or oxidation (LO_2) environment of the cryogenic propellants. It is noted that these high-speed tribo-components are important life-controlling parts in engines [4].

2.2. Self–lubricating bearings

2.2.1. Self–lubrication of retainer

The turbopump bearings are all-steel (AISI 440C) bearings that are self-lubricated by the PTFE transfer film as a lubricant from the reinforced PTFE (polytetra fluoroethylene) retainer. AISI 440C is martensitic stainless steel (with 16-18%Cr) and is one of the most widely used bearing materials in space systems because such high-Cr steel has a high corrosion resistance due to a superficial surface layer of Cr_2O_3. The resin PTFE retainer is reinforced with glass fiber, carbon fiber and laminated glass cloth to reduce wear as well as thermal contraction of the retainer. Although PTFE material has poor mechanical strength at room temperature, it has the best lubricant for use at cryogenic temperature because its mechanical tensile stress drastically increases and reaches to 80 MPa in LO_2 and 130 MPa in LH_2, respectively. In order to reduce wear of the PTFE composite retainer with poor thermal conductivity, sufficient cooling of the retainer is need to eliminate heat generation detrimental to successful bearing operation at high speeds [16].

Since LH_2 and LO_2 are particularly poor as lubricants because of their low viscosity under conditions of reduction or oxidation, hydrodynamic fluid lubrication is less effective. It is noted that the cryogenic pump fluids works to remove severe frictional heat and to prevent the temperature rise in the bearing. At low temperatures, the PTFE transfer film as a lubricant is kept to be hard and to sustain the bearing load, so that softening and rupturing of the transfer film due to a rise in temperature have to be eliminated. Under poor cooling conditions, it appears that the blackened transfer film due to thermal decomposition of PTFE should occur at a high temperature above about 500 K, and the degraded transfer film was not able to sustain the bearing load. Therefore, sufficient cooling by cryogenic fluids, as well as reduction of frictional heat generation, is very important to produce a durable lubricant film transferred from the retainer even in cryogenic fluid [14].

2.2.2. High–speed and load conditions of bearing

For the turbopump bearings, angular-contact bearings are usually used in pairs in duplex mounts (back to back). For example, Table 2 shows main design parameters and internal load conditions for the bearings used in the LH$_2$ turbopumps of the LE-5 and LE-7 [17,18]. In this table, the $SVmax$ value (=$Smax$ x $Vmax$/2) that represents the maximum product of stress times spinning velocity in the contact ellipse zone at the inner race are shown. Here, $Smax$ is the maximum contact stress and $Vmax$ is the maximum spinning velocity. The $SVmax$ value is an important factor related to lubrication and wear at the inner race with ball spinning [13,19]. High $SVmax$ value leads to high frictional heating and to wear of the PTFE transfer film due to spin wear. Under poor cooling condition and large tilted misalignment, the turbopump bearings have an initial contact angel of 15-25 deg. with a large radial clearance to prevent a loss of operating clearance from bearing seizer. As mention later, high-speed bearing has the outer-race ball control that produces high ball spinning at the inner race. In order to reduce the stress level within the spinning contact zone, race curvatures were controlled to be 0.54-0.56 for inner race and 0.52 for the outer race, respectively. The inner race has a counter-bore type to gain sufficient cooling within the bearing.

As the centrifugal force developed on the balls increases at high speeds, the operational contact angle at the inner and the outer races are changed to be different each other. The operational contact angle at the inner race increases rather than the initial contact angle and decreases to near zero at the outer race. This divergence of contact angles tends to increase ball spinning in addition to rolling at the inner race. Its spin velocity due to ball spinning becomes high and results in an occurrence of frictional heat generation. To contrast, rolling contact at the outer race generates differential slip due to curvature of contact ellipse [20].

Under the outer-race control connected with ball spinning at the inner race, heat generation due to ball spin is significantly higher than that of differentia slip, so that sufficient cooling is necessary at the inner race side. Furthermore, sliding velocity of the rolling balls in contact with the outer guide land and the ball pocket is high and resulted in a generation of frictional heating of the retainer. The bearings were effectively cooled by the pump cryogenic fluids circulating in the turbopumps. For example, Fig. 2 shows sliding frictional conditions of the inner and outer raceways for the 25-mm-bore bearing that is at a speed of 50,000 rpm under a thrust load of 980 N [16]. This bearing was used in the LH$_2$ turbopump bearing for the LE-5. In this figure, the distribution of the contact stress, the spinning velocity and the SV value with spin at the inner race are shown. Pattern of spin wear generated by ball spinning becomes similar to the distribution of the SV value. To contrast, for the outer race, the differential slip velocity and the SV value with differential slip are light so that wear due to differential slip is

Parameters	LE-5	LE-7
Bearing		
Dimension [mm]	25 x 52 x 15	40 x 70 x 16
Pitch diameter [mm]	38.5	57
Ball diameter [mm]	7.938	9.525
Number of balls	11	13
Initial contact angle [deg.]	20	25
Initial radial clearance [µm]	57	137
Operating condition		
Rotational speed [rpm]	50,000	46,000
Thrust pre-load [N]	784	1,176
Bearing DN [mm x rpm]	125×10^4	184×10^4
Internal load condition		
Normal load at inner / outer races [N]	157 / 343	176 / 637
Maximum contact stress at inner / outer races (*Smax*) [GPa]	1.58 / 1.49	1.54 / 1.63
Maximum SV at inner race (*SVmax*) [N/mm² x m/s]	2.4×10^3	3.1×10^3

Table 2. Design parameters and internal load conditions for LH_2 turbopump bearings of LE-5 and LE-7

small. Furthermore, for the retainer, the sliding velocity is 50 m/s at the ball pocket and 45 m/s at the outer guide land at a speed of 50,000 rpm, respectively.

For system design of the turbopump high-speed rotor, the thrust load applied on the rotor due to unbalanced fluid pressures is balanced automatically by a balance piston mechanism during operation [17]. As a result, the turbopump bearings can operate only with a spring thrust load to remove internal clearance and control radial stiffness. However, the shaft vibration as well as the fluid action around the impeller should add high dynamic radial load to the thrust load on the bearing. For example, the LH_2 turbopump bearings of the LE-7 had to operate at a speed of 42,000 rpm that was beyond the third critical speed of 32,000 rpm and must support the high shaft-power under high shaft-vibration. Therefore, the bearings must have high combined radial and thrust load capacity at all extremes of the tutbopump operating conditions [14].

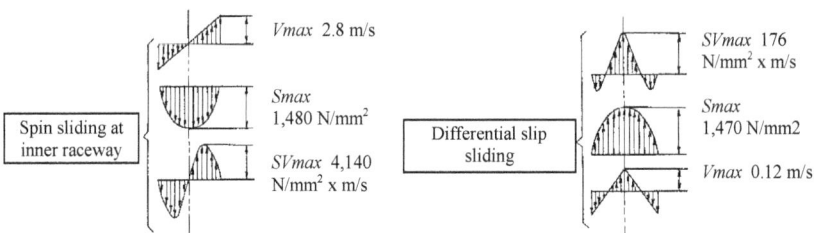

Figure 2. Sliding frictional conditions at inner and outer raceways for LH_2 bearing (25-mm bore, 50,000 rpm, 980 N)

2.3. Shaft seal systems

The required functions for shaft seal systems vary for different engine cycles. Similar to the SSME, the LE-7 has a two-stage combustion cycle. It requires a high pressure seal since the pressure in the pump and turbine is extremely high. To contrast, the pressure of the pump and turbine in the LE-5 with a gas generation cycle is comparatively low. Design parameters (the seal diameter, rubbing speed and seal pressure) for the seal elements used in the LE-5 and LE-7 turbopumps are listed in Table 3 [21]. The seal elements are the LO_2 seal, LH_2 seal, gas helium (GHe) purge seal and turbine gas seal. These shaft seals prevent or minimize the leakage of LO_2 and LH_2 for pump side and hot turbine gas (steam with rich hydrogen gas) for turbine side. In order to make a short length of the shaft, the shaft seals have to be compactly installed between the cryogenic pump and hot turbine.

Parameters	Seal diameter [mm]—Rubbing velocity [m/s] (Rotating speed [rpm])— Seal pressure [MPa]— Seal type	
Engine	LE-5	LE-7
LO_2 seal	46.6—40 (16,500)—0.98—(a)	55—58 (18,000)—4.9—(b)
LH_2 seal	43.2—113 (50,000)—1.4—(a)	50—120 (42,000)—7.1—(c)
GHe purge seal	40—35 (16,500)—0.3—(b)	69—173 (42,000)—0.6—(d)
Turbine gas seal	70—61 (16,500)—0.3—(b)	100—105 (18,000)—0.6—(b)
		55—58 (18,000)—16.7—(c)

(a) Mechanical seal, (b) Segmented seal, (c) Floating ring seal, (d) Lift-off seal

Table 3. Design parameters for seal elements used in LE-5 and LE-7 turbopumps

For the LO_2 turbopumps, when the leakage of LO_2 and hot turbine gas are mixed, an explosion will occur. In order to separate the leakage of LO_2 and hot turbine gas in safety, the system is complicated and requires three types of seal elements (the LO_2 seal, GHe purge seal and turbine gas seal). The GHe purge seal installed between the LO_2 seal and turbine gas seal supplies GHe as a barrier gas. To contrast, for the LH_2 turbopumps, the LH_2 leakage can be discharged to the turbine side so that the seal system is relatively simple. However, the rubbing speed of the seal face becomes considerably high and the contacting seal face is opposite severe tribological condition.

For the low-pressure turbopumps of the LE-5, the LO_2 and LH_2 seals used face-contact mechanical seals to gain small leakage. The GHe purge seal and turbine gas seal used contact-type segmented seal. For the high-pressure turbopumps of the LE-7, the LO_2 seal, LH_2 seal and turbine gas seal used non-contact type, floating-ring seal (annular seal) due to high seal pressure. For example, the shaft seal system of the high- pressure LO_2 turbopump of the LE-7 is shown in Fig. 3 [22]. The shaft seal system was set up between the cryogenic pumps and the hot turbine and prevented the mixing of the leakage of LO_2 and hot turbine gas. The LO_2 seal was composed of a floating-ring seal. The turbine gas seal used two floating-ring seals to seal the low temperature GH_2 that made a barrier to the turbine hot gas. So that the turbine gas seal was kept at a lower temperature against the hot turbine section and the reliability of the shaft-seal system was further increased. Between the LO_2 seal and the turbine gas seal, the segmented circumferential seal (GHe purge seal), that had shrouded Rayleigh step hydrodynamic lift-pads to increase opening force, was paired and purged with GHe to prevent mixing of the leakage of LO_2 and GH_2.

The LH_2 seal system of the high pressured LH_2 turbopump was assembled with the floating-ring seal and lift-off seal. The lift-off seal is similar to a face-contact mechanical seal and is in contact with the mating ring (rotating seal-ring) and its leakage is small when the seal pressure is low. As the rotational speed of the turbopump increases and the seal pressure becomes high, the seal faces are automatically disengaged from contacting and changed to be non-contact seal [21].

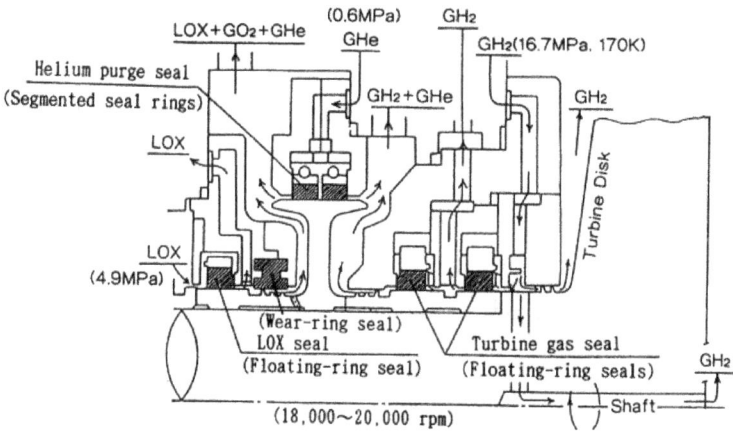

Figure 3. Shaft seal system for high-pressure LO_2 turbopump of LE-7

3. Cryogenic tribological problems [4, 16, 20, 23]

LH_2 is a particularly poor lubricant due to its extremely low viscosity (approximately equal to that of room-temperature air) and chemical reducing effect to remove native oxide film and to make fresh frictional surface, resulting in a severe lubricating condition at the frictional interfaces. Furthermore, at extremely low temperatures in LH_2, the specific heats and thermal conductivities of tribo-materials drop off rapidly rather than those at the liquid nitrogen (LN_2, boiling point 77 K) temperature. At a high temperature in LN_2, the specific heats and thermal conductivities are less changed and same as those at a room temperature. In addition with vaporization of LH_2, it is easy to produce local hot spots at frictional interfaces, so that frictional condition resulted in severe adhesive (welding) wear in LH_2.

LO$_2$ has high oxidization power and forms oxide film at frictional surfaces, so that oxide film produces lower friction compared with that in LH$_2$; however, in boiling of LO$_2$, oxide wear should increase due to high oxidization power. Active cooling is important to prohibit boiling of LO$_2$ at frictional interfaces. Furthermore, violent frictional heating in LO$_2$ can lead to the ignition of tribo-elements due to burn-out phenomenon occurring in nucleate boiling, that is defined by engineering heat transfer. Under burn-out phenomenon in boiling, an extreme rise in surface temperature was experienced because a marked reduction occurred in heat transfer. For example, in boiling of LN$_2$, the sliding surface of Ag-10%Cu alloy (melting point 1,155 K) against Ti alloy (Ti-5Al-2.5Sn) melted due to burn-out wear during friction test [24]. The surface coating of TiN or TiO$_2$ had a high resistance to adhesive welding to the Ti alloy disk was able to protect from burn-out wear. The results were applied to the balance-piston system in the LH$_2$ turbopump of the LE-7.

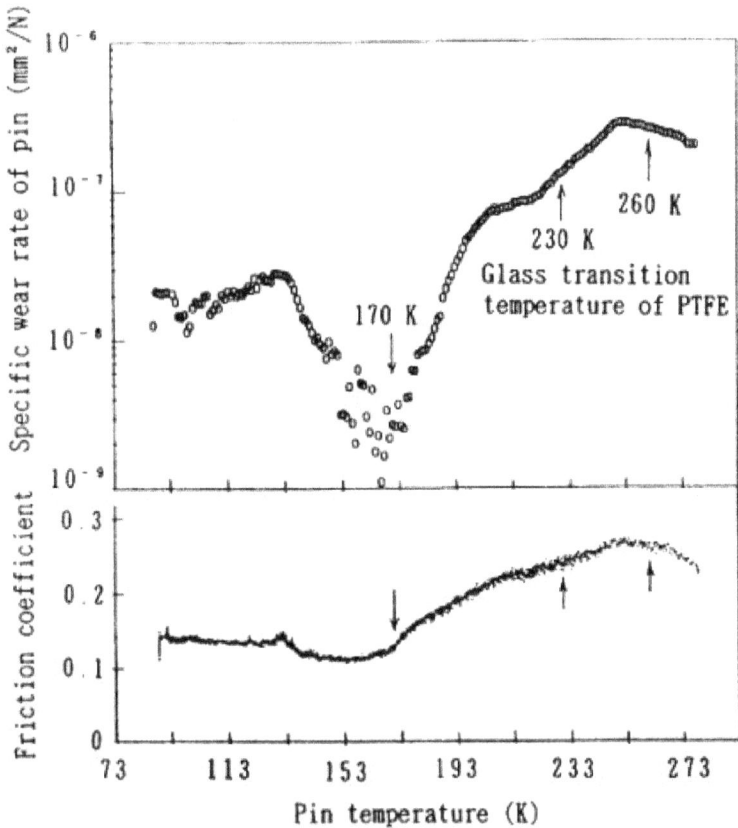

Figure 4. Friction and wear of PTFE pin against 440C disk in cryogenic GO$_2$ as a function of pin temperature

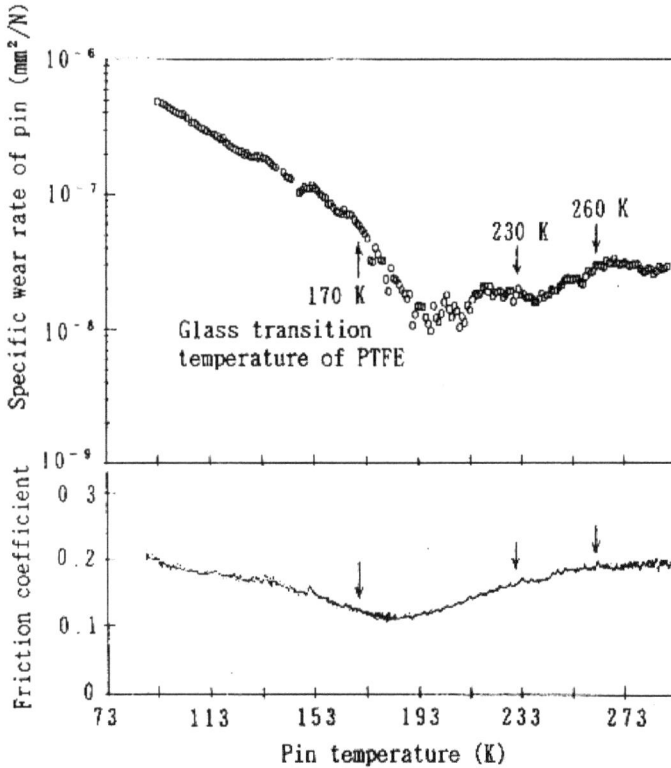

Figure 5. Friction and wear of PTFE pin against oxidized 440C disk in cryogenic GO$_2$ as a function of pin temperature

It is noted that the tribo-characteristics at cryogenic temperatures tend to change complexly. For example, Fig. 4 shows the change of friction and wear of a PTFE pin against a 440C steel disk in cryogenic gaseous oxygen (GO$_2$) as a function of pin temperature [23,25]. This figure denotes the glass transition temperature of PTFE, about 170 K, 230 K and 260 K, those are defined by relaxation of its amorphous layer in the PTFE band structure. When the frictional environment changed from the liquid phase to the gas phase at boiling, the friction coefficient increased drastically and wear began. To the glass transition temperature of 170 K (amorphous layer begins to relax), the friction coefficient remains at a low constant value, but the specific wear drastically decreased at 170 K. In an inert gaseous nitrogen (GN$_2$), there was not such drastically decrease in the specific wear at 170 K. After that, friction and wear begin to increase gradually up to 230 K. The increase of friction and wear above 170 K surely depends on the fact that the strength property of PTFE begins to decrease rapidly above 170 K.

However, when the surface of 440C steel was oxidized, the characteristic curve of friction and wear depended on cryogenic temperatures was changed drastically. Figure 5 shows the change of friction and wear of a PTFE pin in case of using an oxidized 440C steel disk [23,25]. At the pin temperatures above boiling point of LO_2 (90 K), the friction and wear of PTFE pin showed relatively high values as compared with that showed in Fig. 4. As the pin temperature increased from 90 K to near 170 K, the friction and wear of PTFE drastically decreased to low values. The oxidized 440C steel disk was obtained by heating in air at about 623 K for 3 hours. The surface of the oxidized 440C showed an increase of FeO/Fe_2O_3 film in comparison with Cr_2O_3 film. It is noted that the oxidization of 440C steel should result in an increase of friction and wear of PTFE. It seems that PTFE transfer film was less formed due to poor adhesion of PTFE against FeO/Fe_2O_3, and frictional condition became to be severe. Thus, it is very interesting that the friction and wear properties of PTFE changed characteristically at its glass transition temperature, depending on the oxidization of 440C steel.

For other friction tests, wear of PTFE in cryogenic GO_2 was increased as surface roughness of 440C disk was increased; however, in cryogenic GN_2, surface roughness had less effect on wear increase of PTFE. Furthermore, friction and wear of PTFE against Si_3N_4 disk was determined in cryogenic GO_2 and GN_2. In both cryogenic environments, friction coefficient was higher than that of 440C disk. It was noted that wear of PTFE in GO_2 was drastically high compared with that in GN_2. It was assumed that poor formation of PTFE transfer film on the Si_3N_4 disk resulted in an increase of friction and wear in GO_2. This result indicated that the hybrid ceramic bearing with Si_3N_4 ball showed poor self-lubrication in LO_2.

It is interesting to use ceramic material as tribo-materials in cryogenic environments. Friction and wear behavior of typical fine ceramics against 440C disk were evaluated in LO_2 and LN_2. Figure 6 and 7 show wear and friction of five kinds of the ceramic balls in comparison with those in LO_2 and LN_2 [23], respectively. In all the cases of friction tests, the sliding contact surface of ceramic pin was covered by the transfer film of wear debris of 440C steel. The metallic transfer film prevented direct contact between metal and ceramic. As a result, the metal-to-metal contact should control the friction and wear behavior of the sliding pair, and the order of friction seemed to be less affected in the wear resistance of ceramic pins.

In LO_2, Al_2O_3 indicated the lowest wear rate and was followed by SiC, Si_3N_4, Sialon and ZrO_2 in order of the wear resistance. For Al_2O_3 pin, the metallic oxide film of 440C seemed to be strongly adhered onto the ceramic pin and resulted in an increase of protection of the pin wear; however, wear of the 440C disk was prolonged. For SiC, Si_3N_4 and Sialon, sliding friction in oxidized environment made the glassy formation of SiO_2 film due to tribo-chemical reaction. The hardness of SiO_2 is much less than that of ceramic substrate and resulted in an increase in the wear of ceramic pins. It was noted that the wear rate of ZrO_2 was considerably high as similar to that of self-mated 440C steels. Since ZrO_2 has the lowest hardness compared with other ceramics, the hard oxide film of 440C should increase wear of ZrO_2 pin.

To the contrary, in LN_2, Zr_2O_3 indicated the lowest wear rate and was followed by Si_3N_4, Sialon, Al_2O_3 and SiC in order of the wear resistance. The high wear of Al_2O_3 and SiC pins was seemed to be induced by lack of protective film of 440C steel due to weak adhesion to ceramic pin. It is found that the order of wear resistance of ceramics against 440C steel in LO_2 was opposed to that in LN_2 [23].

At cryogenic temperatures, it is noted that sufficient cooling and the restriction of frictional heat generation are essential to prohibit severe tribological conditions. In order to solve these cryogenic tribological problems, it is important that (1) understanding the complex characteristics of tribology at low temperatures, (2) selection of the proper solid-lubricants against the oxidation or reduction power, and (3) active cooling to remove severe frictional heat at local hot spots [4].

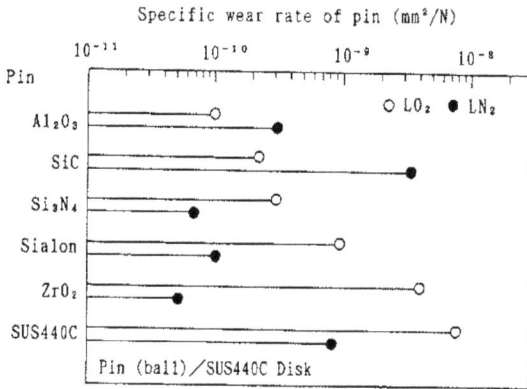

Figure 6. Wear of five kinds of the ceramic balls against 440C disk in LO_2 and LN_2

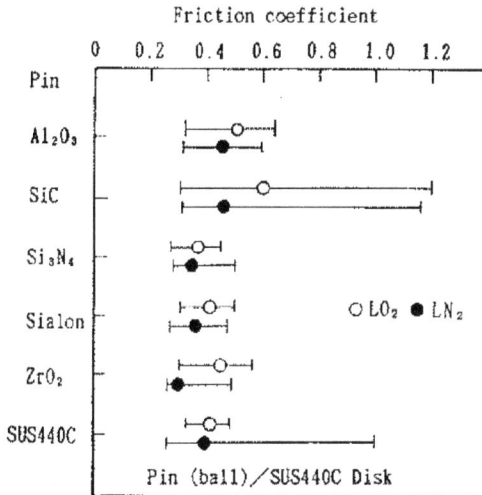

Figure 7. Friction of five kinds of the ceramic balls against 440C disk in LO_2 and LN_2

4. High–speed bearings

4.1. Improvement of self–lubrication of retainer [16, 17, 18, 26, 27]

In the beginning of the development of the turbopump bearing for the LE-5, the bearing had used the composite PTFE retainer reinforced with glass fiber or carbon fiber. The bearing tested in LH_2 by using a bearing tester showed that the glass fiber-reinforced PTFE retainer (24 wt.% glass fiber and additive) could demonstrate stable bearing-torque performance as compared with that of the carbon fiber-reinforced retainer (15 wt.% carbon fiber). From inspection of the ball-pocket surface of the carbon fiber-reinforced retainer, it was found that pile-up of the wear debris of carbon fiber might reduce supply of PTFE transfer film to ball surface. As a result, the LH_2 turbopump bearing selected the glass fiber-reinforced PTFE retainer; however, the real turbopump test showed severe wear of the retainer when the turbopump was operated under poor cooling conditions. This fact indicated low wear resistance of the glass fiber-reinforced PTFE retainer under severe operation of turbopump [16,17].

For the rocket-turbopump bearings, a laminated glass cloth with PTFE binder (laminated glass cloth of 45 wt.% and PTFE of 55 wt.%) was currently used because of its great strength to protect against dangerous retainer rupture [4,17]. This retainer showed poor self-lubrication resulting from abrasion by glass cloth layers exposed on the ball-pocket surface. During the development of the LH_2 turbopumps for the LE-5, the bearing showed unstable high-temperature rise and poor lubrication was observed, resulting in severe wear of the balls. In case of the reusable turbopumps used in the SSME, the bearings similarly experienced a serious wear problem [6]. In order to improve the self-lubricating performance of the retainer, special surface treatment of the retainer was developed [12,18]. The abrasive retainer surface with the exposed glass cloth was chemically etched with hydrofluoric acid (HF) to a depth of 0.10-0.15 mm. Following this treatment, smooth surface for the retainer was obtained. The sliding friction and wear between the ball and ball-pocket surface was reduced, resulting in a sufficient supply of PTFE transfer film from the retainer to the rolling balls.

For the HF etched retainer tested in LH_2, detailed examination of the transfer film on the sound ball surface with hardly any wear was conducted by electron probe microanalysis (EPMA) [12]. The result indicated that F of PTFE of the retainer strongly depended on the Ca concentration on the map and resulted in the tribo-chemical formation of CaF_2 transfer film. The reacted oxide material (49 wt.% of glass fiber) consisted mainly of an oxide of Ca (CaO) remained on the HF etched retainer surface. Therefore, it seems that the formation of CaF_2 transfer film was conducted by tribo-chemical reaction between F of PTFE and CaO remained on the retainer surface in chemical reduction environment in LH_2.

In order to determine the effect of tribo-chemical formation of CaF_2 in transfer film, additional friction tests were conducted. Figure 8 shows the wear of PTFE composite pins with 15 wt.% of various fillers against the 440C disk in cryogenic oxygen gas (GO_2, 123 K) under a high sliding speed (10 m/s) [15]. The PTFE composites with CaO and MgO fillers showed excellent wear resistance (progression of the pin-wear was stopped) due to the formation of good transfer film even in both cryogenic GO_2 and GN_2 (123 K). It seems that alkali-earth-metals

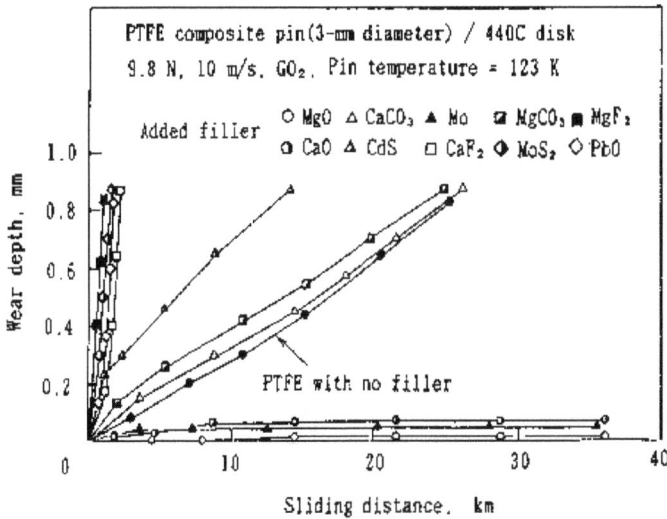

Figure 8. Wear of PTFE composite pins with various fillers against 440C disk in cryogenic GO_2 (123 K) under high-sliding speed (10 m/s)

such as Ca and Mg were able to react easily with F by severe dry sliding friction and resulted in the formation of CaF_2 and MgF_2 within the transfer film [4]. The tribo-chemical formation of CaF_2 and MgF_2 might enhance adhesion of transfer film. When CaF_2 and MgF_2 added as fillers to PTFE, there was no tribo-chemical reaction, resulting in poor wear resistance. Furthermore, oxidation of the Mo filler in GO_2 seemed to be extremely effective except in GN_2.

4.2. Development of elliptical ball–pockets of retainer [13, 14, 26]

During testing of the LH_2 turbopump for the LE-7, the conventional bearings using a retainer with circular pockets showed a significant temperature rise under high shaft vibration. Since high shaft vibration increases the radial load applied to the bearings, ball excursion occurring in the ball pockets of the retainer due to ball-speed-variation (BSV) becomes significantly large. Figure 9 shows the ball excursion due to the BSV *vs.* the radial load for the 40-mm-bore bearing at a speed of 42,000 rpm [13]. The ball excursion tends to increase with increasing of the radial load. At a radial load of about 1.5 times thrust load, the ball excursion reaches a maximum value. When the pocket clearance of the retainer is smaller than the maximum ball excursion, severe contact occurs between the ball and the retainer pocket.

For the 40-mm-bore bearing, a retainer having elliptical pockets with a large pocket clearance was developed. As shown in Fig. 10, this retainer with elliptical pockets is able to allow maximum ball excursion due to BSV in the circumferential direction and to stabilize wobbling of the retainer due to a narrow clearance in the axial direction [13]. The pocket clearance of 1.8 mm was twice as large as that of the conventional circular pocket. Consequently, the LE-7

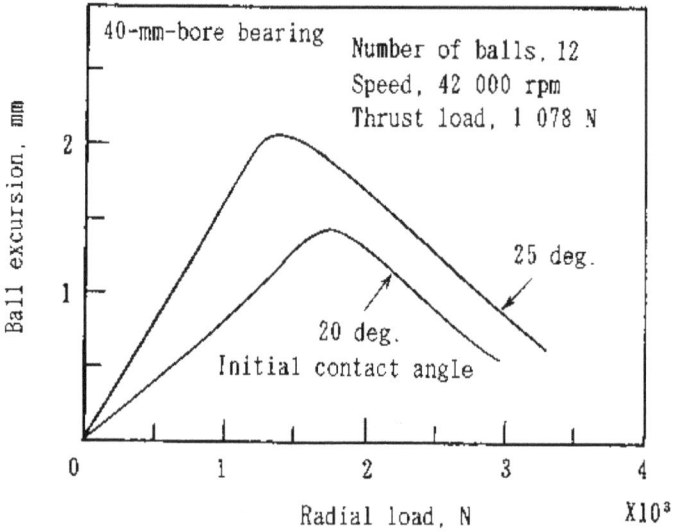

Figure 9. Ball excursion due to BSV vs. radial load for LE-7 LH₂ bearing at 42,000 rpm (40-mm-bore bearing)

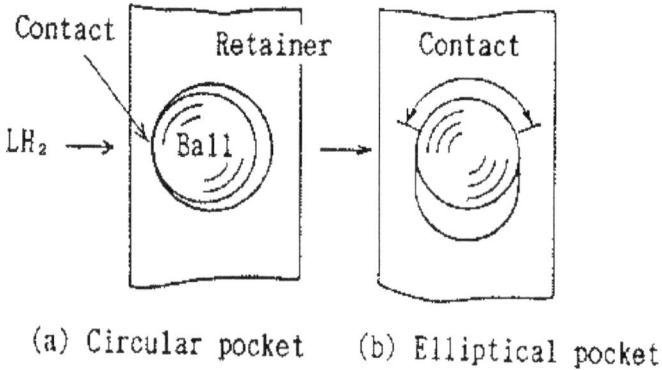

(a) Circular pocket   (b) Elliptical pocket

Ball pocket	Ball-pocket clearance, mm
Circular	0.9
Elliptical	1.8 (Circumferential)
	0.3 (Axial)

Figure 10. Circular and elliptical pockets of retainer and ball pocket clearances for 40-mm-bore bearing

turbopump bearings with the elliptical-pocket retainer exhibited excellent performance by reducing severe frictional heating and high wear of bearing components at a high-speed level of 50,000 rpm (2 million DN). Basic study of the elliptical pocket of the retainer was conducted in the development of the LE-5 turbopump bearing [12,17].

During the development of the LE-7A, the LH_2 turbopump experienced severe operation with high vibration of the rotating shaft. As a result, high vibration of the rotating heavy turbine-disk increased radial load at the turbine-side bearings (40-mm bore) and broke the retainer due to large BSV [26]. It was considered that the ball-retainer contact force due to BSV bent the retainer and hoop stress occurred on the retainer inside, resulting in fracture of the thin (weak) web section of the ball pocket. In order to gain high reliability of the LH_2 turbopump, the retainer using elliptical ball pocket was improved by increasing the pocket clearance to 2.2 mm.

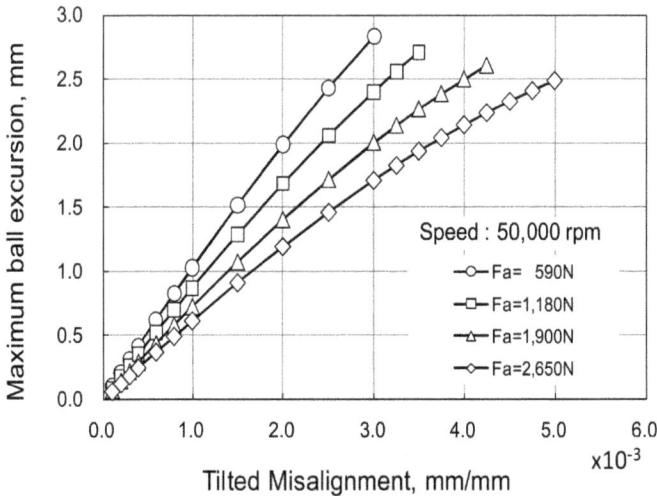

Figure 11. Maximum ball excursion vs. tilted misalignment under various thrust loads at 50,000 rpm (40-mm-bore bearing)

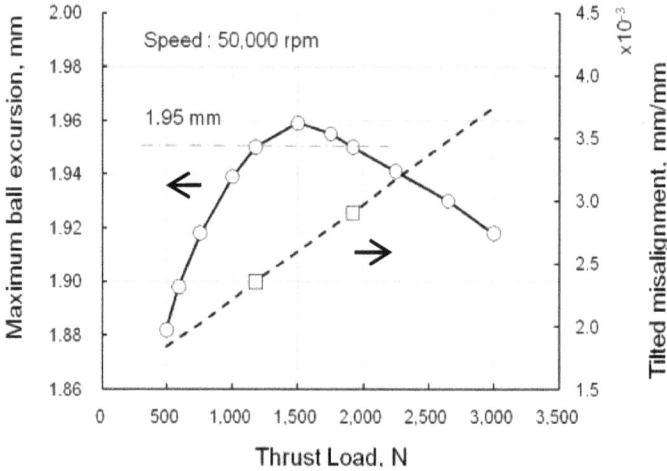

Figure 12. Maximum ball excursion and tilted misalignment *vs.* thrust load at 50,000 rpm (40-mm-bore bearing)

Such BSV was also caused by inclination of the outer race to the shaft (tilted misalignment). The effect of tilted misalignment in a level of 1.9-3.5 x 10^{-3} mm/mm on the tribo-characteristics of 40-mm-bore ball bearing was determined. The bearing used a retainer having various elliptical ball pockets to restrain the ball-retainer contact due to high BSV. The elliptical ball pocket changed the pocket clearance (1.75mm, 1.95 mm and 2.15 mm). Figure 11 shows the relationship of the tilted misalignment and the maximum ball excursion under various thrust loads at a speed of 50,000 rpm [26]. It is understood that maximum ball excursion increased with an enlargement of tilted misalignment.

Figure 12 shows the relationship of the maximum ball excursion and the tilted misalignment *vs.* the thrust load at a speed of 50,000 rpm [26]. The relationship of the maximum ball excursion *vs.* the thrust load was calculated by assuming that the tilted misalignment linearly increased with an increase of the thrust load. As the thrust load increased, the calculated maximum ball excursion tended to increase in a parabolic pattern. It was found that, in case of the pocket clearance of 1.95 mm, ball-retainer contact due to ball excursion possibly occurred within a limited range of thrust loads, resulting in high increase of bearing torque and bearing temperature.

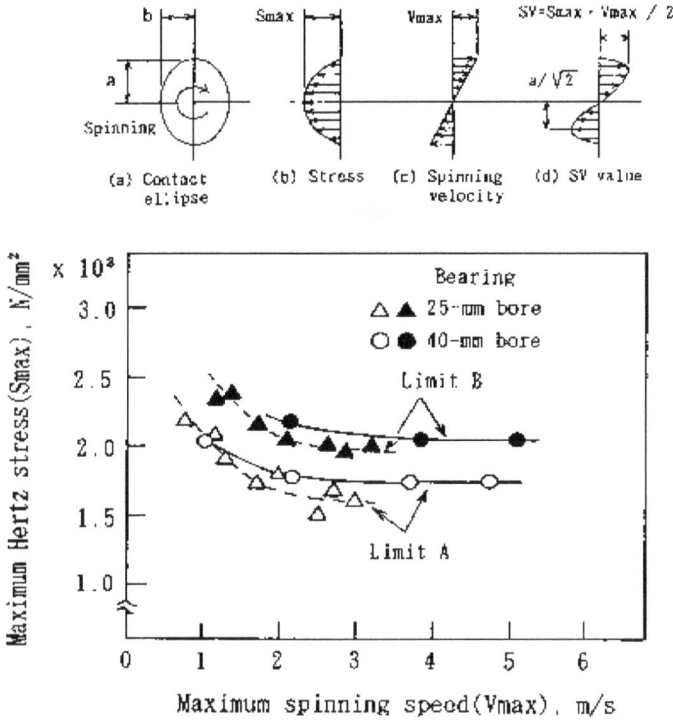

Figure 13. Load capacity of transfer film under inner race ball-spinning in LH$_2$

4.3. Performance of LH$_2$ bearing [12, 13]

Performance of self-lubricating bearing coated with PTFE or MoS$_2$ films was evaluated for the LH$_2$ turbopump bearing of the LE-5. The PTFE and MoS$_2$ films were coated with rf-sputtering. Bearing test was conducted for about 2 hours at a speed of 50,000 rpm in LH$_2$. Frictional heating was estimated from the temperature rise of cooling flow through the test bearing [12]. The coated films are hoped to induce smooth running in the initial operation when the amount of the PTFE transfer film is insufficient. The high self-lubricating performance and durability were experimentally confirmed with the PTFE coated bearing indicating frictional heating of 170-250 W. For the MoS$_2$ coated bearing, the frictional heating was 250-330 W and relatively high. The retainer of the PTFE coated bearing showed less ball-pocket wear than that of the MoS$_2$ coated bearing.

For high-speed bearings, since the bearing was under the outer-race ball control at high speed, the transfer film of the inner raceway was damaged due to the spinning of the ball. In order to evaluate stable operating condition without bearing damage, the load capacity of the transfer film under inner race ball-spinning in LH$_2$ was determined as shown in Fig. 13[13].

This figure shows the critical load capacity, that is, maximum Herze stress (*Smax*) *vs.* maximum spinning speed (*Vmax*). Under high thrust loads, an increasing of the bearing torque and bearing temperature (at limit A) was determined by the bearing tester which could measure the bearing torque in LH$_2$. The film local rupture (at limit B) was also defined by the electrical resistance monitoring between the inner race and outer race. Up to a *Vmax* of 5 m/s at 50,000 rpm, the transfer film was able to sustain a *Smax* up to 2 GPa. It was determined that the load capacity of the transfer film depended more on *Smax* than on *Vmax*. So, in order to increase durability of the bearing, it is important to limit the stress level to a *Smax* of 2 GPa to prevent transfer-film rupture and sufficiently to cool the frictional heat due to high *Vmax*.

4.4. Durability of LO$_2$ bearing [15]

It is noted that violent frictional heating in LO$_2$ can lead to the ignition of tribo-elements due to burn-out phenomenon. Burn out is overheat occurring in a transition from nucleate boiling to film boiling at critical heat flux that is defined by engineering heat transfer. For the LO$_2$ turbopump bearings (32-mm and 45-mm bore) of the LE-7, the durability and fatigue life were evaluated by applying heavy radial loads at a speed of 20,000 rpm in LO$_2$ or LN$_2$. During testing, the bearing-cartridge-acceleration (BCA), i.e., *Gpk* (peak value) and *Grms* (rot-mean-square value), was monitored to detect bearing damage. Testing in LO$_2$ for about 2.2 hours under a system radial load of 5,880 N showed that excellent lubricating conditions without abnormal BCA were obtained for all bearings.

Durability test in LN$_2$ (to keep safety in the experience) under a heavy system radial load of 11,760 N was conducted at a speed of 20,000 rpm for about 5.1 hours [15]. The result detected that the fatigue life of the bearing was about the same as the calculated B$_{10}$ fatigue life. The bearings were operated at steady conditions for 5.1 hours with 20 start-stops. For BCA on bearings A/B, *Gpk* and *Grms* on the chart were abnormally separated from each other in a pattern of abnormal BCA showing an increase of surface roughness due to an occurrence of slight flaking. Then, at a total test time of 3.8 hours, the loaded and unloaded BCA abnormally began to increase concomitantly. Examination of tested bearing B indicted that slight flaking with very shallow depth (about 8.5 μm) was observed on the inner raceway.

4.5. Evaluation of turbopump bearings [14]

The durability of the bearings of the LO$_2$/LH$_2$ turbopumps used in the firing tests of the LE-7 was evaluated based of findings of wear inspection and X-ray photoelectron spectroscopic (XPS) analysis of PTFE transfer film. Inspection of the turbopump bearings used in the engine firing tests is essential for evaluation of their durability under engine operation.

a. *Bearing wear*

After the engine firing test, surface profiles of the raceways of the LH$_2$ turbopump bearings was evaluated [14]. The engine test was conducted for a total time of 31.4 minutes with 20 engine start-stops. The surface profiles included the thickness (1μm) of the initial film coatings of sputtered PTFE film. It is obvious that the wear scars on the raceways of all bearings were flat and spin wear was not observed despite conditions of higher ball spinning on the inner raceway. For the retainer with elliptical pockets, the wear depths in the pockets were smaller than the depth (0.10-0.15 mm) of chemical etching of the glass cloth. The PTFE layer without the abrasive glass cloth sufficiently remained at the bottom of the pocket wear scar.

To contrary, the all inner raceways of the LO_2 turbopump bearings showed typical spin wear with light oxidative wear [14]. These turbopump bearings tested for a total time of 34.6 minutes with 23 engine start-stops. The surface profiles included the thickness of the initial film coatings of sputtered PTFE film (1 μm) on Ion-plated Au film (0.4 μm). The wear depths of raceways seemed to be relatively high; however, smooth surface roughness demonstrated mild wear without severe adhesion due to metal-to-metal. For bearing D that was affected by turbine whirling with radial overload, heavy spin wear with a wear depth of 7 μm was generated on the inner raceway. Furthermore, slight flaking was observed on the inner and outer raceways. This flaking was characterized by a very shallow depth and by fractures on the surface.

For the retainer with conventional circular pockets, the wear depths in the pockets were relatively light compared with those of the LH_2 bearing. The contact area in the retainer pocket and on the ball surfaces was blackened by the thermally degraded transfer film. The degradation of the transfer film seemed to occur at a temperature above about 500 K. This was confirmed by a heating test of the retainer. These facts indicated that the transfer film was severely heated even in cryogenic fluid and the LO_2 turbopump bearings were operated under poor cooling conditions. Thus, to increase the durability of the bearings, it is apparent that sufficient cooling is essential.

b. *XPS analysis of transfer films*

In order to evaluate the excellent lubricating conditions without severe wear, XPS depth analysis of a transfer film on a ball used in the LH_2 turbopump bearing of the LE-7 was conducted. Inspected ball that showed excellent wear condition was from the turbine-side bearing tested for 31.4 minutes in engine tests. The XPS depth analysis with an etching depth of 30 nm (SiO_2 rate) indicated that F(1s) and Fe(2p) spectra show the significant formation of thick CaF_2 and FeF_2 film as shown in Fig. 14 [4]. It seemed that, due to the reduction power of LH_2, the reacted CaO (remained on the retainer surface chemically etched with HF) was tribochemically changed to CaF_2 with the F of PTFE retainer during bearing operation. In addition, due to removing of native oxide film by the LH_2 reducing power, a FeF_2 film was formed by a chemical reaction between the F of PTFE retainer and the Fe of 440C steel. It is noted that the formation of FeF_2 film at the stressed contact area resulted in demonstrating high resistance to metal-to-metal adhesion and in leading to less wear [27].

Thus, the LH_2 turbopump bearings used in the engine firing tests demonstrated excellent performance due to the formation of thick CaF_2 and FeF_2 film. The tribo-chemical formation of CaF_2/FeF_2 film possibly reduced wear at frictional interfaces within the bearings used in LH_2. The basic tribo-chemical reaction was determined as follows [4]:

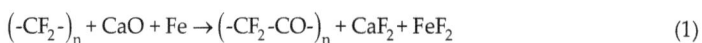

$$\left(-CF_2-\right)_n + CaO + Fe \rightarrow \left(-CF_2-CO-\right)_n + CaF_2 + FeF_2 \tag{1}$$

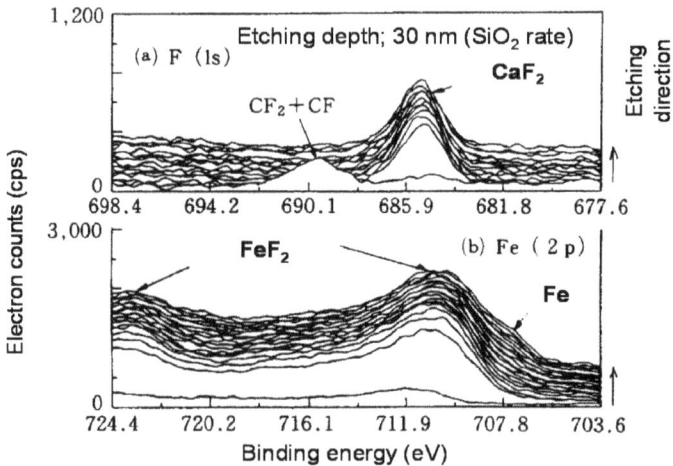

Figure 14. XPS depth analysis of ball for LH_2 turbopump bearing (turbine side)

On the contrary, for the LO_2 turbopump bearings of the LE-7, the inspected ball was from the turbine-side bearing that was tested for 34.6 minutes in engine tests and showed heavy spin wear. Figure 15 shows the XPS depth analysis with an etching depth of 30 nm (SiO_2 rate) for the worn ball due to spin wear. It indicated that the oxidization power of LO_2 prohibited the tribo-chemical formation of CaF_2/FeF_2 transfer film. This bearing was operated under poor cooling conditions, so that the bearing wear was relatively increased and shallow flaking was formed on the raceways. From the F spectrum, it was shown that very thin PTFE/CaF_2 transfer film was formed compared with the thick PTFE/CaF_2 transfer film in the LH_2 bearing. Furthermore, from the Fe spectrum, formation of Fe_2O_3 oxide film was typically shown. Fe_2O_3 oxide film was apt to form at elevated temperature, so that the oxidative mild wear in the bearing was increased due to poor cooling conditions in LO_2 [5]. As mention later (in 6.1.1), for the bearing tested under sufficient cooling condition, the intense formation of Cr_2O_3 film without Fe_2O_3 film was found beneath an extremely thin PTFE film, resulting in high resistance to metal-to-metal adhesion and in a decrease of the bearing wear [28].

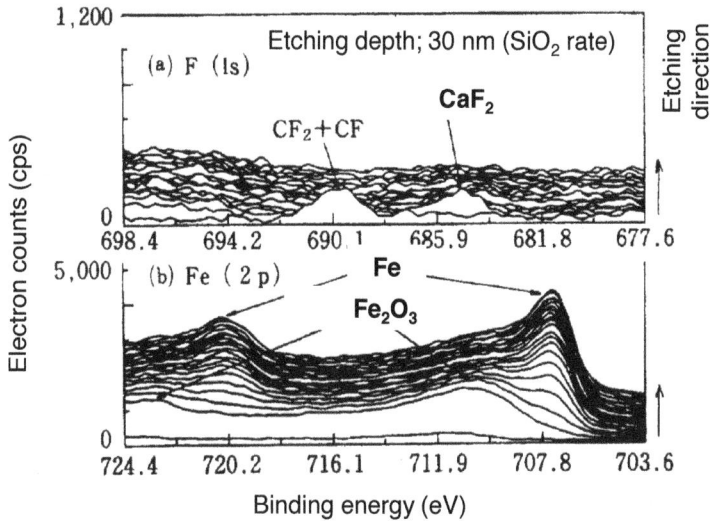

Figure 15. XPS depth analysis of ball for LO$_2$ turbopump bearing (turbine side)

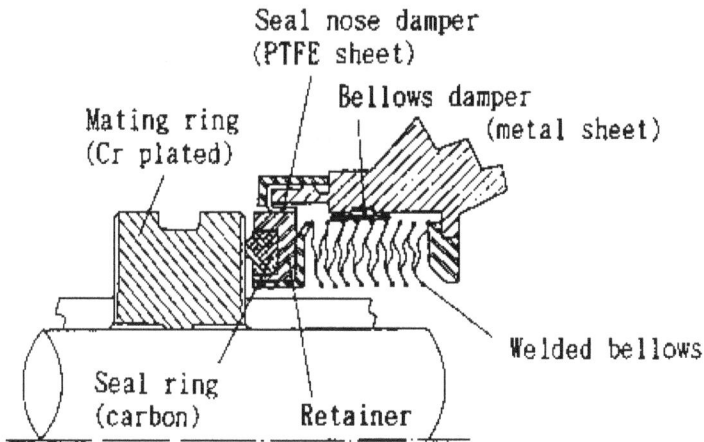

Figure 16. Face-contact mechanical seal for LH$_2$ turbopump of LE-5

5. Turbopump shaft seals

5.1. Mechanical seal [29-34]

For the LE-5 turbopumps operating under the gas generator cycle, the contact-type mechanical seal was able to use for the propellant seals because the pump and turbine pressures were relatively low. Specially, for the LH$_2$ turbopump, a high-speed mechanical seal was required to withstand high rubbing speed (113 m/s) at a speed of 50,000 rpm in LH$_2$. Figure 16 shows the face-contact mechanical seal with a seal diameter of 43.2 mm developed for the LH$_2$ turbopump of the LE-5 [29,30]. In order to reduce seal leakage of LH$_2$, it has a modified seal nose that could reduce the seal face distortion and control the direction of its distortion (to contact at outside of the seal face) under low temperature and high pressure. Furthermore, a modified vibration damper made of PTFE sheets is attached around the seal nose to prevent fluttering during rapid start or stop of the turbopump.

When the closing force to contact seal faces is increased to make seal leakage smaller, wear rate of the seal faces is increased due to the poor lubrication of LH$_2$. If the closing force is set to be smaller than the fluid opening force separating seal face each other, the leakage is considered to be quite large because of the extremely low viscosity and density of LH$_2$. Therefore, to obtain the stable seal performance and the long wear life, it is important that the proper balance between the closing force and the opening force is retained.

Critical value of the seal balance ratio that obtained stable seal performance and reduce wear of the seal faces was experimentally and analytically evaluated [32,33]. In this study, the experimental and analytical study on the friction power loss and seal performance was conducted. It was indicated that the friction power loss fell to a small value after the seal faces were sufficiently run-in. The seal balance ratio [B] that stabilized seal performance was in a range of 0.77-0.82. The seal balance ratio [B] is determined by the following equation;

$$[B] = B + Fsp / (As\Delta P) \tag{2}$$

where, B is the fluid balance ratio, Fsp is the spring force of bellows, As is the seal area and ΔP is the seal pressure. [B] is determined by the initial spring force of the bellows.

When the seal balance ratio was below 0.77, the leakage was apt to increase due to lack of the closing force. In this case, the critical balance ratio [B]c that gains stable seal performance showing small leakage was 0.77. To contrast, its balance ratio above 0.82 increased wear of the seal face by rise of the closing force. This high value of critical balance ratio was due to large opening force that could be explained with leakage flow model, assuming the phase change of leakage (from liquid phase to gas-liquid phase and gas phase) due to viscous frictional heating at high rubbing speed. In this phase change model, a state change of gas was assumed to be irreversibly adiabatic and a curve of gas expansion expressed by the following equation;

$$Pv^m = \text{constant} \tag{3}$$

where, P is the pressure, v is the specific volume and m is the ausfluss exponent. As m decreases with the temperature rise of gas due to viscous friction, the pressure of leakage flow increases particularly in the gas region within gas-liquid phase, and it resulted in the increase of the opening force. The analysis of phase change model of leakage was conducted using the flow and energy equations of liquid and gas leakages.

Figure 17 shows the calculated and experimental results of the relationship between the seal clearance and the opening force ratio $[K]$ at a speed of 50,000 rpm in LH_2. The opening force ratio $[K]$ is expressed by the following equation;

$$[K] = Fo / (As\Delta P) \tag{4}$$

where, Fo is the opening force. It was also shown that the opening force within seal clearance increases linearly as the seal clearance decreases. After the seal faces were sufficiently run-in and the seal clearance was maintained in an average of 0.6 µm, the opening force ratio $[K]$ approaches the critical balance ratio $[B]c$ (= 0.77) that showed critical seal performance. As a result, the difference of $[K]$ and $[B]c$ was decreased and it resulted in the reduction of the load on the seal faces. The frictional loss power was decreased to a small value, resulting in a restrain of wear rate of seal faces. If the seal clearance increases, the leakage becomes large; however, the load on the seal faces is increased with the decrease of the opening force and the seal clearance would become small enough to reduce leakage. Furthermore, the starting torque and static seal performance were markedly affected by the change of the seal face distortion due to wear [31].

Durability of the mechanical seal was evaluated by the long-run test [29]. The long-run test was conducted at a speed of 50,000 rpm with a seal pressure of 1.37 MPaG for 83 minutes. The experimental results showed that the leakage gradually increased until total test time was 50 minutes. During its step, wear of the seal faces was running-in, then the leakage was stabilized. It is noted that an extremely small LH_2 leakage (8-19 cc/min) was kept during test. The seal after the durability test indicated an excellent condition that maximum wear of carbon-ring was 8 µm.

Temperature on the rubbing seal faces was estimated from the reduction rate of the hardness of hard Cr plating on the rotating mating ring [34]. The estimated temperature of rubbing seal face was possibly reached to be about 773 K at a rubbing speed of 113 m/s in LH_2. In an initial stage of running-in, extremely high temperature of the seal faces caused thermal cracks in wear surface of the Cr plating, so that it is necessary to cool the contacting seal faces sufficiently. When the cooling of the sealing unit is insufficient, the surface of the carbon seal ring showed abnormal wear. Furthermore, the Cr plating showed better wear results than the tungsten carbide (WC) coating, because the Cr plating easily forms thin transfer films of graphite contained in the carbon. In the case of the WC coating, the transfer film of graphite was hardly formed in LH_2, resulting in an occurrence of severe seal wear.

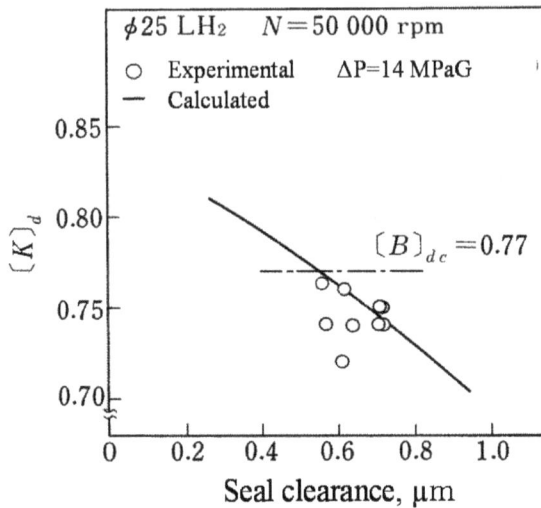

Figure 17. Opening force ratio *[K]vs.* seal clearance at 50,000 rpm in LH$_2$

5.2. Floating ring seal [22, 29, 35, 37]

A floating-ring seal is a type of no-contact annular seal without a rubbing seal surface. It has a simple structure and is able to seal high-pressure fluids, restraining leakage through a small clearance (gap) between the seal ring and the runner. Its gap is in an order of several dozens of μm. The seal ring is free to move in the radial direction, and thus severe contact with the rotating runner can be prevented. Leakage of floating-ring seal is much larger than that of face-contact mechanical seal, but the floating-ring seal shows a high resistance to pressure and a high reliability when used as high-pressure seal. A multi-seal system consisting of several seal rings arranged in series is employed for the high-pressure turbopumps. The floating-ring seals were developed and used in the LE-5 and LE-7.

Figure 18 shows the floating-ring seal with a seal diameter of 50 mm developed for the LO$_2$ turbopump of the LE-7 [22,35]. The carbon seal ring is enclosed with a retainer of the same material as the seal runner. Since the retainer contracts thermally nearly as much as the seal runner at low temperature, the seal gap hardly changes. The seal gap was 50-60μm. When the seal pressure increases, the floating ring is pressed against the secondary seal by the fluid force and its movement in the radial direction is restrained. In order to smooth the radial movement of the floating ring, on the secondary seal of the housing, the PTFE film was coated for the LO$_2$ seal and the MoS$_2$ film was coated for the turbine gas seal (to seal the low temperature GH$_2$). For the GH$_2$ leakage of the floating-ring seal used in the turbine gas seal, leakage rate calculated by the quasi-one-dimensional compressible flow equation agreed quite well with experimental value.

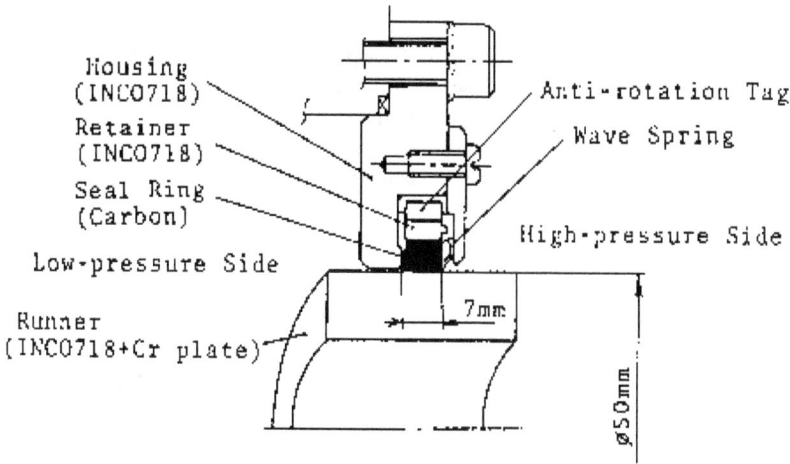

Figure 18. Floating ring seal for LO₂ turbopump of LE-7

The leakage from the floating-ring seal for the LH_2 and LO_2 seal can be calculated from the equation of the incompressible fluid flow in the rotating double cylinders when the leakage is liquid phase flow and the mass flow flux (mass flow/seal area in the flow direction) is large [29,35]. When the seal gap is narrow and the seal pressure is low, the mass flow flux of leakage is reduced, and vaporization of leakage occurred by viscous frictional heating and pressure drop changes liquid phase flow to gas-liquid phase flow (two-phase flow).

Comparison between the experimental and calculated leakage of LH_2 was evaluated by the mass flow flux of leakage for the floating-ring seal with one seal ring or two seal rings [29]. In this study, the LH_2 seal with a seal diameter of 32 mm and various seal gap of 30-86 μm was tested at rotating speeds to 50,000 rpm. It is shown that the leakage of LH_2 is less than the calculated value from incompressible fluid flow equation because the leakage is changed to be tow-phase flow. When the mass flow flux is large, most of leakage flows out in liquid phase. This means that there is not sufficient time to vaporize the leakage to be tow-phase flow within the seal gap.

A flow visualization study of floating-ring seal was conducted to identify the two-phase flow area induced by viscous frictional heating and pressure drop [36]. In order to visualize the two-phase flow in seal gap, the floating ring made of transparent hard plastic (polycarbonate) was tested in a seal fluid of LN_2. It was confirmed that the two-phase flow seemed to be homogeneous mixture of liquid and vapor flow and the two-phase flow area increases with increasing rotational speed and decreases leakage flow rate. When the two-phase flow area was fully prolonged within the seal gap, the leakage rate contrary increased with instability because the inlet flow resistance at the high-pressure side of the seal ring was reduced by two-phase flow.

5.3. Segmented seal [22, 35, 37, 38]

Contact-type segmented seal were used in the GHe purge seals and the low pressured turbine gas seals. The GHe purge seal used in the LO_2 turbopump of the LE-7 is shown in Fig. 19 [22]. Segmented seal has a carbon seal ring cut into three segments. The segmented annular seal ring is pressed on the seal runner with a coil spring and maintains high purge-pressure of GHe as a barrier gas. Wear of the carbon seal ring is reduced by using the shrouded Rayleigh step lift-pads to increase the opening force within the seal clearance. As the rubbing speed increases, the opening force in the Rayleigh step increases, so that the rubbing speed is increased by enlarging the seal diameter using a T-type runner.

Relationship between the purge pressure and the leakage rate of GHe purge seal was evaluated at a steady speed of 20,000 rpm [22]. When the purge pressure is low, the seal face is kept to be non-contact because the Rayleigh step increases the seal opening force. As the purge pressure is set to be high, the seal face condition is changed from the non-contact state to the contact state, it resulted that the dynamic leakage almost equals that of the resting state. Furthermore, for the GHe purge seal combined with the LO_2 floating-ring seal, the environmental temperature around the GHe purge seal was equal to that of LO_2 leakage, so that the carbon seal ring showed severe wear with an appearance of worn-out of the Rayleigh step.

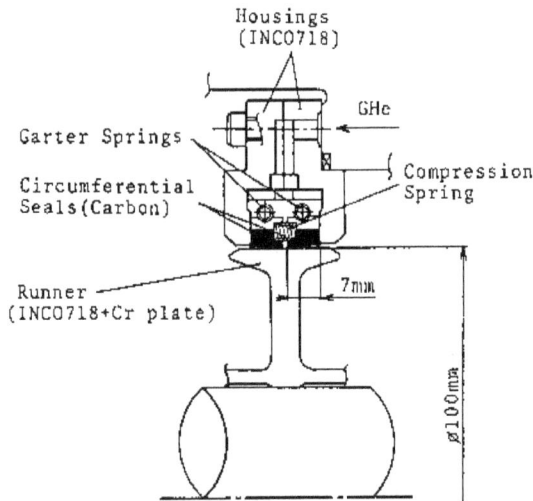

Figure 19. GHe purge seal for LO_2 turbopump of LE-7

Figure 20. Comparison of wear of MoS$_2$ coated and uncoated seal surfaces

Change of the friction and wear of the carbon pin as a function of the pin temperature was determined in the cryogenic GHe environment [23]. Friction test was conducted against the Cr-plated steel disk at a sliding speed of 12 m/s and load of 9.8 N. When the pin temperature is below the solidification temperature of CO_2 (216 K), it is noted that lubricating property of the carbon pin suddenly disappeared and friction and wear became intensive. When absorbed CO_2 gas was changed to be solid phase, lubricity of carbon was lost. This phenomenon resembles that when phase of moisture is transfer to solid phase (ice) below 273 K, lubricity decreases; be well known. From this fact, it seemed that severe wear of the GHe purge seal was generated because the environmental temperature around the seal was lower than 216 K. Spray MoS$_2$ coating on the carbon seal face was drastically able to prohibit progression of wear of the carbon seal ring at low temperature, as shown in Fig. 20.

After a total operating time of 29 minutes for the engine firing test, the GHe purge seal used in the LE-7 indicated that the seal surfaces coated by MoS$_2$ were found to be in excellent condition and wear depth of the carbon seal ring was about 7 μm. It assumes that high opening force produced by the Rayleigh step was kept by prohibit of wear of the Rayleigh step and the GHe purge seal was operated under conditions of nearly no load on the seal surfaces due to balance between the opening and closing forces.

6. Advanced bearings and shaft seals

Future space transport systems require reusable launch vehicles to reduce launch cost and to increase efficiency. The durability of reusable turbopump bearings must be greater than that of currently available (expendable) turbopumps. For the improved high-pressure LO_2 turbopump of the SSME that reduced serious wear of the all-steel bearing, the hybrid ceramic bearing with Si_3N_4 balls was developed and accomplished the required life of 7.5 hours. In this case, to improve self-lubrication of the abrasive retainer made of glass cloth-reinforced PTFE, a new type of the retainer that had PTFE/bronze-powder insert fitted on the ball pocket was developed [7].

It is noted that, at high speeds, the hybrid ceramic bearing that consists of hard, light weight ceramic balls as well as steel rings shows a lower centrifugal force on the ceramic ball. The centrifugal force of the Si_3N_4 ball makes about 60 % lighter than that of the 440C steel ball. This leads to a reduction of bearing load and a smaller contact area with a lower spinning speed, resulting in a low level of heat generation due to ball spin. Additionally, good tribological combinations of the ceramic balls against the steel rings result in a decrease in bearing wear and in instances of seizure, even under insufficient lubricating conditions. Thus, the hybrid ceramic bearing enables higher speed operation rather than the all-steel bearing.

On the other hand, advanced rocket engines that are characterized by high performance (light weight) and high durability (long life) are required today. Ultra-high speed turbopump having a rotational speed level of 100,000 rpm needs to make engine smaller and lighter. Hybrid ceramic bearing is suitable to ultra-high speed turbopump because of lower centrifugal force. In recent years, these advanced research and development on the hybrid ceramic bearing are actively underway.

6.1. Single–guided bearing [27, 28, 39]

In order to increase the durability of self-lubricated bearing, it is apparent that sufficient cooling and restriction of the frictional heat generation in the bearing are essential. Its notification is experimentally identified by a series of studies on the turbopump bearing. In order to improve internal coolant flow through the bearing and to reduce bearing frictional torque, a new type of bearing having a single-guided retainer was developed. Figure 21 shows the 25-mm-bore bearing having a single-guided retainer with elliptical ball pockets [39]. The single-guided retainer is guided only by one side of the outer-ring bore (land) to reduce land friction and to increase the cooling ability within the bearing. However, reduce of retainer guiding is apt to generate unstable wobbling at high speed, so that the elliptical ball pockets with narrow axial clearance is needed to reduce wobbling of the retainer. For the elliptical ball pocket of the single-guided retainer, its circumferential clearance of 1.3 mm was twice as large as that of the conventional circular pocket to reduce ball-to-pocket interaction under high BSV. Furthermore, the axial clearance of 0.1 mm was narrow to stabilize wobbling of the single-guided retainer at high speeds.

Self-lubricating performance, bearing wear and transfer film of two-types of the single-guided bearing, i.e., a hybrid ceramic bearing with Si_3N_4 and all-steel bearing, was evaluated under high thrust loads at speeds up to 50,000 rpm in LH_2, LO_2 and LN_2 [27,39]. Furthermore, to evaluate the durability of the single-guided bearing for long-life bearing, the all steel bearing was tested for total operation times up to 11.7 hours at a speed of 50,000 rpm with high thrust loads in LO_2 [28]. These bearings used the glass cloth-reinforced PTFE retainer which was chemically treated with HF to improve self-lubrication.

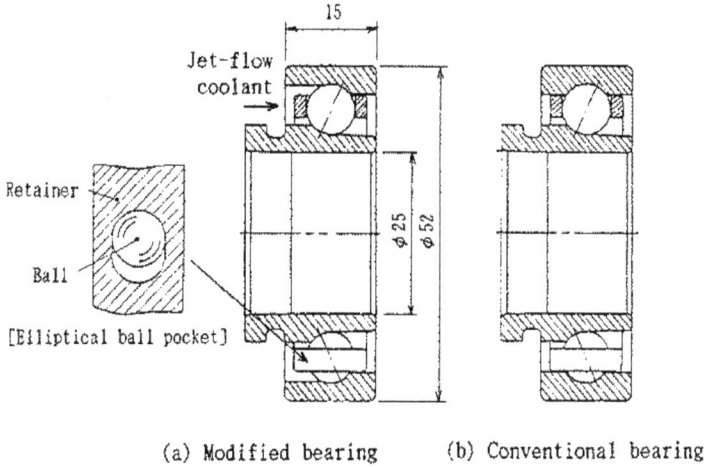

(a) Modified bearing   (b) Conventional bearing

Figure 21. Advanced bearing having single-guided retainer with elliptical ball pocket

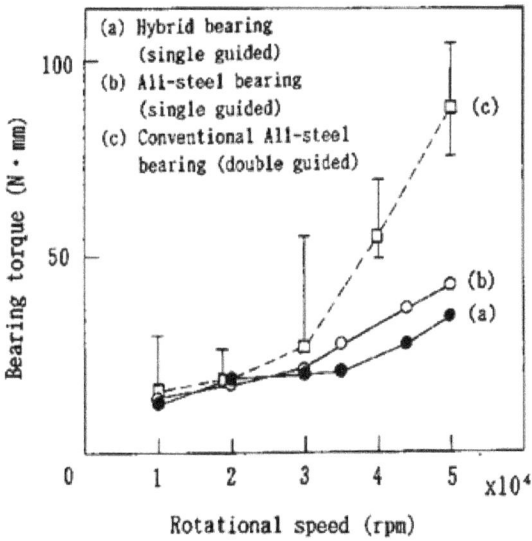

Figure 22. Bearing torque of single-guided bearings and double guided bearing to 50,000 rpm in LH_2

6.1.1. Self–lubricating performance and transfer film [27,39]

a. In LH$_2$

Figure 22 shows the bearing torque of the single-guided bearings (hybrid ceramic bearing and all-steel bearing) and the conventional double-guided bearing at speeds to 50,000 rpm in LH$_2$ [39]. It was observed that the bearing torque of the single-guided bearing effectively decreased to about one-half of that of the double-guided bearing. Its result identified that bearing torque induced by high-speed sliding of the outer land guide of the retainer almost accounted for an overall bearing torque generated at high speeds. In addition, the hybrid ceramic bearing showed lower bearing torque than the all-steel bearing at high speeds.

Critical load capacity of the single-guided bearing without a significant rise of the bearing torque and bearing temperature was evaluated. For the single-guided hybrid ceramic bearing tested in LH$_2$, the critical thrust load was 1,960 N (Smax of inner race, 2.7 GPa) at 50,000 rpm and was two times higher than that of the double-guided all-steel bearing. Furthermore, even when bearing torque increased with a rise of bearing temperature, the hybrid ceramic bearing was able to sustain a thrust load of 2,840 N (Smax, 3.2 GPa) at 50,000 rpm without seizure in LH$_2$. High critical load capacity of the single-guided hybrid ceramic bearing was demonstrated [39].

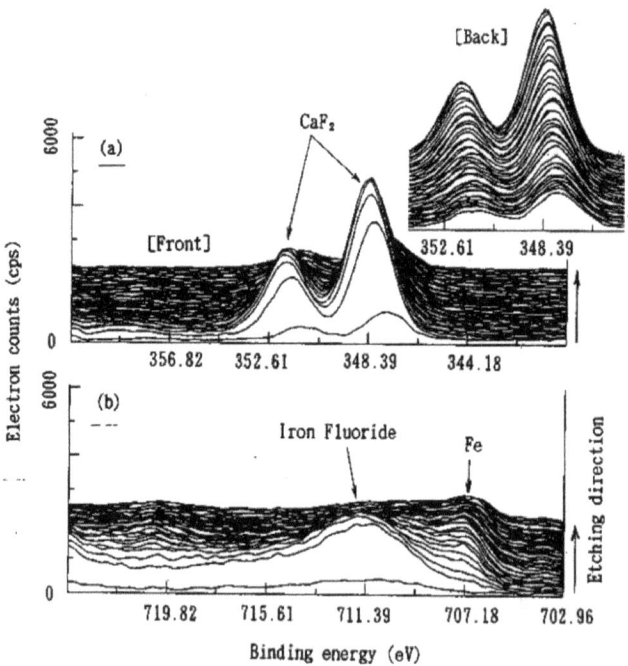

Figure 23. XPS depth analysis of Si$_3$N$_4$ ball of hybrid ceramic bearing tested in LH$_2$

Figure 23 shows the XPS depth analysis of a Si_3N_4 ball taken from the hybrid ceramic bearing tested in LH_2 [27]. Its etching depth was 120 nm (SiO_2 rate). It was found that a considerably thick transfer film consisting of CaF_2/FeF_2 was formed on the ceramic balls. CaF_2 and FeF_2 seemed to be tribo-chemically formed by the reducing power of LH_2. The considerably thick transfer film of CaF_2 and FeF_2 led to exhibit high load capacity. For the all-steel bearing tested in LH_2, a thick CaF_2 film was formed beneath an extremely thin PTFE overlay, but its thickness of CaF_2 transfer film was thinner than that of the hybrid ceramic bearing.

b.   In LO_2

In LO_2, the hybrid ceramic bearing exhibited poor self-lubricating performance even at a low speed of 10,000 rpm. To the contrary, the all-steel bearing indicated excellent load capacity accompanied by a stable bearing and enabled to sustain a thrust load of 2,650 N (*Smax*, 2.7 GPa) at a speed of 50,000 rpm without seizure in LO_2 [39].

For the hybrid ceramic bearing, an extremely thin, weakly adhesive PTFE film was formed on ceramic balls and resulted in a poor load capacity of the bearing. For the all-steel bearing, the intense formation of a Cr_2O_3 film was beneath an extremely thin PTFE film. It is noted that the tribo-chemical formation of Cr_2O_3 film due to high oxidation power of LO_2 could exhibit high resistance to metal-to-metal adhesion leading to seizure [27].

c.   In LN_2

The hybrid ceramic bearing exhibited better load capacity than that of the all-steel bearing in LN_2. The hybrid ceramic bearing enabled to sustain a thrust load of 2,700 N (*Smax*, 3.1 GPa) at a speed of 50,000 rpm without seizure. To the contrary, the all-steel bearing showed unstable change of bearing torque and seized at a relatively light-thrust load of 1,470 N (*Smax*, 2.2 GPa) at a speed of 50,000 rpm [39].

For the hybrid ceramic bearing, the thick transfer film consisting of FeF_2/iron oxide formed on the ceramic balls. To the contrary, the seized all-steel bearing was lubricated by only thin PTFE transfer film, without the tribo-chemical formation of $CaF_2/FeF_2/Cr_2O_3$ films because of its inert environment of LN_2. This fact was determined by that the all-steel bearing once tested in LH_2 or LO_2, whose bearing formed the $CaF_2/FeF_2/Cr_2O_3$ films, showed stable change of the bearing torque without seizure even under high thrust loads above 1,470 N in LN_2 [27].

6.1.2. Long–life bearing [28]

The single-guided all steel bearing was tested for a total operation time to 11.7 hours at a speed of 50,000 rpm with high thrust loads to 2,400 N in LO_2. During long-run test, one-hour operation at a speed of 50,000 rpm was repeated nine times. The test bearing was effectively cooled by the jet-cooling with using nozzles. During the long-run test, the bearing exhibited stable variation of the bearing torque in a range of 93-95 N-mm [28]. The bearing exhibited excellent self-lubrication performance that there was no abnormal change of the bearing torque and bearing temperature.

From the examination of the bearing tested for the long-run test in LO_2, it was observed that sound surface conditions with hardly any wear were determined. The XPS depth

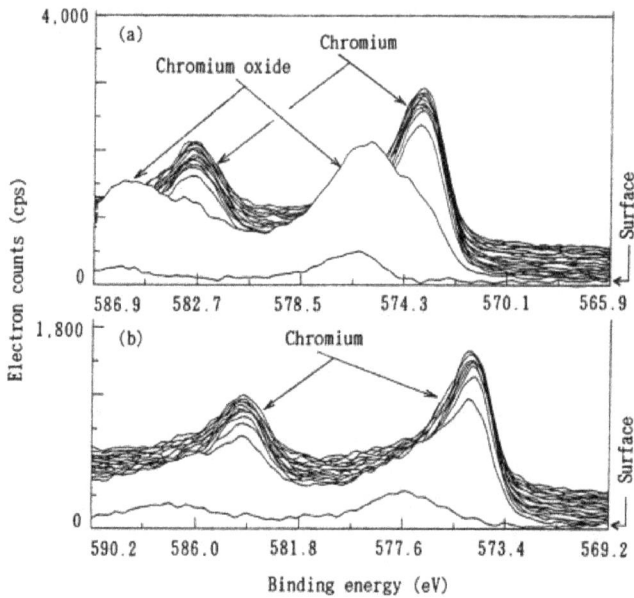

Figure 24. XPS depth analysis of SUS440C ball tested for long run in LO_2 and new ball

analysis of a ball taken from the tested bearing is shown in Fig. 24 [28]. Its etching depth was 30 nm (SiO_2 rate). It is noted that the intense formation of a Cr_2O_3 film was detected and its thickness was thicker than that of the native Cr_2O_3 film on the new ball. Under sufficient cooling conditions in LO_2, the thick Cr_2O_3 film formed by tribo-chemical reaction could provide an extremely high resistance to metal-to-metal adhesion beneath an extremely thin CaF_2 film. To the contrary, under poor cooling conditions in LO_2, the intense formation of oxide film (Fe_2O_3) was mainly produced and led to large mild wear, as discussed in the LO_2 turbopump bearing. Furthermore, the formation of Fe_2O_3 might reduce adhesion of PTFE transfer film, resulting in less lubricant within the bearing. The results indicated that thick formation of a Cr_2O_3 film due to tribo-chemical reaction in LO_2 is important to reduce the bearing wear. Its effect needs sufficient cooling with jet within the bearing components to eliminate the formation of Fe_2O_3 [28].

6.2. Fluorine–passivated bearing [28]

It is experimentally found that the FeF_2 film formed by a tribo-chemical reaction between the F of PTFE and Fe of 440C steel was facilitated by the high reduction power of LH_2 and enhanced to reduce the bearing wear in LH_2. This may suggest that the FeF_2 film has a good solid-lubricant performance to improve the tribological performance of the bearing. Effect of the coated FeF_2 film on the self-lubrication and durability of the all-steel bearing was evaluated. An FeF_2 film was chemically formed by means of a passivating surface treatment of fluoridation in hot pure F_2 gas. The fluorine-passivated bearings coated with FeF_2 film was tested by long run for 11.7 hours at a speed of 50,000 rpm under high thrust loads in LH_2, LO_2 and LN_2. The fluorine-passivated bearings showed excellent self-lubrication in both LH_2 and LN_2 [28].

In a reduce environment of LH_2, even under poor cooling conditions controlled by reducing of the coolant flow, the fluorine-passivated bearing exhibited superior durability for a total test time to 4.4 hours, as compared with signs of seizure for the untreated bearing. The XPS analysis of the transfer film indicated that the fluorine-passivated bearing was tribo-chemically lubricated by a thick CaF_2 film overlaid on a thick FeF_2/Cr_2O_3 films.

In an inert environment of LN_2, the fluorine-passivated bearing showed excellent self-lubrication and wear conditions for the long-run test up to 11.7 hours at a speed of 50,000 rpm. Stable change of the bearing torque (75-80 N-mm) was shown for the passivated bearing during the long-run test in LN_2 [28]. The bearing test was repeated seven times at a speed of 50,000 rpm and a thrust load of 2,600 N in LN_2. From the examination of the fluorine-passivated bearing tested in LN_2, sound surface conditions with hardly any wear were determined. It was found that a thick CaF_2 film was tribo-chemically formed on thick FeF_2/Cr_2O_3 films of the bearing. On the other hand, the untreated bearing was seized at a low thrust load of 1,470 N due to less tribo-chemical reaction in LN_2, as mentioned before. In such inert environment in LN_2, there was less formation of $CaF_2/FeF_2/Cr_2O_3$ films, so that poor self-lubrication and load capacity of the bearing were shown.

To the contrary, in an oxide environment of LO_2, the fluorine-passivated bearing indicated a higher bearing torque with greater unstable change than that of the untreated bearing [28]. The bearing tests were repeated seven times of the bearing test at a speed of 50,000 rpm and a thrust load of 2,450 N in LO_2. Its total test time was 11.7 hours. During long-run test, high bearing torque continued to vary erratically with the variation in a range of 75-120 N-mm. The fluorine-passivated bearing tested in LO_2 showed somewhat high wear. To the contrary, the untreated bearing demonstrated excellent self-lubrication with hardly any wear during the long-run test as mentioned before. It was clearly showed that the FeF_2 film in LO_2 made a typical reduction in self-lubrication.

Inspection of the fluorine-passivated bearing tested in LO_2 indicated that the initial coated film of FeF_2 was worm away. Its result also indicated that oxide power of LO_2 restricted the tribo-chemical formation of FeF_2 film. Such reduction in self-lubrication possibly resulted from that the coated FeF_2 film restricted the tribo-chemical formation of Cr_2O_3 film in LO_2, resulting in an increase of metal-to-metal adhesion. These results indicated that excellent lubrication depended on the tribo-chemical formation of CaF_2/FeF_2 films in LH_2 or Cr_2O_3 film in LO_2, respectively. In order to obtain high self-lubrication and durability of the bearing, it is noted that tribo-chemical reaction is necessary at the frictional interfaces within the bearing [4].

6.3. Ultra–high–speed hybrid ceramic bearing [8,40]

Based on previous bearing tests at high speeds up to 50,000 rpm, the hybrid ceramic bearing (25-mm bore) was tested at ultra-high-speeds up to 120,000 rpm, and results were compared with the all-steel bearing in LH_2. At a ultra-high speed of 120,000 rpm, the inner-race growth of 34μm due to centrifugal force results in a reduction of the radial clearance within the bearing. Table 4 summarizes comparison of the bearing load and speed conditions for the hybrid ceramic bearing and all-steel bearing at a speed of 120,000 rpm with a thrust load of 980 N [8]. At 120,000 rpm, the initial radial clearance of 77 μm was decreased to 43μm. For the hybrid ceramic bearing, the maximum contact stress $Smax$ at the inner race is apt to increase rather than that of the all-steel bearing due to a high elastic modulus. However, the maximum spinning velocity $Vmax$ is reduced and resulted in a lower $SVmax$ value that leads to a reduction of the bearing temperature and spin wear. The maximum contact stress at the outer race becomes higher due to centrifugal force. For sliding conditions of the retainer, the sliding velocity at the outer land and ball pocket reaches to a high level of 110 m/s and the frictional heat generation of the retainer is to be severe. For the cooling system to remove the bearing heat generation at 120,000 rpm, effective jet cooling with nozzles needs to obtain sufficient coolant flow within the bearing. The nozzles were directed to cool the single outer land-guiding side of the retainer where high frictional heat is generated.

Parameters	Hybrid ceramic bearing	All-steel bearing
Bearing		
Rotational speed [rpm]	120,000	
Thrust load [N]	980	
Initial contact angle [deg.]	20	
Initial radial clearance [μm]	77	
Operational radial clearance [μm]	43	
Maximum contact stress at inner/outer races (*Smax*) [GPa]	2.31 / 2.14	2.00 / 2.35
Maximum spinning velocity at inner race (*Vmax*) [m/s]	5.8	7.5
Centrifugal force on ball [N]	454	1,120
Retainer		
Sliding velocity at outer land [m/s]	108	
Sliding velocity at ball pocket [m/s]	116	

Table 4. Bearing load and speed conditions for hybrid ceramic and all-steel bearings at 120,000 rpm with 980 N (25-mm bore)

Figure 25 shows the change of the bearing temperature at a steady speed of 120,000 rpm with a thrust load of 2,160 N [8]. The hybrid ceramic bearing showed excellent performance with a stable condition of the bearing temperature, compared to the seized all-steel bearing showing an irregular change of high bearing temperature. When the thrust load was increased to 3,140 N, the hybrid bearing showed slight damage with a spiky rise of the bearing temperature. It was found that the critical load capacity $Smax$ without seizure at a speed of 120,000 rpm was reached to 3.0 GPa (at a thrust load of 2,160 N) for the hybrid ceramic bearing and 2.0 GPa (980 N) for the all-steel bearing, respectively.

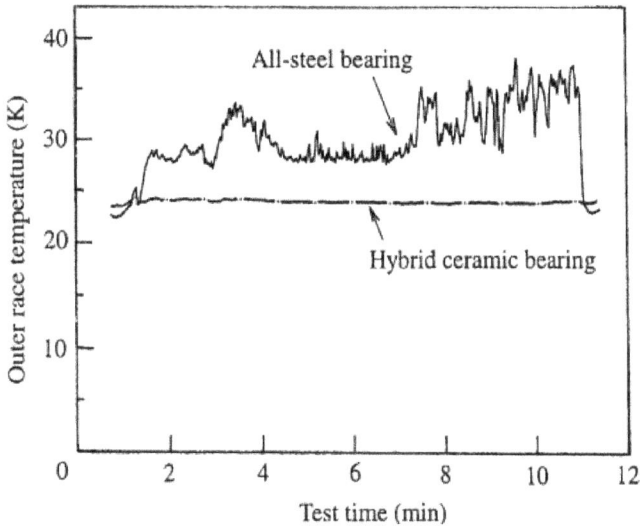

Figure 25. Change of bearing temperature of hybrid ceramic and all-steel bearings at 120,000 rpm with 2,160 N

The power loss around the bearing was estimated based on the heat absorbed by the cooling flow [8]. Figure 26 shows the power loss of the hybrid ceramic and all-steel bearings as a function of rotational speed up to 120,000 rpm in LH_2 under different cooling conditions at a thrust load of 980 N. It was found that the power loss of the bearing significantly increased above 80,000 rpm with increasing cooling flow rate. At 120,000 rpm, the power loss of the bearing that contained the viscous power loss of 2.2 kW at the shaft side was estimated. The power loss was 6.0 kW for the hybrid ceramic bearing and 6.4 kW for the all-steel bearing, respectively. There was not typical difference of the power loss of the bearing because viscous power loss within the bearing almost accounted for an overall power loss generated at ultra-high speeds. It seems that the power loss around the bearing was mainly induced by viscous drag and churning of the cooling flow passing through the bearing.

The components of the hybrid ceramic bearing were in excellent condition with regard to wear at a speed of 120,000 rpm with a thrust load of 3,140 N in LH_2 [40]. On the contrary, the seized all-steel bearing exhibited severe adhesive wear. It was found that the ceramic balls formed superficial micro-cracks on the contact track. Superficial micro-cracks visually extended in a mesh-like pattern on the Si_3N_4 ball tested. It was shown that network of hair crack was propagated along wide-ditch crack. A marked feature of these superficial micro-cracks was that they were very shallow to about 3 μm at maximum and did not extend deeply into the ball. From detailed observation with a scanning electron microscope (SEM), such wide-ditch cracks seemed to be formed by removal of fragments fractured due to contact stress repeated by the rolling balls as shown in Fig. 27. Thus, when the Si_3N_4 balls had lower mechanical strength and fracture toughness, it was clear that wide-ditch cracks were apt to be formed.

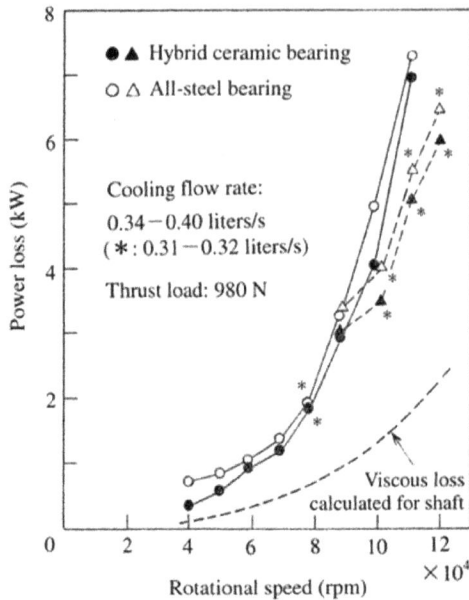

Figure 26. Power loss of hybrid ceramic and all-steel bearings as a function of rotational speed up to 120,000 rpm in LH$_2$

An advanced study was conducted to select a tough Si$_3$N$_4$ ball capable of restraining crack propagation as well as to evaluate the efficient bearing cooling with nozzles. A Si$_3$N$_4$ ball having higher thermal-shock resistance, as well as higher fracture toughness, was found to reduce the propagation of superficial micro-cracks, resulting in a decrease of ball wear. Furthermore, it was observed that the cooling ability of the LH$_2$ jet-flow aimed at the retainer was superior to that aimed at the inner raceway, further reducing the propagation of thermal micro-cracks on the Si$_3$N$_4$ balls. This result also indicated that micro-cracks on the balls were possibly generated at the trace contacting the outer raceway due to a higher centrifugal force under insufficient cooling conditions. Furthermore, under the same cooling rate, the four nozzles achieved a higher cooling ability than the two nozzles with increasing jet speed above 208 m/s. The jet-speed of nozzles reached to the twice of the sliding speed of 108 m/s at the retainer outer-land [40].

In order to prevent the propagation of superficial thermal micro-cracks on the balls, the outer race contact stress was reduced by decreasing the outer race curvature to a limited value of 0.51. Furthermore, sufficient cooling at the outer raceway was gained by a proper clearance of the outer land of the retainer. Decreasing the maximum outer-race stress to 2.0 GPa (thrust load, 1,960 N) in conjunction with sufficient cooling through a narrow outer land clearance could prevent the propagation of superficial micro-cracks even under insufficient cooling conditions [40].

(a) Hair crack   ← 0.01mm →   (b) Wide-ditch crack

Removing of fragments   Wide-ditch crack

Hair crack

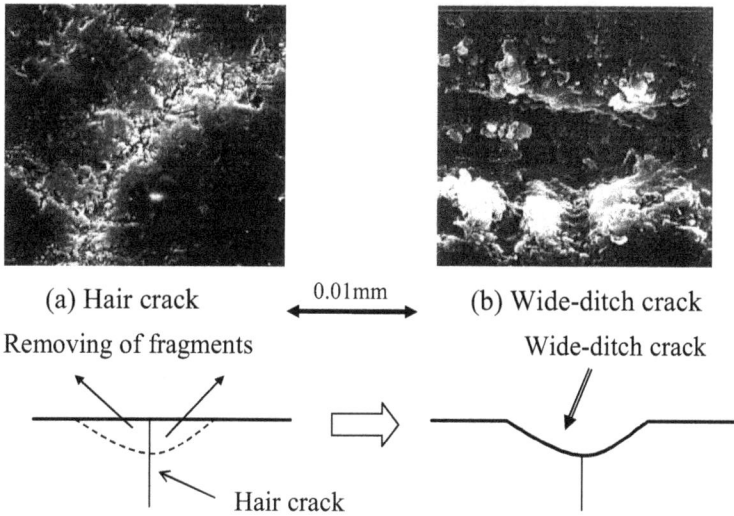

Figure 27. Process model of wide-ditch crack formation on Si_3N_4 ball

6.4. Ultra–high–speed two–phase seal [8]

The floating-ring seal due to noncontact-type is suitable for high-pressure turbopumps; however, conventional seals using carbon seal-rings were weak under high speed and high pressure conditions. Since metal seal-rings have higher mechanical strength and durability, advanced floating-ring seal (with one-seal and two-seal rings) that used Ag-plated metal seal-rings with a seal diameter of 30 mm [8]. This metal seal was studied at ultra-high speeds up to 120,000 rpm in LH_2. Calculated runner growth due to centrifugal force at 120,000 rpm was 29μm, so that the initial seal clearance (gap) was decreased as the rotational speed increased. The test seal had an Ag-plated seal ring made of Inconel 718 that was the same material used for the runner. The runner was coated with a Cr_2O_3 plasma spray, and this coating exhibited excellent friction and wear without adhesion to Ag in LN_2. In order to obtain smooth radial movement of the seal ring, the static seal surface of the housing was coated with a sprayed MoS_2 film.

Figure 28 shows the seal performance of the one-ring seal *vs.* the two-ring seal up to a speed of 100,000 rpm in LH_2 [8]. These seals had a straight bore with a seal gap of 110-120 μm. Figure 29 also shows the phase change models of leakage flow within the seal gap [4]. Seal performance depended on the two-phase flow (gas/liquid phase) of leakage, because the vaporization of leakage was generated by the viscous friction heat and by the seal pressure drop. At lower speeds, the leakage of the one-ring seal was relatively greater than that of the two-ring seal; however, with increasing speed, the leakage of the one-ring seal was drastically decreased and approached the same level of the two-ring seal due to enlargement of the two-phase flow.

For the two-ring seal, the two-phase flow was fully enlarged within the secondary seal ring that was at the downstream of the primary seal ring. Seal leakage was reduced within limits; however, the hydrodynamic force of the liquid phase flow that sustained the seal ring was lost and resulted in seal-ring seizure at a relatively lower speed of 98,700 rpm. Also, shaft vibration for the two-ring seal was likely produced by wobbling of the seal ring under severe rubbing conditions and abruptly increased at speeds of more than 92,000 rpm before resulting in seal-ring seizure at a speed of 98,700 rpm. Furthermore, in the two-ring seal with a seal gap of 70-80 μm, the primary seal-ring seized a speed of 108,600 rpm, because the hydrostatic force decreased due to a low differential pressure.

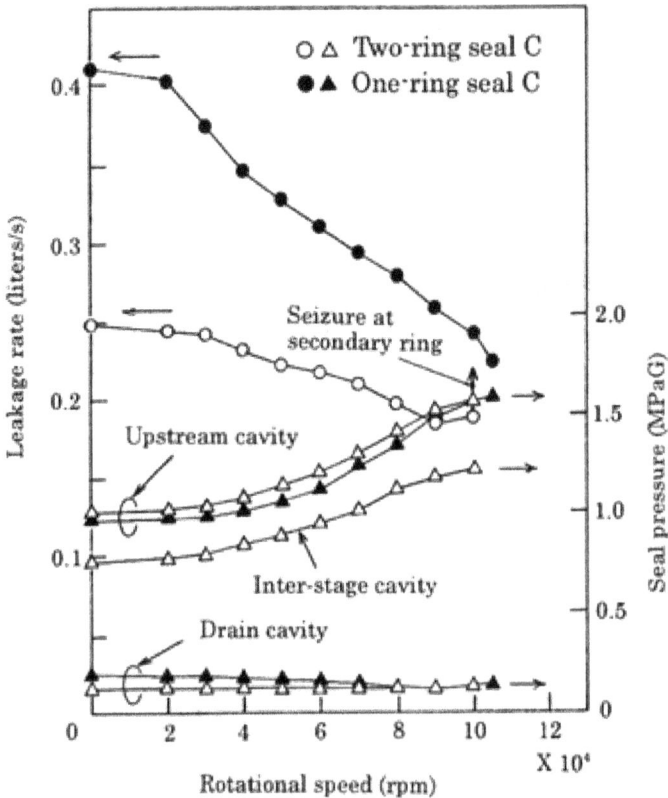

Figure 28. Seal performance of one-ring seal *vs.* two-ring seal up to 100,000 rpm

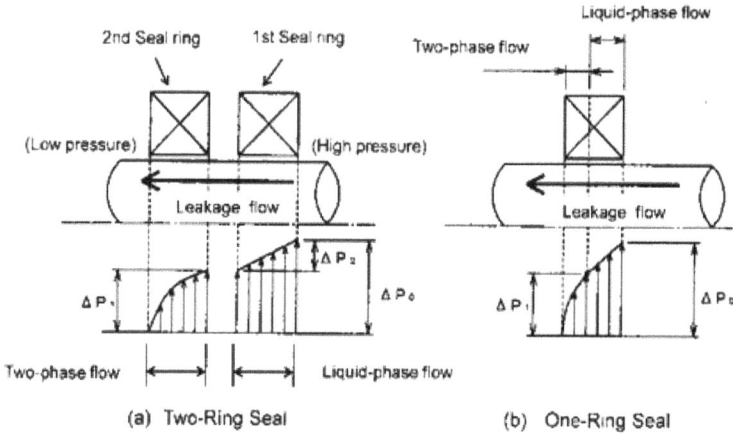

Figure 29. Phase change models of leakage flow within seal gap at ultra-high speed

In contrast, the one-ring seal successfully functioned with no abnormal signs of seizure during tests, because the liquid-phase flow remained within a seal clearance even though the two-phase flow increased. As a result, the hydrodynamic force in the liquid-phase flow as well as the hydrostatic force due to high differential pressure possibly helped to prevent seal-ring seizure. At a steady speed of 120,000 rpm, the one-ring seal exhibited a stable leakage in a range of 0.21-0.24 liters/s that is similar to leakage in the two-ring seal as shown in Fig. 28. Thus, the one-ring seal was superior to the two-ring seal, preventing seal-ring seizure due to an increase of two-phase flow within the sealing clearance.

7. Concluding remarks

For built-up of safe space transport system to achieve high reliability, cryogenic high-speed bearing and shaft seal used in the rocket turbopumps are reviewed historically. These tribo-components have specific lubrication, materials and design requirements in pumping cryogenic liquid propellants in rocket engines. Nowadays, as earth scale issues of energy conservation and environment preservation, a breakaway from the conventional fossil-fuel society becomes a big problem. Clean hydrogen energy is attractive due to its energy efficiency and its smaller impact on the environment, and it is expected to be a key technology in the 21st century. It is proposed that, to build hydrogen infrastructure for LH_2 storage and distribution, development of an industrial tribo-system with long durability and high reliability is essential and advances by supporting of cryogenic tribology studied for LH_2 rocket system.

Acknowledgements

This paper is based on previous cryogenic tribology studies carried out by Japan Aerospace Exploration Agency (JAXA) at Kakuda Space Center. These studies were also supported by IHI Corporation for turbopumps, by NTN corporation for bearings and by Eagle Industry Co., LTD. for shaft seals, respectively. The author is indebted to researchers engaged for their valuable support, to organizations for their enthusiastic cooperation. At last, the author has to thank late Prof. Miyakawa, Y. of Hhosei University, as a pioneer in space tribology in Japan, for his guidance to cryogenic tribology with profound appreciation.

Author details

Masataka Nosaka and Takahisa Kato

Department of Mechanical Engineering, University of Tokyo, Tokyo, Japan

References

[1] Dieter K H & David H H. Modern Engineering for Design of Liquid-Propellant Rocket Engines, *Progress in Astronautics and Aeronautics, Vol. 147*, AIAA, (1992), 155-218.

[2] Liquid Rocket Engine Turbopump Bearing, NASA SP-8048, 1971.

[3] Liquid Rocket Engine Turbopump Rotating- Shaft Seals, NASA SP-8121, 1978.

[4] Nosaka, M. Cryogenic Tribology of High-Speed Bearings and Shaft Seals in Liquid Hydrogen, *Tribology Online*, 6, 2, (2011), 133-141.

[5] Nosaka M, Takada S & Yoshida M. Research and Development of Cryogenic Tribology of Turbopumps for Rocket Engines, *J. of Aeronautical and Space Science Japan*, 58, 681, (2010), 303-313, in Japanese.

[6] Hale J R & Klatt F T. SSME Improvement for Routine Shuttle Operations, AIAA-85-1163, (1985).

[7] Gibson H. Lubriction of Space Shuttle Main Engine Turbopump Bearings, *Lubr. Eng.* 57, 8, (2001), 10-12.

[8] Nosaka M, Takada S, Kikuchi M, Sudo T & Yoshida M. Ultra-High-Speed Performance of Ball Bearings and Annular Seals in Liquid Hydrogen at Up to 3 Million DN (120,000 rpm), *Trib. Trans.*, 47, (2004), 43-53.

[9] Ohta T, Kimoto K, Kawai T, Motomura T, Russ M & Paulus T. Design, Fabrication and Test of the RL60 Fuel Turbopump, AIAA-2003-5073, (2003).

[10] Collongeat L, Edeline E, Frocot M & Dehouve J. Development status of high DN LH2 bearings in Snecma, AIAA-2005-3950, (2005).

[11] Rachuk V & Titkov N. The First Russian LOX-LH₂ Expander Cycle LRV: RD0146, AIAA-2006-4904, (2006).

[12] Nosaka M, Oike M, Kamijo K, Kikuchi M & Katsuta H. Experimental Study on Lubricating Performance of Self-Lubricating Ball Bearings for Liquid Hydrogen Turbopump, *Lubr. Eng.*, 44, 1, (1988), 30-44 .

[13] Nosaka M, Oike M, Kikuchi M, Kamijo K & Tajiri M. Tribo-Characteristics of Self-Lubricating Ball Bearings for the LE-7 Liquid Hydrogen Rocket-Turbopump, *Trib. Trans.*, 36, 3, (1993), 432-442.

[14] Nosaka M, Oike M, Kikuchi M, Nagao R & Mayumi T. Evaluation of Durability for Cryogenic High-Speed Ball Bearings for LE-7 Rocket Turbopumps, *Lubr. Eng.*, 52, 3, (1996), 221-233.

[15] Nosaka M, Oike M, Kikuchi M, Kamijo K & Tajiri M. Self-Lubricating Performance and Durability of Ball Bearings for the LE-7 Liquid Oxygen Rocket-Turbopump, *Lubr. Eng.*, 49, 9, (1993), 677-688.

[16] Nosaka M, Oike M & Kikuchi M. Tribology at Low and High Temperatures, Lubrication in Rocket-Turbopumps, *J. of Japan Society of Lubrication Engineers*, 33, 2, (1988), 90-96, in Japanese.

[17] Nosaka M. Self-Lubricating Performance of High-Speed Ball Bearing for Liquid Hydrogen (1), Design Problems, *J. of Japan Society of Lubrication Engineers*, 32, 10, (1987), 689-695, in Japanese.

[18] Nosaka M. Self-Lubricating Performance of High-Speed Ball Bearing for Liquid Hydrogen (2), Self-Lubricating Performance Improvements. *J. of Japan Society of Lubrication Engineers*, 32, 12, (1987), 833-838, in Japanese.

[19] Winn L W, Eusepi M W & Smalley A J. Small, High-Speed Bearing Technology for Cryogenic Turbo-Pumps, NASA CR-134615, 1974.

[20] Edmond E B & William J A. Advanced Bearing Technology, NASA SP-38, 1965.

[21] Nosaka M & Oike M. Rotating-Shaft Seals in Rocket-Turbopumps, *J. of Japanese Society of Tribologists*, 35, 4, (1990), 233-238, in Japanese.

[22] Oike M, Nosaka M, Watanabe Y, Kikuchi M & Kamijo K. Experimental Study on High-Pressure Gas Seals for a Liquid Oxygen Turbopump, *STLE Trans.*, 31, 1, (1988), 91-97.

[23] Nosaka M, Oike M & Kikuchi M. Cryogenic Tribology of Turbopumps for Rockets, *Cryogenic Engineering*, 31, 10, (1996), 500-511, in Japanese.

[24] Nosaka M. Tribological Burn-Out of Wear, *J. of Japanese Society of Tribologists*, 36, 9, (1991), 689-691, in Japanese.

[25] Nosaka M. Tribology in Low Temperature Environment, *J. of Japanese Society of Tribologists*, 52, 11, (2007), 759-764, in Japanese.

[26] Nosaka M, Takada S, Yoshida M, Kikuchi M, Sudo T & Nakamura S. Effect of Tilted Misalignment of Tribo-Characteristics of High-Speed Ball Bearings in Liquid Hydrogen, *Tribology Online*, 5, 2, (2010), 71-79.

[27] Nosaka M, Kikuchi M, Oike M & Kawai N. Tribo-Characteristics of Cryogenic Hybrid Ceramic Ball Bearings for Rocket Turbopumps: Bearing Wear and Transfer Film, *Trib. Trans.*, 42, 1, (1999), 106-115.

[28] Nosaka M, Kikuchi M, Kawai N & Kikuyama H. Effect of Iron Fluoride Layer on Durability of Cryogenic High-Speed Ball Bearings for Rocket Turbopumps, *Trib. Trans.*, 43, 2, (2000), 163-174.

[29] Suzuki M, Nosaka M, Kamijo K & Kikuchi M. Research and Development of a Rotating- Shaft Seals for a Liquid Hydrogen Turbopump, *Lubr. Eng.*, 42, 3, (1986), 162-169.

[30] Nosaka M, Miyakawa Y, Kamijo K, Suzuki M & Kikuchi M. Study on Sealing Characteristics of High Speed, Contacting Mechanical Seals for Liquid Hydrogen (Part 1), Development of Mechanical Seal for Liquid Hydrogen Turbopump, *J. of Japan Society of Lubrication Engineers*, 29, 1, (1984), 35-42, in Japanese.

[31] Nosaka M, Kamijo K, Suzuki M, Kikuchi M & Miyakawa Y. Study on Sealing Characteristics of High Speed, Contacting Mechanical Seals for Liquid Hydrogen (Part 2), Starting Torque and Static Sealing Performance, *J. of Japan Society of Lubrication Engineers*, 29, 1, (1984), 43-49, in Japanese.

[32] Nosaka M, Kamijo K, Suzuki M, Kikuchi M & Miyakawa Y. Study on Sealing Characteristics of High Speed, Contacting Mechanical Seals for Liquid Hydrogen (Part 3), Friction Power Loss and Dynamic Sealing Performance, *J. of Japan Society of Lubrication Engineers*, 29, 2, (1984), 113-120, in Japanese.

[33] Nosaka M, Kamijo K, Suzuki M, Kikuchi M & Miyakawa Y. Study on Sealing Characteristics of High Speed, Contacting Mechanical Seals for Liquid Hydrogen (Part 4), Characteristics of Running Process and Wear of Rubbing Seal Faces, *J. of Japan Society of Lubrication Engineers*, 29, 2, (1984), 121-128, in Japanese.

[34] Nosaka M, Kamijo K, Suzuki M, Kikuchi M & Miyakawa Y. Study on Sealing Characteristics of High Speed, Contacting Mechanical Seals for Liquid Hydrogen (Part 5), The Formation of Thermal Crack and Wear in Chromium Plate on Rotating Ring, *J. of Japan Society of Lubrication Engineers*, 29, 3, (1984), 187-194, in Japanese.

[35] Oike M, Nosaka M, Kikuchi M & Watanabe Y. Performance of A Shaft Seal System for The LE-7 Rocket Engine Oxidizer Turbopump, *Proc. of The 18th Inter. Symposium on Space Tech. and Sci.*, Kagoshima, (1992), 143-154.

[36] Oike M, Nosaka M, Kikuchi M & Hasegawa S. Two-Phase Flow in Floating-Ring Seals for Cryogenic Turbopumps, *Tribo. Trans.*, 42, 2, (1999), 273-281.

[37] Oike M, Nosaka M, Kikuchi M & Watanabe Y. Performance of a Segmented Circumferential Seal for a Liquid Oxygen Turbopump (Part 1), Sealing Performance, *J. of Japanese Society of Tribologists*, 37, 4, (1992), 339-346, in Japanese.

[38] Oike M, Nosaka M, Kikuchi M & Watanabe Y. Performance of a Segmented Circumferential Seal for a Liquid Oxygen Turbopump (Part 2), Durability, *J. of Japanese Society of Tribologists*, 37, 5, (1992), 389-396, in Japanese.

[39] Nosaka M, Kikuchi M, Oike M & Kawai N. Tribo-Characteristics of Cryogenic Hybrid Ceramic Ball Bearings for Rocket Turbopumps: Self-Lubricating Performance, *Trib. Trans.*, 40, 1, (1997), 21-30.

[40] Nosaka M, Takada S, Yoshida M, Kikuchi M, Sudo T & Nakamura S. Improvement of Durability of Hybrid Ceramic Ball Bearings in Liquid Hydrogen at 3 Million DN (120,000 rpm), *Tribology Online*, 5, 1, (2010), 60-70.

Testing and Modeling

Introduction of the Ratio of the Hardness to the Reduced Elastic Modulus for Abrasion

Giuseppe Pintaude

Additional information is available at the end of the chapter

1. Introduction

Modeling the wear rate is a complex process. The several possibilities of chemical, physical and mechanical changes at the interface are the most probable reasons for this [1]. In this manner, it is reasonable to consider the wear rate as a stochastic process [2, 3], and indeed this approach was taken into account by Archard [4], when he formulated his well-known model. Since then, the majority of available models are based on his proposition, independent on the characteristics of mechanical system. Considering a sharp contact, both Torrance [5] and Yi-Ling & Zi-Shan [6] for sliding and rolling abrasion, respectively, modified Archard's equation based on elastic effects, and the ratio of the hardness (H) to the Young's modulus (E) was the main parameter of the models. In a tribological system with dissimilar materials, for example a ceramic abrading a metal, one material can experience the yielding and the other the brittle failure. This difference in the mechanical behaviors can be decisive for the final performance to wear.

The use of Young's modulus to model the wear resistance was applied for coatings [7]; it appeared in other modifications of Archard's equation [8], and even in empirical relationships between the wear rate and the mechanical properties [9]. Eventually, in all cited cases, the elastic modulus was not the reduced one (E_r), as will be treated here.

A selection of parameters involving hardness and elastic moduli can be summarized (Table 1) [9-11]. All of them have physical meanings that possess some interest for abrasion resistance of materials. The last parameter presented in Table 1, H^3/E^2, is proportional to the load that defines the transition between elastic to plastic contact in a ball-on-plane system, applying the analytical solutions provided by Hertz in Contact Mechanics [12], where the reduced modulus is already taken into account.

Parameter	Physical meaning (taking into account a rigid-plastic material)
H/E	Deformation relative to yielding [9]
$(H/E)^2$	Transition on mechanical contact – elastic to plastic [9]
$H^2/2E$	Modulus of resilience [9]
H/E_r^2	Resistance to the plastic indentation [10]
H^3/E^2	Resistance to the plastic indentation [11]

Table 1. Parameters based on the hardness and elastic moduli, used as indicators of abrasion resistance and their physical meanings

Using some of the abovementioned solutions, a case study will be presented. For two tribological pairs with known wear coefficients, the ratio of hardness to reduced modulus works well than the single property of the worn material (E). The expectation regarding the H/E_r ratio is confirmed also by other aspects used to characterize abrasion, especially the cutting efficiency.

2. Modelling abrasion with E/H ratio

In 1980 Torrance [5] published a model for abrasive wear rate based on the elastic recovery after scratching, supposing that the abrasive particle has a conical geometry. This choice is suitable, because there is an analytical model to describe the changes at the recovered surface [13]. Some years later, another paper [6] adopted the same physical basis but here the application occurred to systems where the abrasive particles roll, instead of slide. The key similarities and differences of both manuscripts will be discussed below.

First of all, it is important to see the main definition presented in [13], because it was the basis for the referred models. This reference presents an equation that relates the reduced modulus with the amount of elastic recovery, h_e (indicated in Figure 1), considering a conical indenter:

$$h_e = h - h_p = \frac{H \times \pi \times a}{E_r} \tag{1}$$

In Equation 1 the term E_r is the reduced modulus, defined as:

$$\frac{1}{E_r} = \frac{1 - v_i^2}{E_i} + \frac{1 - v^2}{E} \tag{2}$$

where,

E_r is the reduced modulus;

E_i is the Young's modulus of conical indenter;

v_i is the Poisson's ratio of conical indenter;

E is the Young's modulus of tested material, and;

v is the Poisson's ratio of tested material.

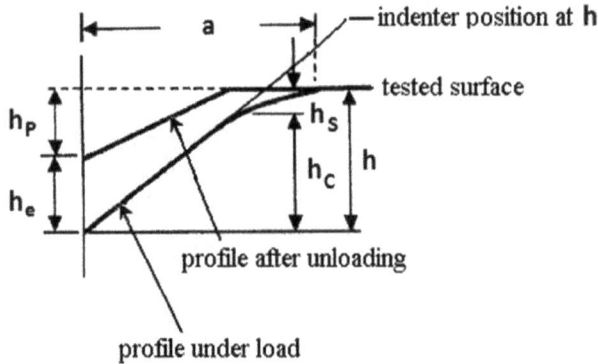

Figure 1. Elastic recovery during an indentation process. Symbology - a: indentation radius, h_e: elastic recovery; h_p: final depth; h: the maximum depth; h_c: contact depth and h_s: deflected depth. Adapted from ISO 14577-1 [14]

A modern definition for the term $E/(1-v^2)$ can be found in ISO/FDIS 14577-1 standard [14], and it is called as 'indentation modulus', using the symbol E_{IT}. Exactly this term was used by references [5] and [6]. In this way, the mechanical properties of abrasive particle were discarded in both cases. It is notable that this aspect has not been ruled out by Stilwell and Tabor [13] in 1961.

A great difference between the Torrance's paper [5] and the Yi-Ling and Zi-Shan one [6] is with respect to the volume of wear. In the former, it was taken as directly proportional to h^2, and for the latter, related to h_p^2, following the symbology of Figure 1. The latter can be considered as more appropriate because it takes into account the elastic effects at a worn surface, so that the final formulation provided by [6] will be presented. Thus, an equation for wear rate, Q (m^3/m), can be written as:

$$Q = C\frac{W}{H}K_P \qquad (3)$$

where,

W is the applied load;

H is the hardness of worn material;

C is a constant and;

K_P will be called here as partial wear coefficient, based on elastic effects during indentation, defined as $(1 + k \times H / E)^2$, being k another constant.

In order to differentiate the use of elastic modulus to the reduced one, another symbology will be considered for the latter case, where K_P' is introduced:

$$K_p' = \left(1 + k \times H / E_r\right)^2 \tag{4}$$

where,

k is a constant.

To define the constant C, Yi-Ling and Zi-Shan [6] made use of nine constants to fit the experimental results. The constant k of term K_P varies with many variables of tribosystem. On the other hand, although Torrance [5] has not make use of constants to fit experimental results obtained by others [15-17], he chose a material as reference (a steel of 401 HV hardness), and all cases were then compared with this material (Equation 5). In this way, he only needed to calculate the constant k (=10), finding a very interesting result, as can be seen in Figure 2.

$$\beta_i^* = \frac{H}{H_{ref}} \frac{\left(1 + 10H_{ref} / E_{ref}\right)}{\left(1 + 10H / E\right)} \tag{5}$$

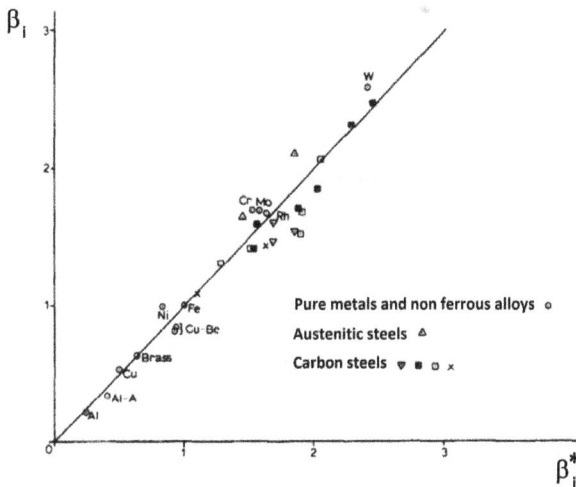

Figure 2. Relative wear resistance (β_i) as a function of relative partial wear coefficient (β_i^*), as defined in [5]. Experimental points derived from [15-17]

Figure 2 shows a linear relationship between the relative wear resistance β_i and the relative partial wear coefficient, defined in [4] as β_i* (Equation 5). The experimental points used to build this curve included pure metals and heat-treated steels. These groups of materials do not present the same behavior when they are abraded by hard particles (Figure 3).

In other words, a pure metal with similar hardness of heat-treated steel presents a higher abrasion resistance. This implies that steels present a different slope on a wear resistance curve when it is put as a function of hardness. Using the ratio of the hardness to the Young's modulus, Torrance [5] put in the same curve these referred groups of materials, which present, a priori, different behaviors. This kind of result was also described in reference [19], but based on a different approach. This research made use of the abrasion factor (f_{ab}) definition (Equation 6 and Figure 4 [20]) to describe the wear resistance. It is a parameter related with the cutting efficiency, i.e., when this factor is equal to unity only the micro-cutting would be the wear mechanism, while for values correspondent to zero no removal of material would be detected. The experimental results obtained in [19] are shown in Figure 5.

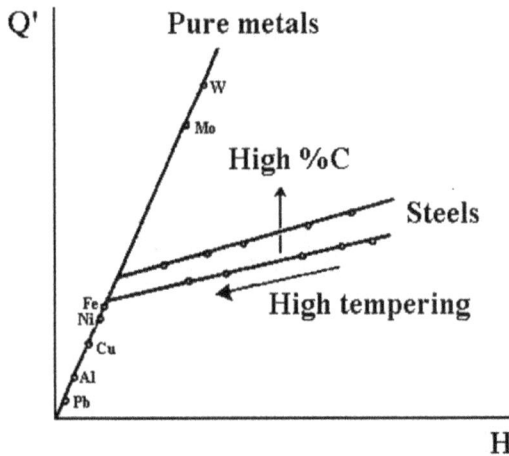

Figure 3. Schematic representation of the wear resistance ($Q' = 1/Q$) produced by hard particles as a function of hardness, considering pure metals and heat-treated steels. Adapted from reference [18]

$$f_{ab} = \frac{A2 - A1}{A2} \tag{6}$$

where,

A1 is the cross section area relative to pileup produced by a single scratch and;

A2 is the cross section area relative to the groove produced by a single scratch.

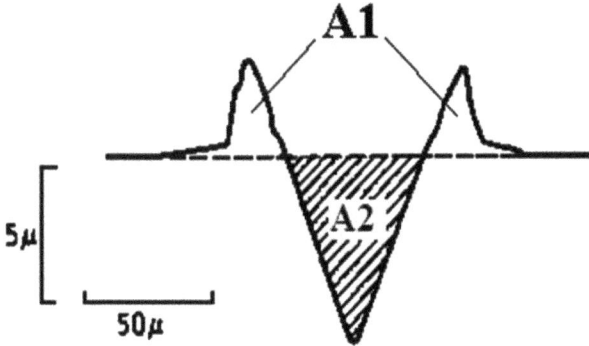

Figure 4. Abrasion factor definition (areas A1 and A2 defined in Equation 6). Real profile adapted from reference [20]

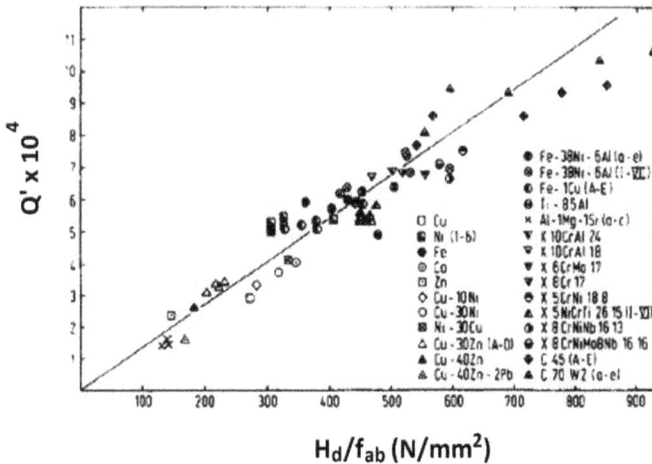

Figure 5. Abrasive wear resistance (Q') produced by flint particles as a function of wear debris hardness-to-f_{ab} ratio. Adapted from reference [19], which defines Q' as a non dimensional parameter

Again, in Figure 5 both pure metals and steels are put in the same curve, showing that the hardness alone, in some cases, is not a complete descriptor of the wear resistance. The similarity between the results presented in Figure 2 and Figure 5 opens a possibility to the abrasion factor to be described as a function of the hardness-to-elastic modulus ratio.

Although in reference [19] the elastic effects have not been incorporated to the model, there are results in the literature relating the pileup formation (Figure 6, [21]) with the mechanical properties [22]. In this case, it is possible to consider that the pileup formation (h_c/h) works for static (hardness test) and kinetic cases (scratch test), being the higher the pileup, the smaller the cutting efficiency.

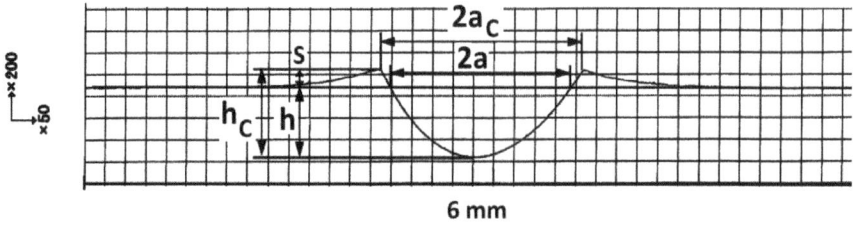

Figure 6. Physical parameters extracted from a residual profile of a spherical indentation. Notation: a_c is the indentation radius at the contact, and s is the height associated to the indentation morphology. Pileup is associated to the h_c/h ratio. Adapted from reference [21]

In reference [22] an equation that correlates the pileup formation with the H/E ratio, based on results obtained in scratch tests conducted with a Berkovich indenter (Figure 7) can be found:

$$h_C\big/h = 0.41498\ln E\big/H - 0.14224 \tag{7}$$

Important evidence was detected in [23] that the application of Equation (7) was dependent on the level of applied load.

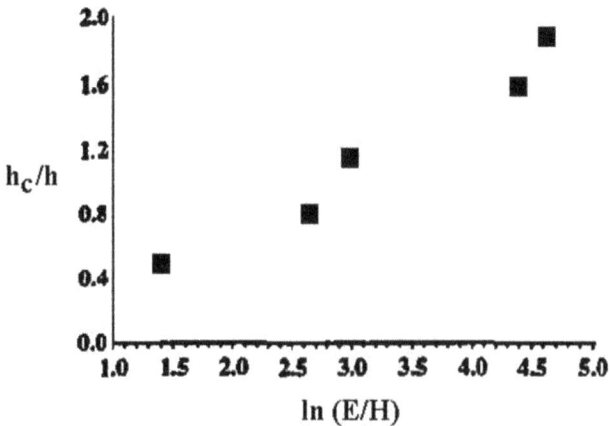

Figure 7. Relation between the pileup (h_c/h) and the $\ln(E/H)$, obtained after scratch tests with Berkovich indenter. Adapted from reference [22]

3. Case study

Two tribological pairs were studied, also considered in a previous investigation [24]. Their wear rates are known: glass abrading a quenched and tempered (Q&T) 52100 steel [25], and alumina wearing a hard metal [26]. The mechanical properties, determined from instrumented indentation testing, used to calculate the partial wear coefficients, K_P and K_P' (Equations 3 and 4), are presented in Table 2. To calculate the reduced modulus (Equation 2), Poisson's ratios were extracted from [27].

As seen in the previous item, due to the similarity of Figures 2 and 5, the partial wear coefficient can be well related to the abrasion factor (cutting efficiency), f_{ab}. An indirect way to know the f_{ab} factor is found in reference [28], which defined it as the ratio between the wear coefficient K (Archard's definition) and the ploughing fraction of friction coefficient (μ_p). The cutting efficiency values, following this definition, are 0.106 for glass abrading bearing steel and 0.079 for alumina wearing hard metal, a difference of 34%. As the constant K_P' can vary along a broad range (Figure 8), we select a value so as to the difference is also 34%, and for this purpose it is 8.25.

Material	H, GPa	E, GPa	E_r, GPa	H/E	H/E$_r$
Soda-lime glass	4.07	69	53.24	0.080	0.103
Q&T steel	5.5	180*		0.023	0.076
Alumina	19.6	376.1	222.43	0.052	0.088
Hard metal (grade K)	11	480		0.023	0.049

Table 2. Mechanical properties of selected tribological pairs. *Obs.: Q&T steel is a wire-drawing, which implies in a reduction of elastic modulus due to the work-hardening effect

An example of variation in K_P' value with k factor is presented in Figure 8, for the tribological pairs considered in Table 2. Another pair was added (calcite-fluorite) in this figure, in order to help the discussion with their differences.

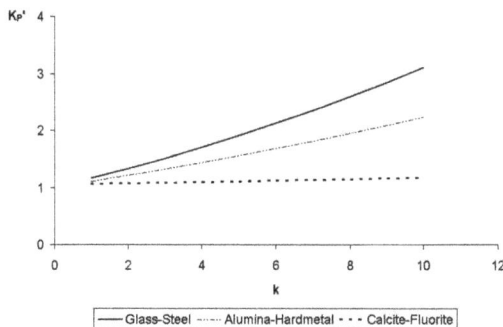

Figure 8. Variation of K_p' wear coefficient with factor k for three cases

The resulting values of partial wear coefficients, using 8.25 as k factor, are presented in Table 3.

Pair	K$_p$ (equation 3)	K$_p$' (equation 4)
Glass - Q&T steel	1.41	2.66
Alumina – Hard metal	1.41	1.98

Table 3. Partial wear coefficient values for selected tribological pairs

Table 3 shows that the K$_P$ values were similar for the studied pairs, while the K$_P$' did not follow this trend. The wear coefficients of tribological pairs (K values) determined using sliding abrasion tests were 0.014 and 0.008, for glass against steel [25] and alumina against hard metal [26], respectively. Visibly this difference in wear rates is only reflected by the K$_P$' values, which is a direct result of the ratio of hardness to the reduced modulus.

Another observation for the K$_P$ values is that the similarity presented in Table 3 is not affected by the variation of the k factor. On the contrary, K$_P$' is strongly dependent on this factor, as can be seen in Figure 8.

One can ask about the good agreement with experimental data found in [5] and [6] when they used only the Young's modulus in wear model. The key point for that is the superiority of abrasive hardness in relation to the worn surface. All experimental results in these cases were performed using abrasives much harder than the tested materials, such that the mechanical properties of them can be considered as unaltered along the tests. An investigation [29] showed significant changes in glass particles when they abraded steel surfaces, even for non heat-treated ones. The same thing was demonstrated in [30] for different abrasives wearing WC-Co thermal sprayed coatings. When the abrasive particle is relatively soft to the abraded material, their mechanical properties play a key role during the wear process, and an extensive discussion on it can be found in [31].

In addition, in reference [24] these tribological pairs were separated using the difference in the plasticity index, δ_H, defined in [32] (Equation 8), being the smallest difference related to the calcite-fluorite, and the largest for glass-steel pair within the materials analyzed in [24]. This aspect seems to be important again when one observes Figure 8. When the difference is insignificant, as in the case of calcite-fluorite, the variation of K$_P$' with k factor is minor, and on the other hand, for the case of the glass-steel pair, for a notable difference in the plasticity index, a great variation of K$_P$' with the k factor occurs.

$$\delta_H = 1 - 14.3\left(1 - v - 2v^2\right){}^H\!\big/_E \tag{8}$$

A brief discussion of the k factor is instructive. Following [6], this factor is especially associated to the particle geometry. As stated, a hypothetical K$_P$' of calcite-fluorite pair would be less affected by the variation on k factor. Therefore, one can imagine a small fragment of mineral used in Mohs scale [33] and, considering the previous assertive, conclude that its geometry would not be important. Probably, this is not the case. The k factor should be understood in a broad manner, i.e., all variables of a system can alter its value. Thus, slight changes in tribological variables could bring higher alterations in the wear coefficient for the pair glass-steel than for the other described cases. At this moment, no experimental result is available to corroborate this hypothesis, but it is an interesting field to be explored further.

Finally, a discussion concerning the Equation (7) and the values presented in Table 2 is valuable. If the values of H/E ratio for steel and alumina were applied in Equation (7), the results would be similar. A similar height of pileup for these materials obviously is not a reasonable result, taking into account the experimental values obtained for wear and friction coefficients for them. Nevertheless, applying the reduced modulus in the place of E in that equation, one can find fewer tendencies to form pileup after abrasion of steel by glass, which means a higher cutting efficiency in this case, meeting with the values described before. Consequently, it is more a case of successful application of reduced modulus to predict changes at the mechanical contact.

4. Conclusions and final remarks

The viability of the use the hardness-to-reduced modulus ratio to model the wear coefficient for abraded materials was demonstrated. Previous models were developed taking into account only the Young's modulus of worn surface, discarding the properties of abrasive material. These cases work only for pairs where the abrasive particle is harder than the abraded material and it was demonstrated that they fail when the abrasive hardness is relatively low.

In addition, other questions were discussed and they open some possibilities to carry out future research. First, the model of wear coefficient treated here involves higher requirements, because a constant is needed. This constant seems to affect more the wear coefficient of some pairs than the others, and the reason for that is not clear. Finally, an extensive work could be made exploring the relation between cutting efficiency (abrasion factor) and H/E_r ratio, computing a large variation of the applied load and the abrasive (indenter) properties. Probably, an investigation based on these aspects should supply answers to the improvement of a wear model containing the H/E_r ratio.

Nomenclature

A1 and A2 Cross section areas used in the definition of abrasion factor

a Indentation radius

a_c Indentation radius at the contact

C Constant

E Young's modulus

E_i Young's modulus of indenter

E_r Reduced modulus

E_{ref} Young's modulus of a reference material

f_{ab} Abrasion factor or cutting efficiency

H Vickers hardness of the worn material

H_d Vickers hardness of wear debris

H_{ref} Vickers hardness of a reference material

h Maximum depth at applied load

h_C Contact depth

h_e Elastic recovery

h_P Final depth

h_S Deflected depth

K Wear coefficient

K_P Partial wear coefficient, defined as $(1 + k \times H / E)^2$

K_P' Partial wear coefficient calculated with Er instead of E

k Constant

Q Wear rate

Q' Wear resistance (= 1/Q)

s Height associated to the indentation morphology

W Applied load

β_i Relative wear resistance

β_i^* Relative partial wear coefficient

δ_H Plasticity parameter

μ_p Ploughing fraction of friction coefficient, taken as 0.2

v Poisson's ratio of the worn material

v_i Poisson's ratio of the indenter

Acknowledgements

Financial support for this study was provided by CNPq under project no. 306727/2011-0.

Author details

Giuseppe Pintaude*

Address all correspondence to: giuseppepintaude@gmail.com

Mechanical Academic Department, Federal University of Technology – Paraná, Curitiba, Brazil

References

[1] Bayer R.G., Comments on Engineering Needs and Wear Models. In: Ludema K.C., Bayer, R.G. (eds.) Tribological Modeling for Mechanical designers. Philadelphia: ASTM International; 1991. p. 3-11.

[2] Silva Jr., C.R.A., Pintaude G., Al-Qureshi H. A., Krajnc M.A. An Application of Mean Square Calculus to Sliding Wear. Journal of Applied Mechanics 2010;77(2) paper 021013.

[3] Silva Jr., C.R.A., Pintaude G. Uncertainty Analysis on the Wear Coefficient of Archard Model. Tribology International 2008;41(6) 473-481.

[4] Archard J.F. Contact and Rubbing of Flat Surfaces. Journal of Applied Physics 1953;24(8) 981–988.

[5] Torrance A.A. The Correlation of Abrasive Wear Tests. Wear 1980;63(2) 359-370.

[6] Yi-Ling W., Zi-Shan W. An Analysis of the Influence of Plastic Indentation on Three-Body Abrasive Wear of Metals. Wear 1988; 122(2) 123-133.

[7] Leyland A., Matthews A. On the Significance of the H/E Ratio in Wear Control: a Nanocomposite Coating Approach to Optimised Tribological Behaviour. Wear 2000; 246(1-2) 1-11.

[8] Liu R., Li D.Y. Modification of Archard's Equation by Taking Account of elastic/pseudoelastic Properties of Materials. Wear 2001; 251(1-12) 956-964.

[9] Finkin E.F. Examination of abrasion Resistance Criteria for Some Ductile Metals. Journal of Lubrication Technology 1974;96(2) 210-214.

[10] Joslin D.L., Oliver W.C. A New Method for Analyzing Data from Continuous Depth-Sensing Microindentation Tests. Journal of Materials Research 1990; 5(1) 123-126.

[11] Tsui T.Y., Pharr G.M., Oliver, W.C., Bhatia C.S., White R.L., Anders A., Brown I.G. Nanoindentation and Nanoscratching of Hard Carbon Coatings for Magnetic Disks. MRS Proceedings 1994;356, p.767.

[12] Brizmer V., Kligerman Y., Etsion I. The Effect of Contact Conditions and Material Properties on the Elasticity Terminus of a Spherical Contact. International Journal of Solids and Structures 2006; 43(18-19) 5736–5749.

[13] Stilwell, N. A., Tabor, D. Elastic Recovery of Conical Indentations. Proceedings of the Physical Society 1961; 78(2) 169-179.

[14] ISO - International Organization for Standardization. ISO/FDIS 14577-1 - Metallic Materials - Instrumented Indentation Test for Hardness and Material Parameter - Part 1: Test method. Geneva, Switzerland, 2002.

[15] Richardson R.C.D. The Maximum Hardness of Strained Surfaces and the Abrasive Wear of Metals and Alloys, Wear, 1967; 10(5) 353-382.

[16] Richardson R. C. D. The Wear of Metals by Hard Abrasives, Wear 1967; 10(4) 291-309.

[17] Richardson R.C.D. The Wear of Metals by Relatively Soft Abrasives, Wear 1968; 11(4) 245-275.

[18] Murray M.J., Mutton P.J., Watson J.D. Abrasive Wear Mechanisms in Steels. In: Ludema K.C., Glaeser W.A., Rhee S.K. (eds.), WOM 1979: Proceedings of International Conference on Wear of Materials, 16-18 April 1979, Dearborn, MI. New York: American Society of Mechanical Engineers, pp. 257–265, 1979.

[19] Zum Gahr, K.H. Modelling of Two-Body Abrasive Wear. Wear 1987; 124(1) 87–103.

[20] Buttery T.C., Archard J.F. Grinding and abrasive wear. Proceedings of the Institution of Mechanical Engineers 1970; 185(1) 537-552.

[21] Pintaude G., Hoechele A.R., Cipriano G.L. Relation between Strain Hardening Exponent of Metals and Residual Profiles of Deep Spherical Indentation. Materials Science and Technology 2012; 28(9-10) 1051-1054.

[22] Jardret V., Zahouani H., Loubet J.L., Mathia T.G. Understanding and Quantification of Elastic and Plastic Deformation during a Scratch Test. Wear 1998; 218(1) 8-14.

[23] Masen M.A., de Rooij M.B., Schipper D.J., Adachi K., Kato K. Single asperity abrasion of coated nodular cast iron. Tribology International 2007; 40(2)170–179.

[24] Pintaude G. An Overview of the Hardness Differential Required for Abrasion, Journal of Tribology 2010;132(3) paper 034502.

[25] Pintaude G., Bernardes F.G., Santos M.M., Sinatora A., Albertin E. Mild and Severe Wear of Steels and Cast Irons in Sliding Abrasion. Wear 2009;267(1-4) 19-25.

[26] Pintaude G., Farias M.C.M., Kohnlein M., Tanaka D.K., Sinatora A. Abrasive Wear of Cutting Tools Used in the Wood Industry (In Portuguese). In: UFRN – Universidade Federal do Rio Grande do Norte (ed.) CONEM 2000: XV Brazilian Congress of Mechanical Engineering, 7-11 August 2000, Natal, Brazil. Rio de Janeiro: Associação Brasileira de Ciências Mecânicas, 2000.

[27] Gercek H. Poisson's Ratio Values for Rocks. International Journal Rock Mechanics and Mining Sciences 2007;44(1) 1–13.

[28] Jacobson S., Wallen P., Hogmark S. Fundamental Aspects of Abrasive Wear Studied by a New Numerical Simulation Model. Wear 1998;123(2) 207-223.

[29] Pintaude G., Tanaka D.K., Sinatora A. The Effects of Abrasive Particle Size on the Sliding Friction Coefficient of Steel using a Spiral Pin-on-Disk Apparatus. Wear 2003; 255 (1-6) 55-59.

[30] Bozzi A. C.; De Mello J.D.B. Wear Resistance and Wear Mechanisms of WC-12%Co Thermal Sprayed Coatings in Three-Body Abrasion, Wear 1999;233-235, 575-587.

[31] Pintaude, G. (2011). Characteristics of Abrasive Particles and Their Implications on Wear, New Tribological Ways, Ghrib, T. (Ed.), ISBN: 978-953-307-206-7, InTech, Available from: http://www.intechopen.com/books/new-tribological-ways/characteristics-of-abrasive-particles-and-their-implications-on-wear

[32] Milman Y.V., Galanov B.A., Chugunova B.I. Plasticity Characteristic Obtained through Hardness Measurement, Acta Metallurgica et Materialia 1993; 41(9) 2523-2532.

[33] Broz M.E., Cook R.F., Whitney D.L. Microhardness, Toughness, and Modulus of Mohs Scale Minerals. American Mineralogist 2006; 91(1) 135–142.

Chapter 7

New Scuffing Test Methods for the Determination of the Scuffing Resistance of Coated Gears

Remigiusz Michalczewski, Marek Kalbarczyk, Michal Michalak, Witold Piekoszewski, Marian Szczerek, Waldemar Tuszynski and Jan Wulczynski

Additional information is available at the end of the chapter

1. Introduction

1.1. Scuffing of gear teeth

In modern machines the problems of the prevention of scuffing of the gear teeth is still very important. One of the reasons is that for many years the technique development is related to increasing the loading of the friction surfaces accompanied by decreasing their size [1]. In the case of gears, the risk of scuffing occurrence rises because of potential design and assembly mistakes, unexpected overloads, as well as extremely different speeds of the rotation of gears, because both very high speeds and very low speeds may cause scuffing [2]. The occurrence of one of the mentioned factors may lead to very serious gear failures.

Apart from the above mentioned factors, the problems of using proper lubricating oils, with high extreme-pressure (EP) properties cannot be neglected.

In gears, the surface destroyed by scuffing appears at the addendum and dedendum of the tooth. This results from the sliding speed of the meshing teeth that reaches the highest values at these places of the gear tooth.

Failures of the gear teeth flanks due to scuffing are shown in Figure 1.

Figure 1. Photographs of failures of the gear teeth flanks due to scuffing: a) "non-symmetric" scuffing observed in gear service, resulting from the incorrect distribution of load along the tooth [3], b) scuffing on the flank of the test gear due to poor extreme-pressure (EP) properties of the tested gear oil during the gear scuffing experiments performed by the authors

Another example of scuffing of gears concerns the rudder speed brake power drive unit of a space shuttle, observed during its inspection after grounding [4]. Figure 2 a) shows the pinion and ring gear of the power drive unit of the space shuttle. Figure 2 b) presents the pinion tooth with wear at the tip and scuffing on dedendum. It was postulated that early shutdown of one of three hydraulic motors driving the gearbox could cause scuffing - in a differential gearbox, early shutdown of one motor could cause the overloading with potential for scuffing.

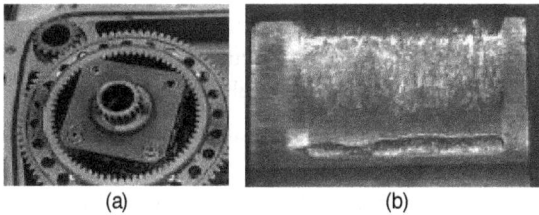

Figure 2. Photographs of the components of the rudder speed brake power drive unit of a space shuttle: a) pinion and ring gear of the power drive unit, b) damaged pinion tooth [4]

From the above example, it is absolutely apparent that the prevention of scuffing is still an important challenge, even in the high-tech sector.

1.2. Scuffing — How is it brought about?

To better understand scuffing, Figure 3 presents the interpretative models of the phenomena in different phases of this process, caused by the continuously increasing load. The models concern the contact between two balls of the four-ball tribosystem (the rotating upper ball with one of the three stationary lower balls) during the testing of the automotive gear oils of API GL-4 and GL-5 performance levels. Such oils contain chemically active extreme-pressure (EP) lubricating additives to prevent scuffing. API GL-4 oils are used to lubricate synchronised manual transmissions of European cars and contain up to 4% of EP additives. API GL-5 oils containing up to 6.5% of EP additives are employed to lubricate automotive gears especially susceptible to scuffing, i.e. hypoid gears, in axles operating under various combinations of high-speed/shock-load and low-speed/high-torque conditions.

It should be emphasised here that a four-ball tribosystem is very often used for tribological testing of the performance of automotive gear oils.

The lower graph in Figure 3 presents the friction torque curve (M_t) obtained at continuously increasing load (P). The brackets over the graph indicate particular phases of the scuffing process. In these phases, the friction coefficient values (μ) were determined, and they are given in the red rectangles in the graph area. The thick red line below the graph denotes the time from the beginning of the run until the occurrence of the scuffing initiation reflected by a sharp rise in the friction torque.

The interpretative models of phenomena related to scuffing are presented over the graph in Figure 3. Because the models concern the contact zone between two balls of the four-ball tribosystem where the upper ball rotates and the lower ball is stationary, the direction of the movement was indicated in the upper part of the models by an arrow. If there is no arrow, the given model illustrates no movement of the balls, i.e. at the beginning of the run (before the motor of the tribotester starts).

For the phase "scuffing initiation," the upper model in Figure 3 illustrates the surface that did not exhibit very rough topography typical of scuffing (shown in the surface topography image), while the lower one concerns the surface already destroyed by scuffing.

In the models, three characteristic zones in the wear scar surface layer were identified: a chemically modified zone through the action of the lubricating additives and the steel surface, a zone of plastic deformation, and a zone of elastic deformation. All of these zones are described in the legend above the models in Figure 3.

1.2.1. Phase: "Beginning of run"

After immersing the test balls in the tested gear oil and applying the initial load close to 0, a phenomenon known as physical adsorption or physisorption appears. In this phase, adsorbed molecules constitute the boundary layer on the friction surface, which protects the surface asperities against direct contact. The model with the heading "Beginning of run" in Figure 3 illustrates this, reflecting the situation before the start of the relative movement of the test balls.

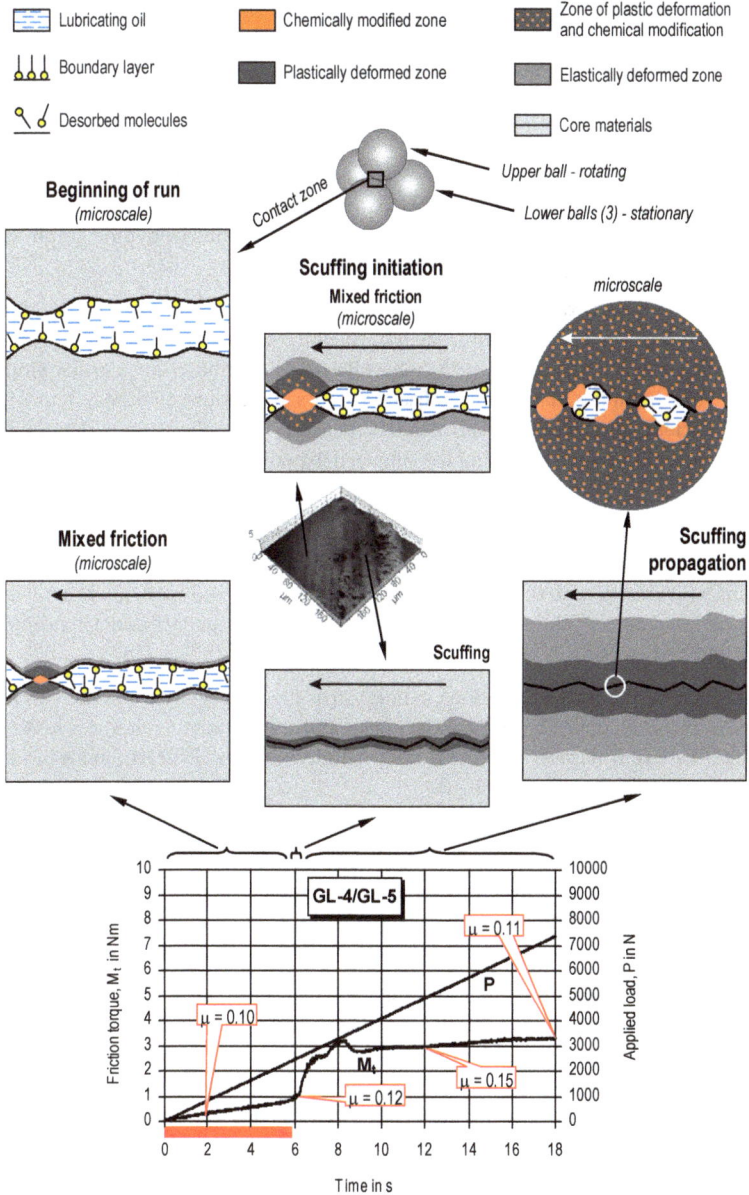

Figure 3. Models of scuffing in different phases for the automotive gear oils of the API GL-4 and GL-5 performance levels

1.2.2. Phase: "Mixed friction"

In publications, the terms "mixed friction" and "mixed lubrication" are often used equivalently and concern the same phenomena. For the purpose of this chapter, one can assume that occurrences during the regime of the mixed lubrication result in the mixed friction with its specific friction coefficient.

The phase "Mixed friction" concerns the first stage of the run from the moment of the start of the relative movement between the test balls to the scuffing initiation reflected by a sharp rise in the friction torque. Its duration is denoted by the thick red line below the graph with the friction torque (M_t) and applied load (P) - Figure 3.

In this phase the mixed friction occurs. This can be stated on the basis of the fundamental criterion that is the friction coefficient value. The friction coefficients typical of particular types of friction were adopted from the work [5], where the four-ball tribosystem was also employed. From that work, it implies that the mixed friction occurs in the four-ball tribosystem when the friction coefficient is in the range between 0.07 and 0.1. Thus, the authors determined the friction coefficient at the 2nd second of the four-ball experiment, being 0.1, denote the mixed friction.

It is worth noting here that the idea of the occurrence of the mixed friction regime (instead of EHL, i.e. elastohydrodynamic lubrication) at the very start of the relative movement between the test balls (load is close to 0) is also supported in the mentioned work [5]. From that work it is apparent that "pure" EHL occurs in the four-ball tribosystem only under conditions of a low load and high speed.

At mixed friction, the micro-EHL films mainly carry the load and the mating surfaces are protected from direct contact by the boundary layer. But at some micro-zones, due to the failure of the micro-EHL film surface, asperities locally collide, which is illustrated in the model with the heading "Mixed friction" in Figure 3.

Due to collisions of surface asperities, the temperature in the micro-contact rises. At a higher temperature, physically adsorbed molecules may be attracted to the surface with greater forces, and chemical adsorption or chemisorption appears. The decomposition of the active compounds in the lubricating additives catalyses the transformation of some chemically adsorbed molecules into chemical compounds at higher temperatures.

The collision of the surface asperities and the local high pressure of the oil induced by the approaching asperities bring about elastic (reversible) and plastic (irreversible) deformations of the contacting surface. Due to the thermal (temperature rise) and mechanical activation (plastic deformation causing surface defects), the conditions exist for the initiation of the diffusion of "active" atoms from the lubricating compounds (e.g. sulphur atoms) into the surface layer.

The described phenomena lead to the formation of inorganic chemical compounds of iron with sulphur, phosphorus, and oxygen, coming from EP lubricating additives in the tested gear oil. Such additives (based on organic S-P compounds) form e.g. iron sulphide FeS [6]. FeS compounds, apart from hampering the creation of adhesive bonds with their shear strength being 1/5th that of steel and their hardness being 1/4th that of steel, facilitate shearing of the chemically modified surface asperities, and the shear plane is transferred to the thin FeS layer, which protects the surface from tearing out the material from deeper layers, reducing the wear intensity.

For the tested oil, containing EP lubricating additives, the surface asperities are covered by the protective layer of the above mentioned chemical compounds. This is illustrated in the respective model in Figure 3. Due to this, for the gear oils with EP lubricating additives, the scuffing initiation is delayed to appear at much higher loads than in the case of oils without lubricating additives (e.g. API GL-1 ones, not presented here).

1.2.3. Scuffing phase: "Scuffing initiation"

In this phase, scuffing initiates - the friction torque (M_t) sharply increases and measured friction coefficient values exceed the maximum value assumed for the mixed friction, i.e. 0.1 [5].

The scuffing initiation occurs at a load called the scuffing load, which is characteristic for each tested lubricating oil. At this load, the lubricating film collapses, the number of colliding surface asperities drastically increases, and the destruction changes its occurrence from the micro- to macro-scale and scuffing appears. Initially only part of the friction surface undergoes scuffing. It can be observed in the surface topography image of the border between the surface that did not exhibit very rough topography typical of scuffing (left side) and the surface destroyed by scuffing (right side) - Figure 3.

The described phenomena leading to scuffing are illustrated in the models with the heading "Scuffing initiation" in Figure 3. The upper model concerns the surface that did not exhibit very rough topography typical of scuffing, where still the mixed friction exists, while the lower one refers to the surface already destroyed by scuffing.

The upper model shows that the micro-scale phenomena in the zone intact by scuffing are similar to those described in the phase "Mixed friction" apart from the thickness of elastic and plastic deformations which increased due to rising load. Probably, in view of plastic deformation that causes surface defects, the reactive diffusion of "active" atoms from the EP lubricating additives (e.g. sulphur atoms) into the surface layer takes place and iron sulphides form, which is confirmed by other researchers, e.g. in the work [7]. The diffusively modified micro-zones inside the highest asperities are plastically deformed and are indicated in the respective model as orange spots - Figure 3.

By observing phenomena in the part of the friction surface that undergoes scuffing, one can indicate that the situation changes radically. The lower model illustrates that, in the first phase of scuffing, the lubricating film no longer exists, nor is there any boundary layer. This leads to a rapid intensification of the material destruction. Much plastic deformation appears, turning into the transfer, flowing and mingling of the material of the rubbing test balls. For the tested oils with EP lubricating additives, much of the surface layer starts to be chemically modified. This will be decisive for the scuffing propagation character.

1.2.4. Scuffing phase: "Scuffing propagation"

This phase refers to the scuffing process, after its initiation. It is reflected by a sharp increase in the friction torque (M_t), accompanied by a high intensity of the lower test balls wear - Figure 4 a, b). This situation is illustrated in the models with the heading "Scuffing propagation" in Figure 3.

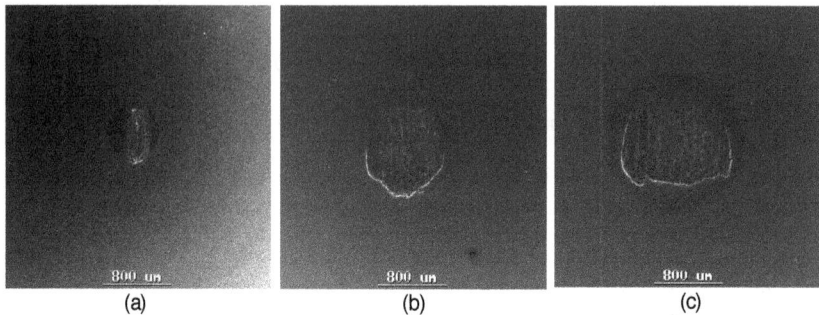

(a)   (b)   (c)

Figure 4. Development of the wear of the lower test balls due to scuffing: a) at scuffing initiation, b) at 12th seconds of the run (scuffing propagation), c) at the end of the run; images obtained at the same magnification

For the tested gear oils, after the scuffing initiation due to rapid chemical reactions of their EP additives with the surface, a rise in the friction torque is mitigated to quickly stabilise at relatively low value - Figure 3. It is accompanied by continuously evolving wear of the lower test balls that is not intensive - Figure 4 b, c). A drop in the pressure in the contact zone due to wear, brings about the possibility of oil introduction into the contact zone and the regeneration of the boundary layer on much of the friction surface. Such an action is indicated by the friction coefficient within the range 0.11 to 0.15, typical of boundary friction. On the basis of the work [5], which also concerns four-ball experiments, it was assumed that the boundary friction occurs in the four-ball tribosystem when the friction coefficient is in the range between 0.09 and 0.15. The determined values of the friction coefficient being in the middle and upper limit typical of boundary friction denote that some part of the friction surface must have undergone scuffing; It can be assumed from [5] that "full scuffing" occurs when the friction coefficient exceeds 0.3. The specific state of the surface layer in this phase is called the "Secondary Boundary Layer" (SBL) in the work [8]. The round model in the micro-scale concerning the scuffing propagation (Figure 3) illustrates the places of oil appearance in the contact zone. Let us call them "the micro-pockets." One can presume that inside the oil micro-pockets the following phenomena take place: the intensive adsorption and desorption of the base oil and lubricating additives molecules on/from the steel surface, chemical reactions of the lubricating additives with the surface, and - in view of plastic deformation that causes surface defects - the diffusion of "active" atoms from the lubricating compounds (e.g. sulphur atoms) into the surface layer. In view of the transfer and mingling of the material of the rubbing test balls, the chemically modified zones appear across the entire zone of plastic deformations - orange spots.

For the API GL-4 and GL-5 gear oils, the effective chemical modification of the surface mitigates the increase of the wear scar diameter - Figure 4 b, c) - in the phase of the SBL formation, accompanied by a mitigated rise in the friction torque and a decreasing friction coefficient (Figure 3).

1.3. Gear tests of scuffing

Nowadays, two manners of the improvement of the resistance to scuffing of gears are in use in the world. One is focused on the improvement of extreme-pressure (EP) properties of gear oils. The other one is related to the improvement of the properties of gear materials, e.g. by the deposition of thin hard coatings onto the tooth flank surface.

The verification of the quality of gear oils and new techniques of surface engineering of the tooth surface of gears requires that gear testing should be used. The most known is a unique complex of gear test methods developed in the Gear Research Center (FZG) at the Technical University of Munich. Approximately, 500 FZG gear test rigs are used around the world [9].

The most often used and popular gear tests for lubricating oils are performed using the FZG A/8.3/90 scuffing test method. Unfortunately, this method makes it impossible to differentiate between gear oils having very good extreme-pressure (EP) properties, from the point of view of the resistance to gear scuffing [10]. This is why various scientific centres have developed their own test methods [10-13].

Recognising the problem of the low resolution of A/8.3/90 scuffing test, the FZG has developed two new scuffing methods denoted as A10/16.6R/90 and S-A10/16.6R/90 (S - *shock*). The new test methods are described in detail in the literature, e.g. [14-19]. They are carried out under much severer conditions compared to the A/8.3/90 test. This is a result of the reduced face width of the small gear (pinion), doubling rotational speed, and reversing the direction of rotation. Additionally, according to the S-A10/16.6R/90 method, the test is started at once with a load at which the failure is expected, hence the name "scuffing shock test." Shock loading prevents the test gears from running-in and in turn increases their susceptibility to scuffing, which further increases the method resolution.

Nowadays, one of the research directions in numerous scientific centres in the world is an improvement in the scuffing resistance of toothed gears, achievable by the deposition of thin, hard, low-friction coatings onto the gear teeth, e.g. the a-C:H:W or MoS_2/Ti coatings [20-22]. For the last several years, intensive research work has also been performed on this subject in the Tribology Department of ITeE-PIB. Until now, the FZG A/8.3/90 gear scuffing test method has been used most often in various scientific centres, which, like in the case of testing gear oils, exhibits a resolution that is too low to differentiate between the coated gears from the point of view of their resistance to scuffing [23-25] - Figure 5. It should be explained here that a-C:H:W and a-C:H coatings are DLC (diamond-like carbon) coatings, and the a-C:H:W coating has an outermost DLC layer doped with W (tungsten).

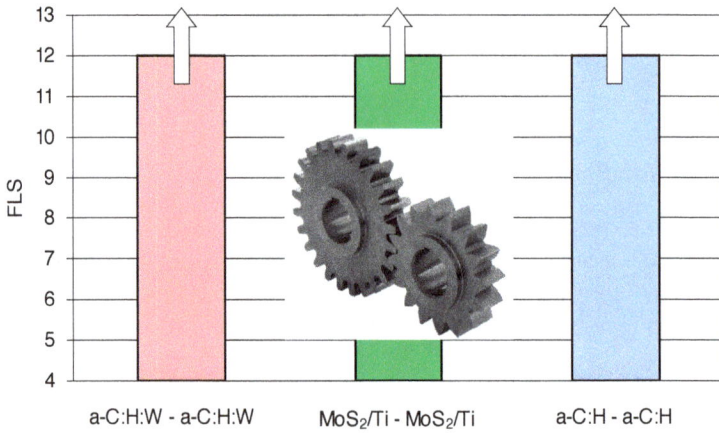

Figure 5. Failure load stages (FLS) obtained for the tested coatings (both gears coated) - FZG A/8.3/90 test method; data compiled from [23-25]

It is apparent from Figure 5 that the failure load stages (FLS), indicating the gear resistance to scuffing exceed the maximum number 12, so that the it is impossible to differentiate between the coated gears using the FZG A/8.3/90 test method.

To solve this problem, in the Tribology Department of ITeE-PIB, research was undertaken to apply the new FZG scuffing tests for coated gears to differentiate between their resistance to scuffing. Because the FZG test methods are dedicated exclusively to lubricating oils, their application for testing coated gears required introducing significant modifications - unique test methods have been developed, being the subject of this chapter. They are called the "Gear Scuffing EP Test for Coatings" and "Gear Scuffing Shock Test for Coatings."

2. New test methods

2.1. Idea of the methods

The main difference between the test methods designed by the authors and the gear scuffing tests A10/16.6R/90 and S-A10/16.6R/90, developed by FZG, is a rise in the initial oil temperature to 120 °C, adoption of a failure criterion related to wear of the wheel (big gear), and resigning from the criterion of invalidation of the test results when wear of the wheel exceeds 20 mg.

The tests are performed on a pair of lubricated test gears with a coating (it can be applied on one or both the gears) at a constant rotational speed, and at the initial temperature of the lubricating oil identical for all the runs - until a failure load stage (FLS) is determined, i.e., such a load at which at least one of the failure criteria is met. In the Gear Scuffing EP Test for Coatings, based on the FZG S-A10/16.6R/90 test, the load is increased stepwise, from the lowest to the highest value. According to the Gear Scuffing Shock Test for Coatings, based on the FZG S-A10/16.6R/90 test the load is not increased in stages from the lowest value, but the expected failure load is applied to an unused gear flank (hence, the name "shock test"). In the shock test, each change of the load requires an unused gear flank; therefore, before subsequent runs, the test gears should be disassembled and reversed or replaced with new ones.

Although the authors have introduced some significant changes to the FZG gear scuffing tests, the core procedures of performing the tests are the same as in the FZG tests, and they can be found in the relevant publications, e.g. in [14].

To better explain the differences between the "old" FZG gear scuffing test A/8.3/90 and the new test methods designed by the authors, the test conditions according to each method and the failure criteria are specified in Table 1.

After starting the run, the oil in the test chamber is heated by the heaters and friction. The oil temperature is allowed to rise freely. No cooling system is used in the tests.

Like in the FZG gear scuffing tests, if the failures are observed only within 1 mm from the tooth addendum, they are only scratches, or the failures are so small that the original criss-cross-grinding pattern (Figure 6) is still intact, they should be neglected when calculating the total area of the failures.

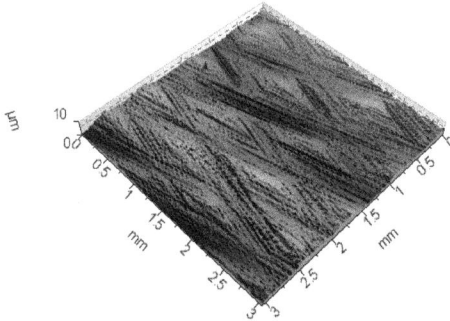

Figure 6. Original criss-cross-grinding pattern on the test gear teeth - stylus profilometry image

The failure load stage (FLS) is the main measure of the resistance of the test gears to scuffing. According to the Gear Scuffing EP Test for Coatings, the FLS is such a load at which the main failure criterion specified in Table 1 has been met. According to the Gear Scuffing Shock Test for Coatings, the FLS is such a load at which at least one of the failure criteria has been met and, when at the load stage lower by 1, neither of the failure criteria has been met.

When there is significant decohesion of the coating due to poor adhesion to the surface, the run should be invalidated.

	Gear scuffing test FZG A/8.3/90	Gear Scuffing EP Test for Coatings based on FZG A10/16.6R/90	Gear Scuffing Shock Test for Coatings based on FZG S-A10/16.6R/90
Purpose of test	Testing lubricating oils	Testing coatings deposited on gears	Testing coatings deposited on gears
Test gear type	FZG A-type (pinion and wheel width 20 mm)	FZG A10-type (pinion width 10 mm, wheel width 20 mm)	FZG A10-type (pinion width 10 mm, wheel width 20 mm)
Test materials	20MnCr5	20MnCr5, but at least one gear coated	20MnCr5, but at least one gear coated
Motor rotational speed	1500 rpm	3000 rpm	3000 rpm
Circumferential speed	8.3 m/s	16.6 m/s	16.6 m/s
Direction of motor rotation	"Normal"	"Reversed" (R)	"Reversed" (R)
Run duration	15 min.	7 min. 30 s	7 min. 30 s
Maximum load stage	12	10	12
Maximum loading torque	535 N·m	373 N·m	535 N·m
Maximum Hertzian pressure	1.8 GPa	2.2 GPa	2.6 GPa
Loading type	Stepwise, from load stage 1	Stepwise, from load stage 1	Shock (i.e. starting with a load at which the failure is expected)
Initial lubricating oil temperature	90 °C	120 °C	120 °C
Temperature stabilisation during the run by cooling	No	No	No
Type of lubrication	Dip lubrication	Dip lubrication	Dip lubrication
Main failure criterion for FLS determination	$A_p \geq$ area of one pinion tooth ($\approx$200 mm^2)[a]	$A_p >$ area of one pinion tooth ($\approx$100 mm^2)	$A_p >$ area of one pinion tooth ($\approx$100 mm^2), or $W_w >$ 200 mg[b]
Additional criteria of failure assessment	None	Failures on the pinion teeth	Failures on the pinion teeth
Criterion of invalidation of the run	None	Significant decohesion of the coating	Significant decohesion of the coating

[a] A_p - total area of failures on the pinion

[b] W_w - wear (mass loss) of the wheel

Table 1. Comparison of the FZG gear scuffing test and the methods designed by the authors

After run completion at a given load stage, the failures on the pinion teeth should be noted using the symbols from Table 2. These data are used for additional failure assessment, complementarily to FLS.

Mode of wear	Symbol	Appearance
Polishing	W	
Scratches	R	
Scoring	B	
Scuffing	Z	

Table 2. Modes of wear of the test pinion (small gear)

Polishing can be identified when the "mirror-like" surface on the tooth flank is observed with the disappearing criss-cross-grinding pattern shown in Figure 6.

Scratches appear as shorter or longer fine lines in the sliding direction of the tooth flanks.

Scoring marks run in the same direction as scratches. On the basis of CEC L-07-95 standard [26], it can be adopted that they occur singly or in zones as light, medium or deep grooves continuing towards the tip of the tooth and having a rougher appearance than the criss-cross-grinding pattern (Figure 6).

Scuffing marks occur as single, fine marks or strips, or areas covering a part or all of the flank width. According to CEC L-07-95 standard, they appear as dull areas with the roughness much greater than the original criss-cross-grinding pattern shown in Figure 6. In this case, the grinding pattern is no longer visible.

The difference between scuffing and scoring is that scuffing originates from the adhesive bond creation between the mating surfaces, which are then sheared, and scoring results from mechanical abrasion of the surface by the very hard wear particles under conditions of a very high load. Like scuffing, scoring is one of the most dangerous modes of gear wear.

When both test gears are uncoated, a respective standardised test method A10/16.6R/90 or S-A10/16.6R/90, developed by FZG should be used. However, to compare the results with the new test methods, it is necessary to start a run at the initial oil temperature of 120 °C rather than 90 °C.

2.2. Test gears

A photograph of the FZG A10 scuffing test gears employed in the tests according to the developed methods is shown in Figure 7.

Figure 7. Photograph of the FZG A10 scuffing test gears

The A10 test gears are made of 20MnCr5 steel. They are carburized, case hardened, tempered and Maag criss-cross ground. The surface hardness is HRC = 60 + 2 and the case hardness depth (CHD) is 0.6 to 0.9 mm (Eht). The effective face width of the pinion is 10 mm, and the wheel is 20 mm. The number of pinion teeth is 16, and wheel 24. The gears are identical to the ones used to perform tests according to the FZG A10/16.6R/90 and S-A10/16.6R/90 methods.

2.3. Test equipment

For the complex testing of gears, a back-to-back gear test rig, denoted as T-12U, was designed in the Tribology Department of ITeE-PIB in Radom. Its photograph is presented in Figure 8 and kinematic schemes are presented in Figure 9.

The T-12U test rig is equipped with a control-measuring system, which consists of measuring transducers (thermocouple, speed transducer) and the controller (Figure 8).

Figure 8. Photograph of the T-12U gear test rig

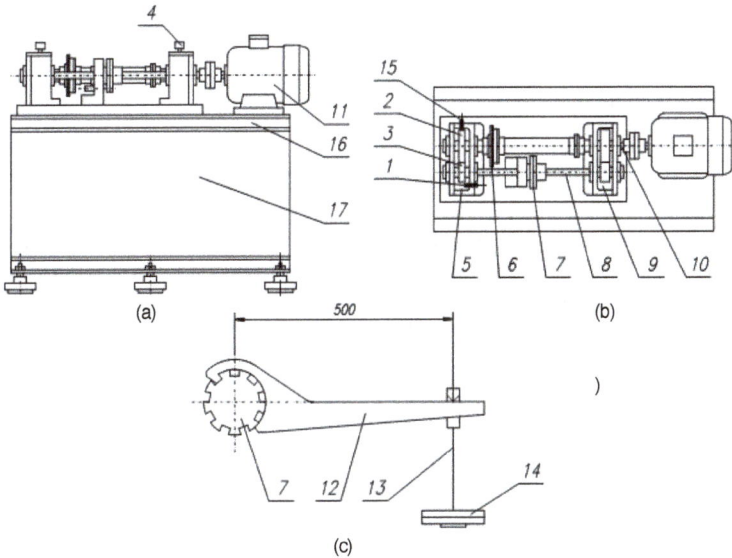

Figure 9. Kinematic schemes of the T-12U gear test rig: a) front view, b) top view, c) loading equipment; 1 - thermo-couple, 2 - test wheel, 3 - test pinion, 4 - vent, 5 - test chamber, 6 - shafts torsion angle indicator, 7 - load clutch, 8 - front shaft, 9 - slave chamber, 10 - drive clutch, 11 - electric motor, 12 - loading lever, 13 - weight hanger, 14 - weights, 15 - heaters, 16 - frame, 17 - concrete base

During runs, the following quantities are measured: rotational speed, lubricating oil temperature, motor current load, time, and the number of motor revolutions. The measured values are displayed on the controller.

The test rig is mounted on the concrete base equipped with vibration-dumping feet.

The T-12U gear test rig is a back-to-back rig (Figure 9) where the test gears (2) and (3), located in the test chamber (5), are connected by two shafts to the slave gears, located in the chamber (9). The front shaft (8) has two parts. Between them there is the load clutch (7). To apply the loading torque between the meshing gears, before the run, one part of the shaft (the left part of the front shaft (8) is fixed to the base with the lock-pin via the clutch and its support. A round-shaped loading lever (12) is placed on the right part of the clutch (7), and then the weight hanger (13) is suspended and the appropriate number weights (14) put on it. They give a static loading torque by twisting the shafts, which is measured indirectly using the torsion angle indicator (6). When the load has been applied, the two halves of the clutch (7) are firmly fixed against each other with the bolts. Then, the lock-pin is removed to close the safety cover. During the run, this loading torque "circulates" between the gears. In the back-to-back solution the motor (11) must overcome only the friction between gears, rolling bearings, and some minor components of friction (friction against seals, internal friction in the oil). Thus, the whole design is very simple and compact.

An AC squirrel-cage motor (11) of the nominal rotational speed of 3000 rpm is used to drive the rig. It is controlled by the frequency converter, which enables to change the rotational speed within a wide range.

In the gear scuffing tests the test gears are dip lubricated. In the test chamber where the test gears are located, there are heaters (15) to heat up the lubricating oil. The thermocouple (1), with the measuring point inserted in the lubricating oil, is to measure the oil temperature. A PID regulator is used to protect against overheating of the lubricating oil.

The motor (11) of the machine is automatically stopped when the preset time elapses. The required time is set on the controller panel. Additionally, the operator can read out the number of motor revolutions to confirm the correct duration of the run. The number of motor revolutions is displayed on the controller panel (Figure 8) connected to the speed transducer.

In the T-12U machine, the friction torque can be measured indirectly by measurement of the motor current load, which can be assumed to be proportional to the friction torque.

The test rig has a special support on the side cover of the test chamber (5) for mounting vibration transducers (accelerometers) to enable the operator to monitor the level of vibrations along different axes. However, now there is no possibility to automatically stop the motor when the vibration level is very high. This feature (together with other features like direct measurement of the friction torque) will be included in a new test rig, denoted as T-12UF, being developed at present.

Additional equipment includes a mass comparator for a very precise determination of the mass loss (wear) of the wheel.

2.4. Test materials

The gears coated with the low-friction a-C:H:W coating (trade name: WC/C) of DLC type and composite low-friction MoS_2/Ti coating (trade name: MoST) were tested. All material combinations were tested: coating-coating (both gears coated), coating-steel, steel-coating, and steel-steel for reference (both gears without the coating). In all cases, mineral, automotive gear oil of API GL-5 performance level and of SAE 80W-90-viscosity grade was used for lubrication.

2.5. Statistical analysis

To check statistical differences between the results obtained (FLS values), the uncertainty of measurement was assessed for the both developed test methods. This was done according to the procedures specified in the document EA-4/16 G:2003, which are binding in the accredited laboratories meeting the requirements of ISO/IEC 17025:2005.

Once the uncertainty of measurement has been calculated, the test result "y" and the uncertainty of measurement "U" should be reported as "y ± U."

As a normal practice, the uncertainty of measurement is given in relation to the average value of the measurement. For example, in the case of the gear scuffing shock tests, the respective formula derived by the authors is expressed as follows:

$$U = 0.45 + 0.06 * FLS \qquad (1)$$

where:

U - uncertainty of measurement,

FLS - failure load stage.

According to ILAC-G8:03/2009, if the uncertainty intervals expressed by U do not overlap each other, one can say that the compared results are statistically different.

3. Results and discussion

3.1. Gear Scuffing EP Test for Coatings

3.1.1. Material combinations with the a-C:H:W coating

Failure load stages (FLS) obtained for the tested material combinations with the a-C:H:W coating are presented in Figure 10. The coated gear is dark grey coloured, and the uncoated one is light grey.

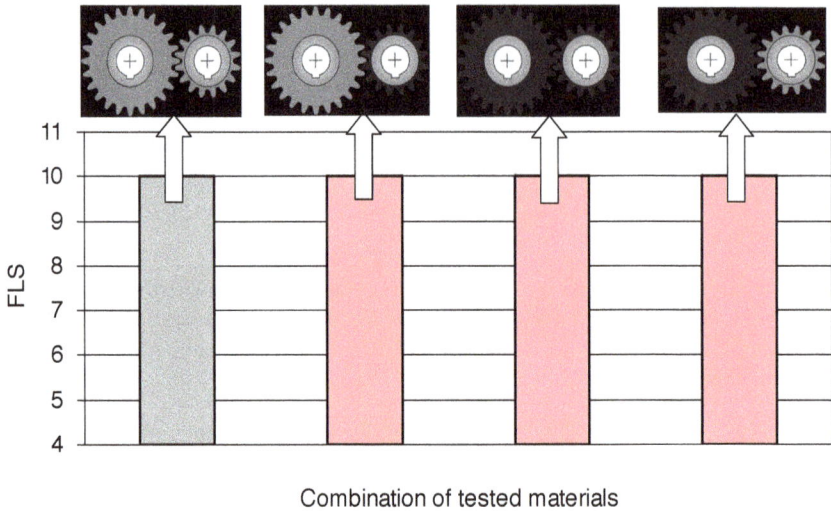

Combination of tested materials

Figure 10. Failure load stages (FLS) obtained using the Gear Scuffing EP Test for Coatings for the tested material combinations with the a-C:H:W coating

Figure 10 shows that the Gear Scuffing EP Test for Coatings is unable to differentiate between the tested material combinations from the point of view of the main criterion - FLS. All the FLS values exceed the maximum load stage, i.e. 10th. Thus, the additional criteria of failure assessment, related to the wear of the pinion after runs at particular load stages, were taken into account - Table 3. The table presents the symbolic modes of the wear of the test pinion at particular load stages for the tested material combinations with the a-C:H:W coating, and the mode of wear that appeared most often on the pinion teeth was considered. Below are the symbols of the wear modes, the total area of failures on the pinion (A_p) are given. The used symbols of wear were presented earlier in Table 2.

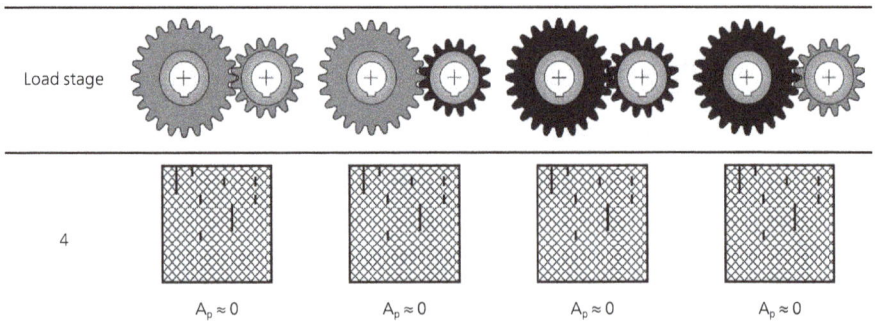

Load stage				
5	$A_p \approx 0$	$A_p \approx 0$	$A_p \approx 0$	$A_p \approx 0$
6	$A_p \approx 0$	$A_p \approx 0$	$A_p \approx 0$	$A_p \approx 0$
7	$A_p \approx 0$	$A_p \approx 0$	$A_p \approx 0$	$A_p \approx 0$
8	$A_p \approx 0$	$A_p \approx 0$	$A_p \approx 0$	$A_p \approx 0$
9	$A_p \approx 0$	$A_p \approx 0$	$A_p \approx 0$	$A_p \approx 0$
10	$A_p \approx 0$	$A_p \approx 0$	$A_p \approx 0$	$A_p \approx 0$

Table 3. Modes of the wear of the test pinion at particular load stages for the tested material combinations with the a-C:H:W coating, together with the total area of failures on the pinion (A_p); Gear Scuffing EP Test for Coatings

As can be observed in Table 3 for the tested material combinations, the three modes of wear that appear most often on the pinion teeth are scratches, polishing, and scoring. The uncoated pinion undergoes the process of polishing through the rubbing by the hard a-C:H:W coating deposited on the meshing wheel. Similar action was observed on the uncoated wheel meshing the coated pinion (results not shown here). The role of such polishing is to be explained in further experiments planned by the authors.

To sum up this part of the experiment, the Gear Scuffing EP Test for Coatings gives minor differences between the tested material combinations with the a-C:H:W coating, observed only when the pinion is uncoated and the wheel is coated. From the point of view of the practical applications of the a-C:H:W coating in gears, the situation when the both gears are coated seems to be better than in the case of one of the gears uncoated, because it is exposed to the abrasive action of the meshing coated gear, which results in polishing and scoring.

3.1.2. Material combinations with the MoS_2/Ti coating

Failure load stages (FLS) obtained for the tested material combinations with the MoS_2/Ti coating are presented in Figure 11. The coated gear is dark grey coloured, and the uncoated one is light grey.

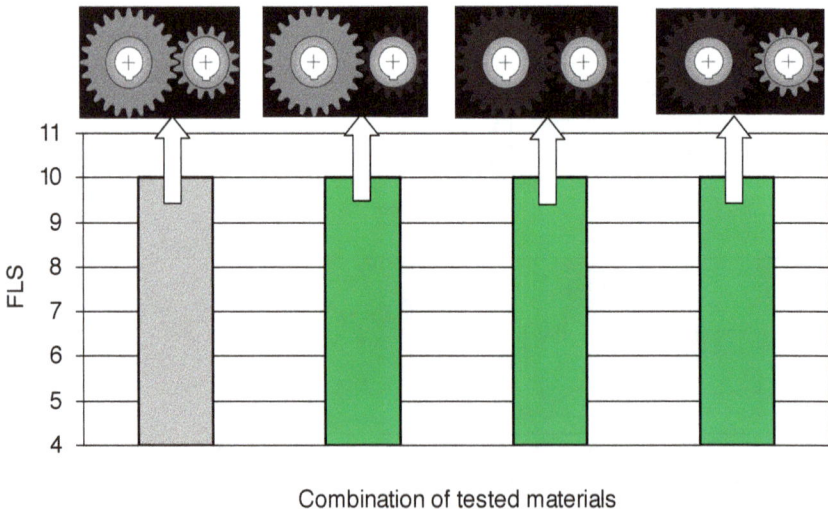

Figure 11. Failure load stages (FLS) obtained using the Gear Scuffing EP Test for Coatings for the tested material combinations with the MoS_2/Ti coating

Figure 11 shows that the Gear Scuffing EP Test for Coatings is unable to differentiate between the tested material combinations from the point of view of the main criterion - FLS. As in the case of testing the a-C:H:W coating, all the FLS values exceed the maximum load stage, i.e. 10th. Thus, the additional criteria of failure assessment, related to the wear of the pinion after runs at particular load stages, were taken into account - Table 4. The table presents the symbolic modes of the wear of the test pinion at particular load stages for the tested material combinations with the MoS_2/Ti coating, which is the mode of wear that appeared most often on the pinion teeth was considered. Below are the symbols of the wear modes, and the total area of failures on the pinion (A_p) are given. The used symbols of wear were presented earlier in Table 2.

Load stage				
4	$A_p \approx 0$	$A_p \approx 0$	$A_p \approx 0$	$A_p \approx 0$
5	$A_p \approx 0$	$A_p \approx 0$	$A_p \approx 0$	$A_p \approx 0$
6	$A_p \approx 0$	$A_p \approx 0$	$A_p \approx 0$	$A_p = 5\ mm^2$
7	$A_p \approx 0$	$A_p \approx 0$	$A_p \approx 0$	$A_p = 10\ mm^2$
8	$A_p \approx 0$	$A_p \approx 0$	$A_p \approx 0$	$A_p = 10\ mm^2$

Load stage

9

$A_p \approx 0$ $A_p \approx 0$ $A_p \approx 0$ $A_p = 10\ mm^2$

10

$A_p \approx 0$ $A_p \approx 0$ $A_p \approx 0$ $A_p = 10\ mm^2$

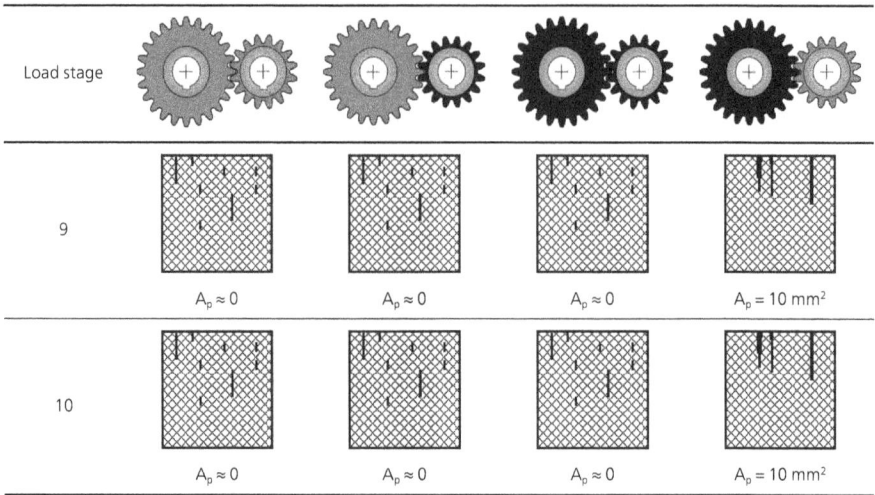

Table 4. Modes of the wear of the test pinion at particular load stages for the tested material combinations with the MoS$_2$/Ti coating, together with the total area of failures on the pinion (A_p); Gear Scuffing EP Test for Coatings

As can be observed in Table 4 for the tested material combinations, the two modes of wear that appear most often on the pinion teeth are scratches and scoring. In the material combination of the uncoated pinion meshing the coated wheel, the pinion bears the mark of scoring caused by the rubbing by the hard coating deposited on the meshing wheel.

Thus, the Gear Scuffing EP Test for Coatings gives minor differences between the tested material combinations with the MoS$_2$/Ti coating, observed only when the pinion is uncoated and the wheel is coated. From the point of view of the practical applications of the MoS$_2$/Ti coating in gears, the situation when the both gears are coated seems to be better than in the case of one of the gears uncoated as it is exposed to the abrasive action of the meshing coated gear, which results in scoring. However, when one of the gears needs to remain uncoated, using the a-C:H:W coating is more preferable than MoS$_2$/Ti, because a-C:H:W causes less wear of the uncoated gear.

3.2. Gear scuffing shock test for coatings

3.2.1. Material combinations with the a-C:H:W coating

Failure load stages (FLS) obtained for the tested material combinations with the a-C:H:W coating are presented in Figure 12. The coated gear is dark grey coloured, and the uncoated one is light grey. The assessed uncertainties of measurement for each result obtained are also shown in the Figure.

Figure 12 shows that the Gear Scuffing Shock Test for Coatings makes it possible to differentiate between the tested material combinations. The best resistance to scuffing (highest FLS) is observed when both gears are coated.

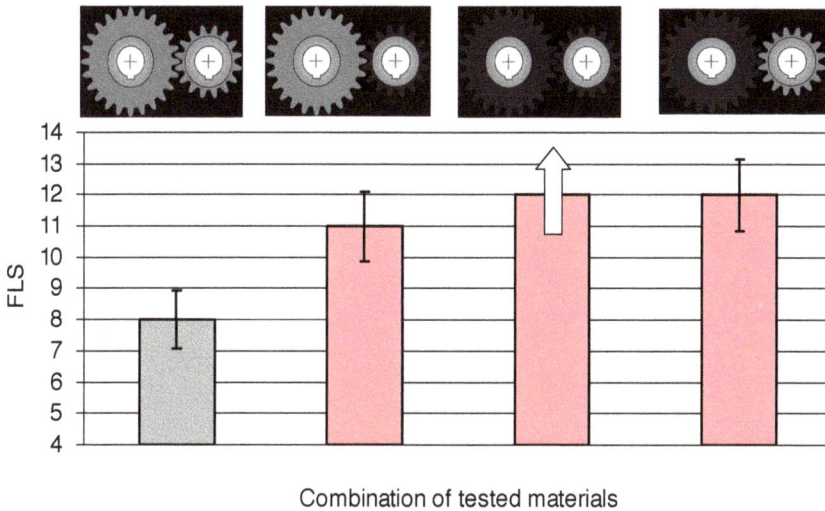

Figure 12. Failure load stages (FLS) obtained using the Gear Scuffing Shock Test for Coatings for the tested material combinations with the a-C:H:W coating

Under "shock" conditions, when the pinion is uncoated and the wheel is coated with the a-C:H:W coating, the resistance to scuffing is slightly higher than in the case when the pinion is coated and the wheel is uncoated. Hypothetically, there is a transfer of graphite (solid lubricant) from the a-C:H:W coated gear to the teeth of the uncoated one, which is more effective for the wheel coated than in the opposite situation, because the area of the coated steel surface of the wheel (larger gear with 24 teeth) is greater than in the case the coating is deposited on the pinion (small gear having only 16 teeth). However, one must have in mind that the difference in the scuffing resistance of the two material combinations is not statistically significant, because the measurement uncertainties overlap each other.

In comparison to the case of the both gears uncoated, when the a-C:H:W coating is deposited on one or two gears, much higher resistance to scuffing is observed. This is a result of a high surface energy for metals (here for steel) promoting adhesive bonding in the steel-steel contact, and smaller affinity in the different materials than when both of them are identical (i.e. steel-steel), which protects the surface from adhesive bonding. Yet another phenomenon can be attributed to it. When one of the mating materials (coating) is much harder than the other one (steel), or when two very hard materials are in contact (coating-coating) there is a reduction in the tendency to adhesive bonding, hence scuffing.

The additional criteria of failure assessment, related to the wear of the pinion after runs at particular load stages, were also taken into account - Table 5. The table presents the symbolic modes of the wear of the test pinion at particular load stages for the tested material combinations with the a-C:H:W coating, which is the mode of wear that appeared most often on the pinion teeth was taken into account. The photographs of the most often appearing mode of wear of the pinion at the highest load stage are shown also shown in the table. Red-shadowed cells in the table denote the failure load stage (FLS). Below are given the symbols of the wear modes, the total area of failures on the pinion (A_p), and wear of wheel (W_w). The used symbols of wear were presented earlier in Table 2.

As can be observed in Table 5 for the tested material combinations, the three modes of wear that appear most often on the pinion teeth are scratches, scuffing, and scoring. When one or both gears are a-C:H:W-coated, only scratches and scoring predominate on the pinion teeth.

What was observed also during the Gear Scuffing EP Test for Coatings, and what seems to by typical of "the action" of the a-C:H:W coating, the uncoated gear undergoes the process of polishing or scoring through the rubbing by the hard coating deposited on the meshing gear. The polishing on the wheel teeth flanks can be seen in Figure 13.

Figure 13. Photograph of the tooth flank of the uncoated wheel, polished by the a-C:H:W-coated pinion; Gear Scuffing Shock Test for Coatings

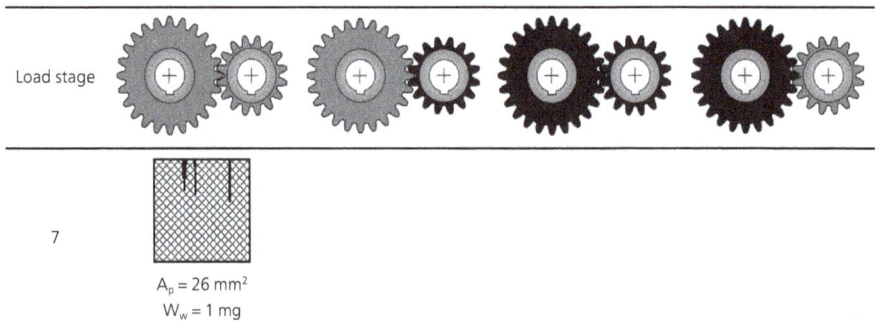

Table 5. Modes of the wear of the test pinion at particular load stages for the tested material combinations with the a-C:H:W coating, together with the total area of failures on the pinion (A_p), and wear of wheel (W_w), obtained in the Gear Scuffing Shock Test for Coatings; red-shadowed cells - the failure load stage (FLS)

Scuffing is observed only when both gears are uncoated - Table 5. This is one of the most dangerous modes of gear wear. As mentioned earlier, scuffing marks occur as single, fine marks or strips, or areas covering a part or all of the flank width. They appear as dull areas with the roughness much greater than the original criss-cross-grinding pattern shown in Figure 6. In this case, the grinding pattern is no longer visible.

To sum up this part of the experiment, the Gear Scuffing Shock Test for Coatings gives a much better resolution than the Gear Scuffing EP Test. However, one needs to have in mind that the cost of the former is about four-times higher than in the case of the latter, because the "shock tests" require more test gears to be used. From the point of view of the practical applications of the a-C:H:W coating in gears, the situation when the both gears are coated seems to be better than in the case of one of the gears uncoated, because it is exposed to the abrasive action of the meshing coated gear, which results in polishing or scoring. This positively verifies the observations taken during performing the Gear Scuffing EP Test for Coatings.

3.2.2. Material combinations with the MoS$_2$/Ti coating

Failure load stages (FLS) obtained for the tested material combinations with the MoS$_2$/Ti coating are presented in Figure 14.

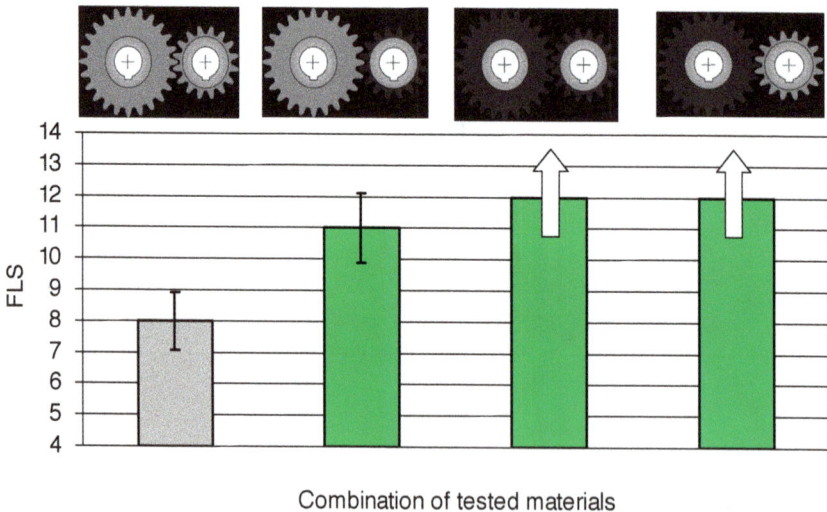

Figure 14. Failure load stages (FLS) obtained using the Gear Scuffing Shock Test for Coatings for the tested material combinations with the MoS$_2$/Ti coating

Figure 14 shows that the best resistance to scuffing (highest FLS) is observed when both gears are coated with the MoS$_2$/Ti coating, or when the uncoated pinion meshes the coated wheel.

As in the case of the a-C:H:W coating, when only the wheel is MoS$_2$/Ti-coated, under "shock" conditions the resistance to scuffing is higher than in the situation when only the pinion coated.

Hypothetically, there is a transfer of MoS_2 (solid lubricant) from the teeth of the coated gear to the uncoated one. The transfer is more effective in the case of the MoS_2/Ti-coated wheel meshing the uncoated pinion than in the opposite situation, because the area of the coated steel surface of the larger gear (wheel) is greater than in the case when the coating is deposited on the small gear (pinion).

In comparison to the case when both gears are uncoated, a much higher resistance to scuffing is observed when the MoS_2/Ti coating is deposited on one or two gears. The respective mechanisms of this behaviour were described earlier.

Table 6 presents the symbolic modes of wear of the test pinion at particular load stages for the tested material combinations with the MoS_2/Ti coating. As in the case of the a-C:H:W coating, the mode of wear that appeared most often on the pinion teeth was taken into account. The photographs of the most often appearing mode of wear of the pinion at the highest load stage are also shown in the table. Red-shadowed cells in the table denote the failure load stage (FLS). Below are given the symbols of the wear modes, the total area of failures on the pinion (A_p) and wear of wheel (W_w). The symbols of wear were presented earlier in Table 2.

As can be observed in Table 6 for the tested material combinations, the two modes of wear that appear most often on the pinion teeth are scuffing and scoring. When one or both gears are coated, only scoring predominates on the pinion teeth.

When the pinion is coated and the wheel is uncoated, and when both the gears are coated, identical results were obtained for the two investigated coatings - FLS values are respectively 11 and higher than 12 (Figures 12 and 14). Therefore, the main criterion of assessment of the resistance to scuffing (FLS) makes it impossible to differentiate between these two situations. Under these circumstances, the analysis of additional criteria of failure assessment, related to the modes of wear at particular load stages, like in the previous cases can give additional, valuable information. In the case of the material combinations with the a-C:H:W coating, the predominating mode of wear of the pinion were only scratches. For the material combinations with the MoS_2/Ti coating, the pinion wear was much more sever, and scoring instead of scratches could be met most often. Thus, the a-C:H:W coating provides better protection against severe wear than MoS_2/Ti, especially when it is deposited on the both gears. This positively verifies the observations taken during performing the Gear Scuffing EP Test for Coatings.

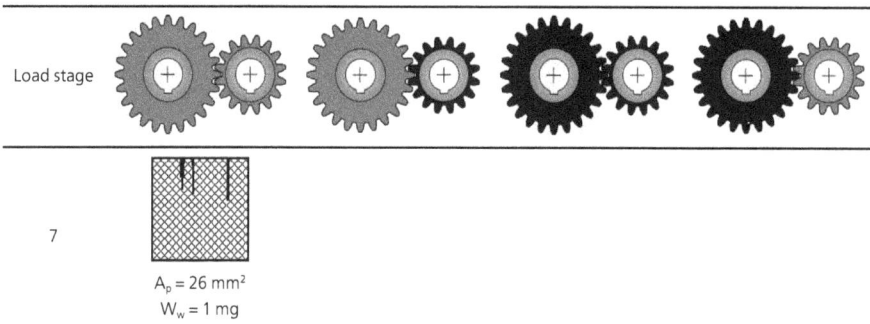

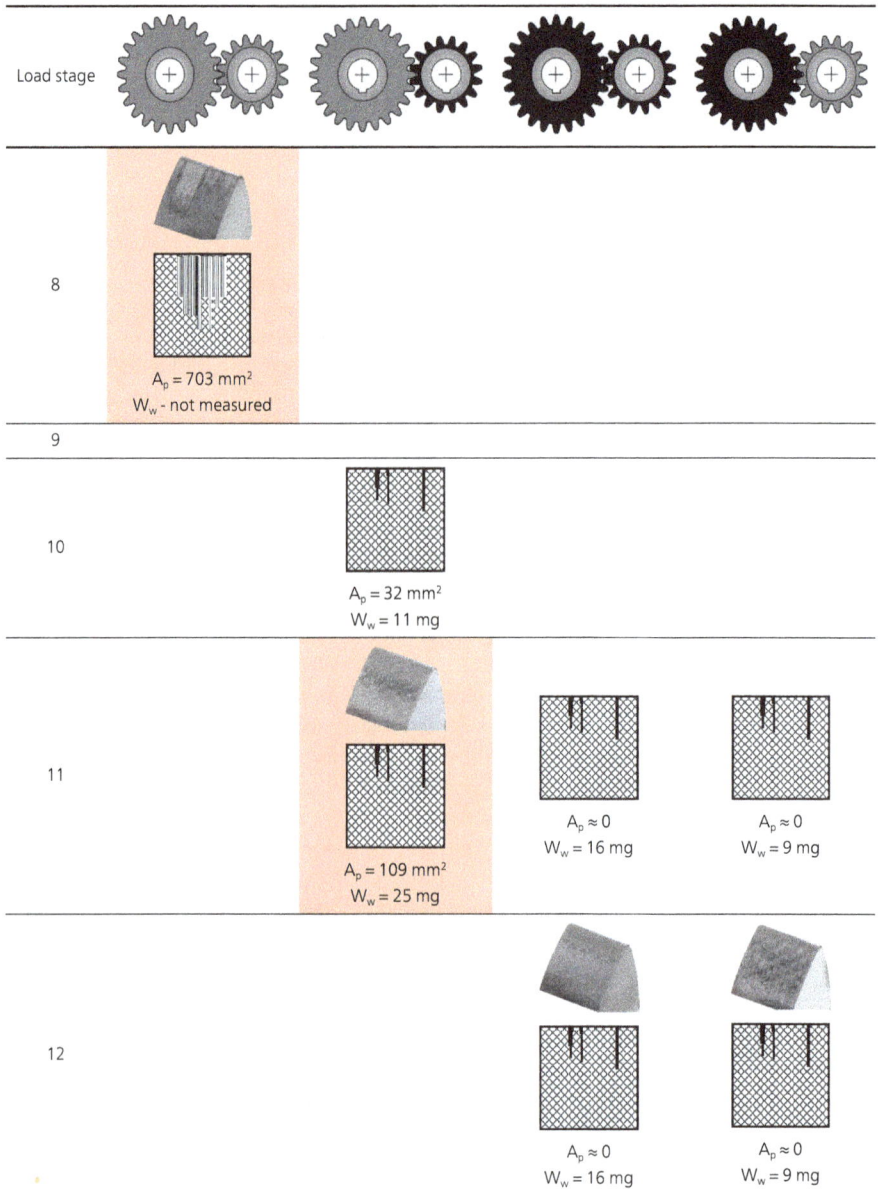

Table 6. Modes of wear of the test pinion at particular load stages for the tested material combinations with the MoS$_2$/Ti coating, together with the total area of failures on the pinion (A$_p$), and wear of wheel (W$_w$), obtained in the Gear Scuffing Shock Test for Coatings; red-shadowed cells - the failure load stage (FLS)

4. Summary and conclusions

The authors have developed unique test methods, being the subjects of this chapter. They are called the "Gear Scuffing EP Test for Coatings" and "Gear Scuffing Shock Test for Coatings."

The analysis of the values of the failure load stage (FLS), reflecting the resistance to scuffing, shows that the developed Gear Scuffing EP Test for Coatings has too little resolution to differentiate between the tested material combinations - coating-coating (both gears coated), coating-steel, steel-coating, and also steel-steel (both gears without a coating). Additional criteria of failure assessment need to be employed to reveal minor differences between the tested material combinations observed only when the pinion is uncoated and the wheel is coated.

In comparison, Gear Scuffing Shock Test for Coatings makes it generally possible to differentiate between the tested material combinations from the point of view of the main criterion of assessment of the gear resistance to scuffing, i.e. FLS. Thus, this test method has a sufficient resolution. However, as in the case of the Gear Scuffing EP Test for Coatings, apart from the analysis of only FLS values, analysis of the additional criteria of failure assessment related to predominating modes of wear at particular load stages is recommended and may give additional, valuable information. For the two coatings tested (a-C:H:W and MoS_2/Ti), the best resistance to scuffing/scoring (FLS > 12) is observed when both gears are coated; however, the a-C:H:W coating gives a slightly better protection against severe wear than MoS_2/Ti - only scratches instead of scoring are observed for a-C:H:W.

Although the Gear Scuffing Shock Test for Coatings gives a much better resolution than the Gear Scuffing EP Test, one needs to have in mind that the cost of the former is about four-times higher than in the case of the latter, because the "shock tests" require more test gears to be used.

In the both tests, when one or both gears are coated, three modes of wear occur most often on the pinion teeth - polishing, scratches, or scoring. Scuffing is observed only when the two gears are uncoated.

The following conclusions can be drawn:

- The developed Gear Scuffing Shock Test for Coatings has been successfully verified by the testing of thin, hard coatings deposited on the gears; therefore, it can be implemented in the laboratories of the R&D centres devoted to surface engineering and the engineering of advanced materials intended for modern toothed gears, having in mind that this test is rather expensive.

- If the coating is intended for application on gears, from the point of view of the highest achievable resistance to scuffing/scoring, it is recommended that both meshing gears are a-C:H:W-coated.

- Although the T-12U gear test rig has been effectively employed in the performed research, it is suggested that its research capacities should be extended by the measurement and data acquisition of the friction torque, which will make it possible to investigate, as postulated by gear transmissions manufacturers, the possibility of the reduction of friction between the meshing teeth by the application of a low-friction coating. At present, a new version of the T-12U test rig, denoted as T-12UF, is being developed within the framework of the Strategic Programme executed at ITeE-PIB in Radom, and the planned deadline of this work is in 2013.

• Both the differentiation between the tested objects (lubricating oils, material combinations) and the predictability of gear failures in real applications (transmissions) are important when assessing gear tests. This is why the authors plan to verify the results obtained by application of coated gears in transmissions (speed reducers) of different devices manufactured by one of the Polish producers. What is more, at present another test rig - a back-to-back bevel gear test rig, denoted as T-30 - is being developed in the Tribology Department of ITeE-PIB in Radom with the deadline in 2012. The reason is that until now widely used test devices and methods have allowed researchers to perform runs on only spur gears having the tooth geometry significantly different than the geometry of bevel gears. The new tribotester will allow researchers to better predict the failures of bevel gears.

Acknowledgements

Scientific work was financed:

• From the means of the Minister of Science and Higher Education, executed within the Strategic Programme "Innovative Systems of Technical Support for Sustainable Development of the Country's Economy" within Innovative Economy Operational Programme.

• By the National Centre for Research and Development (NCBiR) within the scope of the R&D project No. N R03 0019 06.

The authors wish to express their thanks also to Dr. Maksim Antonov from Tallinn University (Estonia) for his support with the gear scuffing tests and helpful discussions, within the framework of Marie Curie RTN (6th EU FP); Contract No MRTN-CT-2006-035589.

Author details

Remigiusz Michalczewski[*], Marek Kalbarczyk, Michal Michalak, Witold Piekoszewski, Marian Szczerek, Waldemar Tuszynski and Jan Wulczynski

[*]Address all correspondence to: remigiusz.michalczewski@itee.radom.pl

Institute for Sustainable Technologies - National Research Institute (ITeE-PIB), Tribology Department, Radom, Poland

References

[1]   Pytko S., Sroda P. Classification and Evaluation of Machines for Investigation of Materials for Production of Gear Wheels. ZEM 1975; 1 39-58 (*in Polish*).

[2]   Tomaszewski J., Drewniak J., editors. Scuffing of Gears. Gliwice: CMG KOMAG; 2007 (*in Polish*).

[3]   ARTEC Machine Systems. Failure by scuffing due to poor distribution of load. http://www.artec-machine.com/images/failure_due_to_poor_distribution_of_load_1.pdf (accessed 22 September 2012).

[4]   Kratz S.H., Wedeven L.D., Black W.F., Carlisle D.J., Wedeven G.G. Simulation of Space Shuttle Gear Performance. In: proceedings of the III World Tribology Congress, 2005, Washington, USA.

[5]   Kuo W.F., Chiou Y.C., Lee R.T. A Study on Lubrication Mechanism and Wear Scar in Sliding Circular Contacts. Wear 1996;201 217-226.

[6]   Ku P.M., editor. Interdisciplinary Approach to Friction and Wear. Washington D.C.: Southwest Research Institute; 1968.

[7]   Makowska M., Matuszewska A., Gradkowski M. Migration of Active Elements from a Lubricant to the Material of the Friction Pair. Tribologia 2011;4 163-176 (*in Polish*).

[8]   Wachal A. Analysis of Boundary Layer Estimating Criteria in Lubricating Oils Investigations. ZEM 1983;3 325-332.

[9]   Höhn B.-R., Oster P., Schedl U. Pitting Load Capacity Test on The FZG Gear Test Rig with Load-Spectra and One-Stage Investigations. Tribotest journal 1999;5-4 417-430.

[10] Szczerek M., Tuszynski W. A Method for Testing Lubricants under Conditions of Scuffing. Part I. Presentation of the Method. Tribotest journal 2002;8-4 273-284.

[11] Piekoszewski W., Szczerek M., Tuszynski W. The Action of Lubricants under Extreme Pressure Conditions in a Modified Four-Ball Tester. Wear 2001;249 188-193.

[12] Tuszynski W., Michalczewski R., Piekoszewski W., Szczerek M. Effect of Ageing Automotive Gear Oils on Scuffing and Pitting. Tribology International 2008;41 875-888.

[13] Bisht R.P.S., Singhal S. A Laboratory Technique for the Evaluation of Automotive Gear Oils of API GL-4 Level. Tribotest journal 1999;6-1 69-77.

[14] Method to Assess the Scuffing Load Capacity of Lubricants with High EP Performance Using an FZG Gear Test Rig. FVA Information Sheet No. 243 Status June 2000.

[15] Höhn B.-R., Michaelis K., Eberspächer C., Schlenk L. A Scuffing Load Capacity Test with the FZG Gear Test Rig for Gear Lubricants with High EP Performance. Tribotest journal 1999;5-4 383-390.

[16] Michaelis K., Höhn B.-R., Graswald C. Scuffing Tests for API GL-1 to GL-5 Gear Lubricants. In: proceedings of the 13th International Colloquium Tribology, 2002, Ostfildern, Germany.

[17] Michaelis K., Höhn B.-R., Oster P. Influence of Lubricant on Gear Failures - Test Methods and Application to Gearboxes in Practice. Tribotest journal 2004;11-1 43-56.

[18] Höhn B.-R., Oster P., Tobie T., Michaelis K. Test Methods for Gear Lubricants. Goriva i maziva 2008;47-2 141-152.

[19] Tuszynski W. Performance Classification of Automotive Gear Oils Using the Gear Scuffing Shock Test. Tribologia 2009;2 259-274 (*in Polish*).

[20] Kalin M., Vižintin J. The Tribological Performance of DLC-Coated Gears Lubricated with Biodegradable Oil in Various Pinion/Gear Material Combinations. Wear 2005;259 1270-1280.

[21] Martins R.C., Moura Paulo S., Seabra J.O. MoS_2/Ti Low-Friction Coating for Gears. Tribology International 2006;39 1686-1697.

[22] Martins R., Amaro R., Seabra J. Influence of Low Friction Coatings on The Scuffing Load Capacity and Efficiency of Gears. Tribology International 2008;41 234-243.

[23] Szczerek M., Michalczewski R., Piekoszewski W. The Problems of Application of PVD/ CVD Thin Hard Coatings for Heavy-Loaded Machine Components. In: proceedings of the ASME/STLE International Joint Tribology Conference, 2007, San Diego, USA.

[24] Michalczewski R., Piekoszewski W., Szczerek M., Tuszynski W. The Lubricant-Coating Interaction in Rolling and Sliding Contacts. Tribology International 42;2009 554-560.

[25] Michalczewski R., Piekoszewski W., Szczerek M., Tuszynski W. Scuffing Resistance of DLC Coated Gears Lubricated with Ecological Oil. Estonian Journal of Engineering 2009;15-4 367-373.

[26] CEC L-07-95: Load Carrying Capacity Test for Transmission Lubricants (FZG Test Rig).

Chapter 8

Artificial Slip Surface: Potential Application in Lubricated MEMS

M. Tauviqirrahman, R. Ismail, J. Jamari and
D.J. Schipper

Additional information is available at the end of the chapter

1. Introduction

1.1. Background

For the last years, there has been a tremendous effort towards the development of Micro-Electro-Mechanical System (MEMS) for a wide variety of applications in aerospace, automotive, biomedical, computer, agricultural industries, electronic instrumentation, industrial process control, biotechnology, office equipment, and telecommunications. MEMS devices integrate chemical, physical, and even biological processes in micro-scale technology packages.

Stiction (a subtraction of 'static friction') in micro-system technology has been a problem ever since the advent of surface micromachining in the eighties of the last century. As the overall size of the machine is reduced, the capillary and surface tension force of liquid become large, which induce stiction rendering the devices to fail or malfunction. In particular, stiction forces created between moving parts that come into contact with one another, either intentionally or accidentally, during operation are a common problem with micro-mechanical devices. Stiction-type failures occur when the interfacial attraction forces exceed restoring forces. Consequently, the surfaces of these parts either temporarily or permanently adhere to each other, causing device malfunction or failure.

Several approaches to address the stiction between two opposing surfaces have been presented in the various literatures [1-4]. The basic approaches to prevent stiction include increasing surface roughness (topography) and/or lowering solid surface energy by coating with low surface energy materials. This includes self-assembled molecular (SAM) coatings, hermetic packaging and the use of reactive materials in the package [5].

Other attractive technique to tackle the stiction problem is by inserting a lubricant into the region around the interacting devices to reduce the chance of stiction-type failures. As is well-known, many MEMS devices include moving (sliding/rolling) surfaces and thus it is necessary to apply a lubricant between the contacting surfaces to reduce friction and wear. However, a significant barrier to the development of MEMS lubrication is the problem of achieving effective tribological performance of their moving parts. This is because the lubricant behavior is different at micro-scale compared to macro-scale. At the macroscopic level, it is well accepted that the boundary condition for a viscous fluid at a solid wall is no-slip, i.e. the fluid velocity matches the velocity of the solid boundary. While the no-slip boundary condition has been proven experimentally to be accurate for a number of macroscopic flows, it remains an assumption that is not based on physicals principles. At micro-scale level, certain phenomena must be taken into account when analyzing liquid flows such as a slip condition at solid wall boundaries.

As a consequence of the MEMS technology revolutionary application to many areas, it is possible for scientists to observe the boundary slip on micro/nano-meter scale. A variety of techniques are now available that are capable of probing lubricant flow on micro-scales and are therefore suitable for the investigation of boundary conditions. There are three techniques so far for detecting the boundary slip: nano-particle image velocimetry (NPIV) [6], atomic force microscope (AFM) [7-9] and surface force apparatus (SFA) [10]. The NPIV technique is a direct observation method with a measurement precision depending on the size of the nano-particles but with poor moderate accuracy. The AFM and SFA are indirect observation techniques based on the assumption that boundary slip takes place precisely on the interface of liquid and solid. These methods need a high accuracy boundary slip model to infer the slip velocity. Boundary slip has been observed not only for a hydrophobic surface [6, 7, 10] but also for a hydrophilic surface [8, 9]. Therefore, the slip evidence has been generally accepted and for certain cases the no-slip boundary condition is not valid.

There is a large body of literature dealing with the analysis of lubricant slip flow based on the analytical and numerical solution of molecular dynamic simulations [11, 12], Lattice-Boltz-man [13, 14], and the Reynolds equation [15-23]. The accurate description of slip at the wall is very difficult and still remains a subject of intensive research. The so-called Navier slip model and the critical shear stress model are usually used to describe a boundary slip. In fact, nearly two hundred years ago Navier [24] proposed a general boundary condition that incorporates the possibility of fluid slip at a solid boundary. Navier's proposed boundary condition assumes that the velocity, u, at a solid surface is proportional to the shear stress at the surface. It reads: $u = b \, (du/dz)$ where b is the slip length and du/dz is the shear rate. The slip length b, which is defined as the distance beyond the solid/liquid interface at which the liquid velocity extrapolates to the velocity of the solid, is used to quantify a boundary slip. If $b = 0$ then the generally assumed no-slip boundary condition is obtained. If $b = $ finite, fluid slip occurs at the wall, but its effect depends upon the length scale of the flow. The Navier-slip boundary condition is the most widely used boundary condition with the methods based on the solution of continuum equations.

In micro-scales such as MEMS, the boundary condition will play a very important role in determining the lubricant flow behavior. Control of the boundary condition will allow a degree of control over the hydrodynamic pressure in confined systems and be important in lubricated-MEMS. To prevent a stiction, in a controlled way, one is able to enhance, a hydrophobic/hydrophilic behavior of surfaces. If one surface is hydrophobic (slip) and the other is hydrophilic (no-slip) the sliding velocity or displacement between the surfaces is accommodated by shear at the hydrophobic surface (the lubricant is kept in the contact by the hydrophilic surface). In this way wear of the surfaces is prevented and the surfaces are able to move because stiction is prevented.

The slip situation, however, can be controlled to obtain a positive effect by surface technology. Coating and texturing technologies can be used to engineer large slip. In practice, a large slip can be made using super-hydrophobic surfaces. Such surfaces can be manufactured by grafting or by deposition of hydrophobic compounds on the initial surface at a certain zone. Super-hydrophobic surfaces were originally inspired by the unique water-repellent properties of the lotus leaf. It is the combination of a very large contact angle and a low contact-angle hysteresis that defines a surface as super-hydrophobic. Implementing the slip property (hydrophobicity) on a surface in a wide range of application for the mechanical components is of great challenge by numerous authors recently. In published works [15-23], both experimentally and numerically, slip surface is able to reduce friction force at the contacting surfaces and finally reduce energy consumption, increase component's life-time and reduce economic and environmental costs.

1.2. Problem statement

In classical liquid lubrication it is assumed that surfaces are fully wetted and no-slip occurs between the fluid and the solid boundary. In MEMS, this wetting is actually an unwanted process because it can encourage the occurrence of stiction and as a result micro-parts can not be moved [25]. It is expected that slip can reduce the friction and improve the load support. However, with respect to the engineered slip pattern, the choice of slip zone on a certain surface must be taken carefully in relation to such tribological performances. In other words, an inappropriate slip zone pattern on a certain surface or the election of inappropriate surface containing a slip situation may lead to the deterioration of the lubrication performance. How to control the boundary slip in the application of a lubricated-MEMS is one of the challenging tasks in the future. This chapter will explore the provision of a new lubrication model based on the continuum approach for moving parts in MEMS in order to improve the tribological performance of lubricated contacts. In MEMS, by lubrication, low friction force and high load support are the goals which want to be achieved. The artificial slip surface will be introduced as one of the solutions to improve the lubrication performance of MEMS so as MEMS with a longer life-time can be obtained. The term "artificial slip surface" is used to address a non-homogeneous engineered slip/no-slip pattern, i.e. a surface consisting of a slip zone and a no-slip zone.

2. Research methods

Full film lubrication of lubricated contact is often described by the Reynolds theory [26]. According to the classical Reynolds theory, no-slip boundary is assumed and the convergent geometrical wedge is one of the most important conditions to generate hydrodynamic pressure. Therefore, the lubrication model for lubricated MEMS will be an extension of the classical lubrication theory. This means that modeling the lubricant through very narrow gap, normally modeled by assuming no-slip at the boundaries will be modified by introducing a boundary slip.

2.1. Modified reynolds equation

The classical Reynolds equation that is valid under no-slip condition can be generalized for taking into account slip conditions. It is then possible, for any film height distributions, to calculate the pressure distribution and the shear rate profile. The model of lubrication presented here is based on the fact that slip of the lubricant will exist at the interface of a lubricated sliding contact. Thus, a boundary slip is employed both on the moving and stationary surface, see Figure 1. The proposed lubrication model with slip leads to a modified Reynolds equation as presented in Eq. (1).

$$
\frac{\partial}{\partial x}\left(\frac{h^3}{12\mu} \frac{h^2 + 4h\mu(\alpha_t + \alpha_b) + 12\mu^2\alpha_t\alpha_b}{h(h + \mu(\alpha_t + \alpha_b))} \frac{\partial p}{\partial x} \right) = \frac{u_w}{2} \frac{\partial}{\partial x}\left(\frac{h^2 + 2h\mu\alpha_t}{h + \mu(\alpha_t + \alpha_b)} \right)
$$
$$
-u_w \frac{\alpha_t\mu}{h + \mu(\alpha_t + \alpha_b)} \frac{\partial h}{\partial x} + \frac{h}{2\mu} \frac{\partial p}{\partial x} \frac{\partial h}{\partial x} \frac{h\alpha_t\mu + 2\alpha_t\alpha_b\mu^2}{h + \mu(\alpha_t + \alpha_b)}
$$

(1)

The physical meanings of the symbols in Eq. (1) are as follows: h the lubrication film thickness (gap) at location, p the lubrication film pressure, μ the lubricant viscosity, α the slip coefficient (subscripts t and b denote the top (stationary) and bottom (moving) surface, respectively) and u_w the velocity of the moving surface. It can be seen in that if the slip coefficient α is set to zero (no-slip condition), Eq. (1) reduces to the classical Reynolds equation. It should be noted that the product of multiplication of the slip coefficient by the viscosity, $\alpha\mu$, is usually called as 'slip length' b.

Eq. (1) is derived by following the usual approach to deduce the Reynolds equation from the Navier-Stokes system by assuming classical assumptions except that boundary slip is applied both on the stationary surface and moving surface as depicted in Figure 1.

Eq. (1) can be derived by considering the equilibrium of an element of fluid.

$$
\frac{\partial^2 u}{\partial z^2} = \frac{1}{\mu} \frac{\partial p}{\partial x}
$$

(2)

where z lies along the direction across the thickness of the film, and u is the velocity field. To obtain the velocity profile, Eq. (2) can be integrated twice.

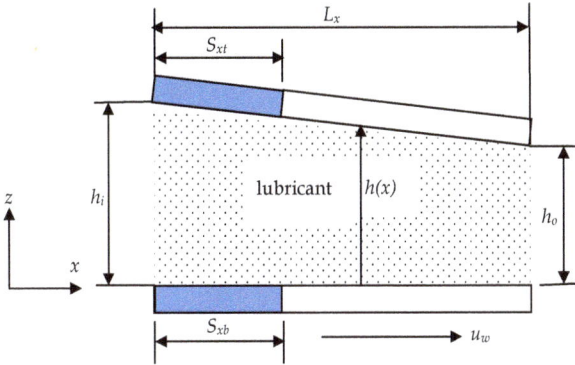

Figure 1. Schematic of a lubricated sliding contact with artificial slip surface applied both on the stationary and the moving surface. The boundary slip zones (S_{xt} and S_{xb}) are located at the leading edge of the contact. (Note: u_w is the sliding velocity, L_x is contact length, h_i and h_o are inlet and outlet film thickness, respectively, $h(x)$ is variable film thickness).

$$u = \frac{1}{2\mu}\frac{\partial p}{\partial x}z^2 + C_1 z + C_2 \tag{3}$$

C_1 and C_2 in this case are either constants or functions of x and can be solved by applying a boundary condition for u. The bottom and the top surfaces have the slip condition based on slip length model.

$$\left. \begin{array}{l} at\ z=0,\ u = u_w + \alpha_b \mu \dfrac{\partial u}{\partial z}\bigg|_{z=0} \\[3mm] at\ z=h,\quad u = -\alpha_t \mu \dfrac{\partial u}{\partial z}\bigg|_{z=h} \end{array} \right\} \tag{4}$$

This gives

$$u = \frac{1}{2\mu}\frac{\partial p}{\partial x}z^2 - \left(\frac{u_w}{h + \mu(\alpha_t + \alpha_b)} + \frac{h}{2\mu}\frac{\partial p}{\partial x}\frac{h + 2\alpha_t \mu}{h + \mu(\alpha_t + \alpha_b)} \right)z + u_w \frac{h + \alpha_t \mu}{h + \mu(\alpha_t + \alpha_b)} \tag{5}$$

$$- \frac{h}{2\mu}\frac{\partial p}{\partial x}\frac{\alpha_b \mu (h + 2\alpha_t \mu)}{h + \mu(\alpha_t + \alpha_b)} \tag{6}$$

This velocity is used to compute the flow rate, q by integrating across the fluid film thickness, h. When q is differentiated to fulfill the continuity of flow, assuming μ is constant; this gives a modified Reynolds equation as stated in Eq. (1).

When full film lubrication is assumed, the entire load w is carried by the lubricant film and the calculation is simply an integration of the lubricant film pressure distribution over contact area, i.e.

$$w = \int_0^{L_x} p(x)dx \tag{7}$$

The friction force f generated by the lubrication system is due to the fluid viscous shear. It is calculated by integrating the shear stress over the surface area. These shear stresses are given by

$$\tau(x,z) = \left(\mu \frac{\partial u}{\partial z} \right)_{z=h} \tag{8}$$

The simulation results will be presented in dimensionless form, i.e. $P = ph_o^2 / \mu L_x u_w$ for the dimensionless pressure, $W = wh_o^2 / (u_w \mu L_x^2)$ for the dimensionless load support (where w is the load per unit width), $F = fh_o / \mu u_w L_x$ for dimensionless friction force (where f is the friction force per unit width), and $m = F / W$ for dimensionless friction coefficient. In the present study, the dimensionless slip length A is determined by normalizing the slip length b with the outlet film thickness h_o. For slip analysis in the following computations, the dimensionless slip length A varies from 3 to 300, which are reasonable values of the slip length based on the results published in literature [17, 18, 20].

2.2. Solution method

The modified Reynolds equation, Eq. (1) is discretized over the flow using the finite volume method, and is solved using the tridiagonal matrix algorithm (TDMA), [27]. By employing the discretization scheme, the computed domain is divided into a number of control volumes using a grid with uniform mesh size. The grid independency is validated by various numbers of mesh sizes. An assumption is made that the boundary pressures are zero at both sides of the contact.

A numerical simulation is conducted to investigate the possible application so as a boundary slip can be beneficial to achieve a high load support and low friction force. In order to maximize the performance of lubrication, the boundary conditions (slip zones, S_{xt} and S_{xb}, see Figure 1) of the model are optimized through a parametric analysis. The object of optimization is to maximize the hydrodynamic load support. The load support satisfies two main functional purposes: (a) carry the applied external load, and (b) to minimize the contact between the opposing solids, and thus wear. The optimization analysis attempts to satisfy both functional requirements with a single design parameter, the slip zone.

The parametric analysis is performed using a developed computer code to investigate the effect of various slip parameters on the lubrication performances (load support, friction force, and friction coefficient). A parametric study is conducted with the variation of slip parameters (slip zone and slip length) over a large range of values considering different performance parameters. The design variables and the objective function are referred to as the optimization variables. The design variables are independent quantities which are varied in order to achieve the optimum design. The objective function is the dependent variable that is maximized, i.e. the load support. In the present study, the design variables are slip zones as indicated in Figure 1. The algorithm used in the present study is depicted on Figure 2.

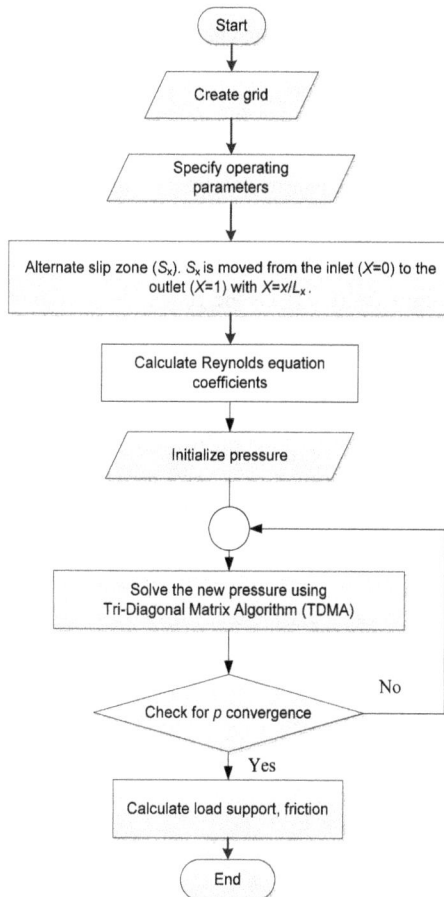

Figure 2. Flow chart for numerical method.

3. Key results

The behavior of traditional (no-slip) hydrodynamic lubrication between the opposing surfaces can be estimated by a classical form of the Reynolds equation. The derivation of the classical Reynolds equation is based on the assumption of no-slip between the lubricant and the surfaces. In the classical Reynolds lubrication, the mechanism to generate a pressure is due to the convergent wedge effect. An artificial slip surface presented here is designed to be able to carry the external load during lubrication even if the wedge effect is not present. This situation is very beneficial in designing lubricated-MEMS which exhibits parallel gaps.

In this chapter, there are two main investigations. At first, the study is conducted in order to validate the developed numerical scheme. It assures that the numerical method used can be employed for solving other hydrodynamic characteristics. The no-slip case of lubricated contact is of main interest due to the availability of the analytical solution. Secondly, the study will be extended to explore the effect of the slip zone of the artificial slip surface on pressure, load support, friction force, and friction coefficient. The comparison between the modified sliding contact containing an artificial slip surface and the traditional one is conducted in order to describe the benefit of the use of an artificial slip pattern quantitatively.

3.1. No-slip condition

The modified Reynolds equation (Eq. (1)) is the governing equation for the fluid lubrication system containing a boundary slip. If the slip coefficient, α, is set to zero, Eq. (1) reduces to the classical Reynolds equation. In this section, in order to validate the developed computer code containing a numerical scheme using finite volume method combined with tridiagonal matrix algorithm (TDMA), the classical Reynolds equation (no-slip condition) is solved numerically for calculating the pressure distribution, and finally the friction in a lubricated sliding contact as depicted in Figure 1 and Table 1, respectively. These results are compared with the analytical solution based on the work of Cameron [28] as follows:

$$p = \frac{6u_w\mu L_x}{h_o^2} \frac{K\dfrac{x}{L_x}\left(1-\dfrac{x}{L_x}\right)}{(2+K)\left(1+K-K\dfrac{x}{L_x}\right)^2} \tag{9}$$

for the pressure distribution where $K = (h_i/h_o) - 1$, and

$$f = \frac{L_x\mu u_w}{h_o}\left(\frac{4\ln(1+K)}{K} - \frac{6}{(2+K)}\right) \tag{10}$$

for the friction force per unit width.

In Figure 3 the numerical results obtained with TDMA as well as analytical results for the dimensionless pressure distribution along the bottom wall of the contact are shown alongside those obtained from the Reynolds approximation. The wedge ratio h^* of 2.2 was considered based on the fact that a maximum load support for a no-slip contact occurs when $h^* = 2.2$ [28]. In the present study the wedge ratio h^* is defined as the inlet film thickness over the outlet film thickness, h_i/h_0, and sometimes quoted as slope incline ratio in other literature. It is observed from Figure 3 that the maximum error is within 0.01% between the pressure obtained from the analytical solution and the numerical result.

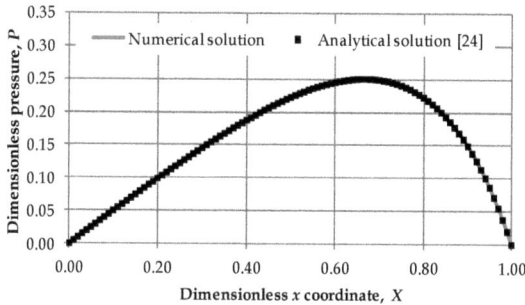

Figure 3. Normalized pressure distribution along the bottom wall of the linear wedge with no-slip boundary condition for an optimal wedge ratio h^* of 2.2.

The comparison between the dimensionless friction force F obtained with the numerical prediction and the analytical solution are presented in Table 1. Like the result of the pressure distribution, the predicted dimensionless friction force shows very good agreement with the analytical solution. In general, the numerical solution of the classical Reynolds equation is matched well with the analytical solution. It assures that the numerical method used is valid and thus can be extended for analysing other hydrodynamic characteristics.

	Dimensionless friction force, F
Analytical solution [28]	0.77
Numerical prediction [present study]	0.77

Table 1. The comparison between the analytical solution performed by Cameron [28] and the numerical simulation code.

3.2. Artificial slip surface

In MEMS, liquid lubrication has generally been omitted due to high hydrodynamic friction force that occurs in fluid film. Compared with a solid coating, stiction prevention using liquid lubrication is less practical. However, recent studies have demonstrated that it is possible for

Newtonian liquids to slip along very smooth solid walls [20] and this result may make liquid lubricants for MEMS devices feasible. The main advantage of a liquid lubricant over a solid lubricant is that they generally produce no-contact shear stresses. Unfortunately, a stiction-type failure due to a large shear and capillary forces occurs. In [15, 16], a lubrication model of low load contacts was proposed to reduce such stiction or friction. The idea behind that work was how to use a lubricant that does not wet one of the solid surfaces. It was found that a half-wetted bearing generates a significant friction reduction compared to a traditional bearing.

In order to reduce stiction, two principal methods are available, chemical and physical modification of the surfaces. To generate wall slip, in the chemical approach, the chemical composition of the surface is altered. In the physical approach, the surfaces are roughened to decrease the effective contact area [29].

In practice, the slip zone of the artificial slip surface can be prepared from (super)hydrophobic surface which uses chemical properties as well as micro- and nano-structures in order to achieve a high level of friction force reduction. The main characteristic of (super)hydrophobic surfaces is the slip length. Extensive studies have confirmed that the chemical treatment of the surfaces generates a slip length in the order of 1 μm [30], while longer slip length up to 100 μm can be obtained through a combination of a hydrophobic surface with textured structure [20, 31, 32]. In the present study, it will be shown by the computational analysis that a longer slip length applied on the slip zone of the artificial slip surface leads to a greater friction force reduction in combination with an improved load support.

3.2.1. Beneficial surface of slip

Recently, the use of an engineered slip surface has become popular with respect to lubrication, since this type of surface enhancement would give a better tribological performance. The great challenge for an engineered slip surface from the perspective of a numerical simulation is choosing the optimal slip zone geometry with respect to the lubrication performance. Two engineered slip surface modes were used currently: homogeneous slip surface (i.e. slip applied over the whole surface) and artificial slip surface (i.e. surface consisting of slip zone and no-slip zone). It can be noted that term "artificial slip surface" was sometimes also called as heterogeneous slip/no-slip surface [17, 18] and mixed slip surface [19]. The first study to mention using a homogeneous slip was dedicated by Spikes [15, 16] who numerically studied the effect of slip profiles on friction. The author pointed out that by introducing the half-wetted bearing having a homogeneous slip boundary on one of the surfaces, a reduced friction can be obtained. Subsequently, an experimental study was published in [20] confirming the finding of [15, 16]. However, in addition to the friction reduction, it was shown that a homogeneous slip surface usually has a negative effect, i.e. the decrease in the load support. If the lubricated contact exhibits a perfect slip property, it was found that the fluid load support was only half of that without slip [115, 19, 21, 23]. Clearly, this is unwanted effect with respect to the lubrication. Therefore, to date, an artificial slip surface has become of great interest by some researchers [17-19, 21, 23] with the focus of how to balance the slip effect on the load support and friction.

The big question with respect to the tribological performance of lubricated-MEMS emerges in accordance with at which wall boundary slip must be applied, at the stationary surface, moving surface, or both of them. Besides that, the types of slip zone pattern become also great issue. Therefore, a series of simulations were conducted with such boundaries to find the best possibility of slip boundary application in terms of load support. Investigations were made for four kinds of slip boundaries to find the best boundary slip in terms of tribological performance, i.e. (1) slip applied on both the stationary and moving surfaces is referred as 'condition 1', (2) slip applied on the stationary surface is referred as 'condition 2', (3) slip applied on the moving surface is referred as 'condition 3', and (4) no-slip condition applied on the both of surfaces is referred as 'condition 4'. Here, a homogeneous slip surface is employed for all slip conditions.

Figure 4 presents the effect of the wedge ratio h^* on the dimensionless load support W. It is shown that the contact with homogeneous slip condition of 1, 2, and 3 have a negative effect, i.e. a reduced load support. The highest achievement of a load support W is obtained when the slip is applied on the stationary surface (condition 2). However, the value of the predicted load support is much lower than the conventional lubricated contact for all values of wedge ratio. It is only half of what the conventional Reynolds theory predicts for an optimal wedge ratio of a traditional slider contact. Fortunately, the direct trend of homogeneous slip to decrease the load support W is counterbalanced by the fact that such surface also reduces friction significantly. This is indicated in Figure 5 which shows the comparison between the dimensionless friction F for condition 2 and the condition 4 for the range of wedge ratio, h^*. The reduced friction as an advantageous effect in the lubrication can be explained by the fact that the boundary slip tends to reduce the wall shear rate at a prescribed film thickness, and then the wall shear stress and friction. Therefore, in the following design for the maximal lubrication performance, the lubricant has a no-slip boundary condition at the moving solid surface but can slip along the stationary surface. In the next section, the geometry of the boundary slip zone making an artificial slip surface will be investigated in order to achieve a higher load support as well as a low friction force.

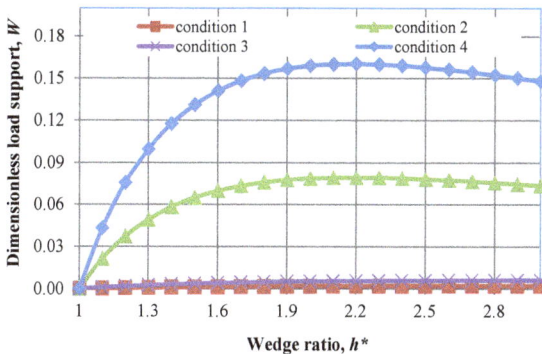

Figure 4. Dimensionless load support versus wedge ratio for several homogeneous boundary slip conditions. (Note: slip on stationary and moving surfaces (condition 1); slip on stationary surface (condition 2); slip on moving surface (condition 3); traditional no-slip (condition 4)). For the slip cases, the dimensionless slip length A of 20 is assumed.

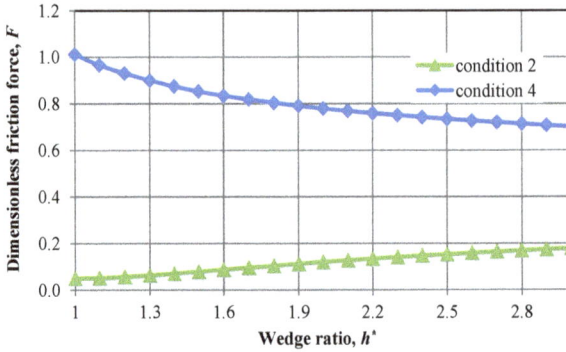

Figure 5. Dimensionless friction force versus wedge ratio at boundary condition in which slip applied on the station-ary surface (condition 2) compared to the no-slip condition (condition 4).

3.2.2. The optimum slip zone of the artificial slip surface

This section is intended to investigate the optimum slip zone of the artificial slip surface for several values of wedge ratios h^* with respect to the load support.

In the following computations, as discussed in the previous section, for a high load support, it is considered that the artificial slip surface will take place on the stationary surface, whereas no-slip condition occurs on the moving surface. The parameter S_{xD} (in which $S_{xD} = S_x/L_x$) is introduced in order to completely define the dimensionless slip zone. A parametric analysis is conducted by varying the dimensionless slip zone S_{xD} from zero (i.e. no-slip boundary) to one (i.e. homogeneous boundary slip). The effect of slip zone geometry on the load support is presented in Figure 6. It can be shown that the load support has a maximum value when S_{xD} = 0.65 and h^* = 1 (i.e. parallel sliding surfaces). It should be noted that no hydrodynamic pressure can be built up in parallel sliding surfaces for traditional no-slip contact. From Figure 6, it is also shown that with the increase in wedge ratio h^*, a clear shift of the maximum load support toward to outlet can be observed in this work. In addition, the maximum load support decreases with the increase in h^*. A first conclusion is that for the best artificial slip surface with respect to the load support performance, the slip zone must be employed on the leading edge of the parallel sliding contact.

Figure 7 shows the normalized representation of lubrication film pressure distributions as a function of wedge ratio which are predicted by the modified Reynolds equation (Eq. (1)). For slip configuration, the optimal slip zone S_{xD} of 0.65, is employed. It can be shown that compared with the no-slip boundary condition, the artificial slip surface yields a positive fluid pressure. However, the performance improvement obtained through artificial slip surface is rather mild for the high wedge ratio. Obviously, the lower the base geometry wedge ratio (thus leads to the parallel sliding contact), the larger improvement that boundary slip can induce. In other words, using the artificial slip surface considered here, the maximum pressure occurs not at a

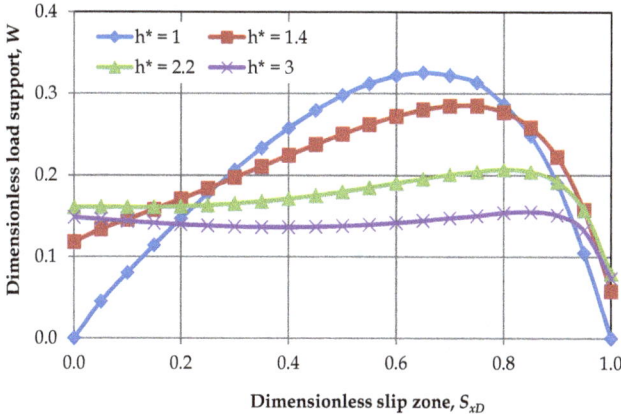

Figure 6. Effect of the dimensionless slip zone of the artificial slip surface S_{xD} on the dimensionless load support W for several wedge ratios h^*.

convergent wedge as predicted by the classical Reynolds assumption ($h^*_{opt} = 2.2$ [28]), but at a parallel surface. The predicted maximum pressure for a parallel gap is over three times as large as the maximum pressure obtained from a no-slip contact when the wedge ratio $h^* = 2.2$. From this perspective, the well-chosen artificial slip surface pattern can be considered as a potential application which is possible for lubricated-MEMS based devices with respect to the load support. In this way, liquid lubricants on the modified opposing surfaces (i.e. artificial slip surface) can prevent the lubrication to break down during device operation to the point where they no longer give proper lubrication.

Figure 8 shows the dimensionless surface friction force F at the bottom surface. It can be seen that the friction of the artificial slip surface becomes smaller than that of a traditional no-slip lubricated contact especially when S_{xD} is larger than about 0.6. This agrees with the numerical analysis of Wu *et al.* [19] eventhough the slip model and numerical method used are different. One can remark that when there is no wedge effect (i.e. $h^* = 1$) and $S_{xD} = 0.65$, the artificial slip surface gives the minimum dimensionless friction force of 0.65, while the no-slip lubricated contact gives its dimensionless minimum friction of 0.77 at $h^*_{optimal} = 2.2$, see Table 1. It means that the optimized artificial slip surface of lubricated-MEMS can produce a lower friction force than a no-slip contact.

3.2.3. Effect of dimensionless slip length on lubrication performance

The hydrophobicity of a solid surface, as discussed in the previous section, is usually expressed in terms of a slip length, which quantifies the extent to which the fluid elements near the surface are affected by the surface energy and the surface geometry. The surface energy is an intrinsic property of a material that can be controlled by chemical treatment, such as etching approach and/or coat-on/cast approach. The surface roughness of a hydrophobic solid material can be

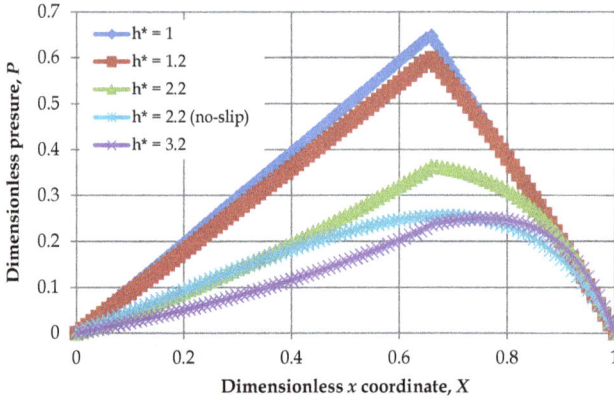

Figure 7. Lubrication film pressure distributions for several values of wedge ratio. The slip profiles are calculated for optimized dimensionless slip zone $S_{xD} = 0.65$ and dimensionless slip length $A = 20$.

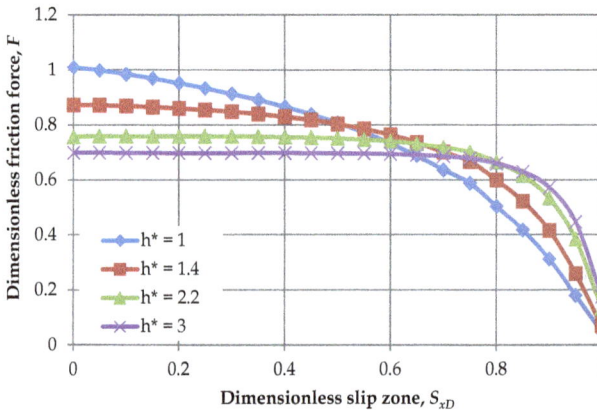

Figure 8. Effect of the dimensionless slip zone of the artificial slip surface S_{xD} on the dimensionless friction force F at bottom surface for several values of wedge ratio h^*. The slip profiles are evaluated for dimensionless slip length $A = 20$.

tuned in order to increment its hydrophobicity and obtain a super-hydrophobic solid surface [33, 34]. In this section, from the numerical point of view, the effect of slip length on the lubrication behavior is studied. The dimensionless slip length is varied from 3 to 300.

Figure 9 shows the effect of slip zone of the artificial slip surface for several dimensionless slip length values on the dimensionless load support. As indicated in Figure 9, the increase in the slip length leads to an increase in the predicted load support. Generally, the larger the slip length at the optimized slip zone of the artificial slip zone, the higher the load support.

However, when dimensionless slip length is larger than 30, the dimensionless load support is not affected significantly with the increase in the dimensionless slip length. So, the increase of the load support is not infinitely large. It can be deduced that there is no fluid load support for a lubricated-MEMS when it contains no-slip condition ($A = 0$). It indicates that the absence of the wedge effect on the pressure generation at parallel lubricated sliding contact has been counterbalanced by the influence of the ariticial slip surface application. Again, this condition is very advantageous in engineering a lubricated-MEMS which demonstrates parallel gaps.

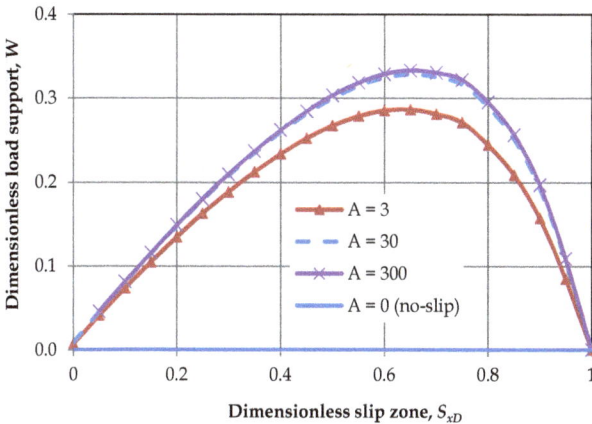

Figure 9. Effect of the dimensionless slip zone of the artificial slip surface S_{xD} on the dimensionless load support W for several values of dimensionless slip length A. All profiles are calculated for parallel sliding surfaces ($h^* = 1$).

In Figure 10 the effect of slip zone of the artificial slip surface for several dimensionless slip length values on the dimensionless friction force is presented. As is well-known, the ability to control and manipulate friction force during sliding is extremely important key to prolong a life-time of lubricated-MEMS. Better understanding of the friction force phenomena at micro-scales is needed to provide designers and engineers the required tools and capabilities to control friction force and predict failure of lubrication in MEMS.

As can be seen in Figure 10 that the artificial slip surface leads to a reduction of the friction force for all dimensionless slip length. The friction force decreases with increasing the slip zone S_x. It can be seen that when the slip zone covers over the whole surface, i.e. homogeneous slip surface, the friction forces has a minimum value, especially for a high slip length. If the reduction in friction force is of only particular interest, the homogeneous slip surface ($S_{xD} = 1$) is very beneficial. But if the performance is also related to the load support, homogeneous slip is not recommended because when $S_{xD} = 1$, the predicted load support is very small for all wedge ratios, see Fig. 6. With respect to the influence of the dimensionless slip length, opposite to the hydrodynamic load support, the dimensionless friction force becomes smaller for higher dimensionless slip length. Therefore, the optimized artificial slip surface is a very promising way to increase the hydrodynamic performance and the stability of the lubricated MEMS system because it gives an advanced load support in combination with a reduced friction force.

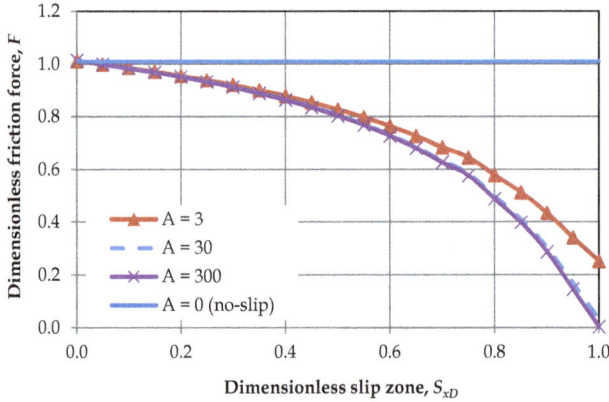

Figure 10. Effect of the dimensionless slip zone of the artificial slip surface S_{xD} on the dimensionless friction force F for several values of dimensionless slip length A. All profiles are calculated for parallel sliding surfaces ($h^* = 1$).

The combined effect of slip zone parameter on load support and friction force can be better analyzed using the dimensionless friction coefficient. In the present study the dimensionless friction coefficient m is defined as the ratio of the dimensionless friction coefficient F to the dimensionless load support W. Figure 11 shows the variation of the dimensionless friction coefficient m as a function of the dimensionless slip zone S_{xD} for various slip lengths A. It appears that when the dimensionless slip zone is smaller than 0.2, the friction coefficient increases significantly. After $S_{xD} = 0.2$, increasing the slip zone will be less significant to the reduction of the friction coefficient. It can also be deduced from Figure 11 that dimensionless friction coefficient m will decrease with the increase of the dimensionless slip length A, especially for small S_{xD}. It implies that the minimum dimensionless friction coefficient is accord with the the highest dimensionless load support (when $S_{xD} = 0.65$).

4. Concluding remarks

Numerical results show that the hydrodynamics of a lubrication film confined between a moving no-slip surface and a stationary with an artificial slip surface differ significantly from that of a film confined between two no-slip surfaces. It is found that a homogeneous slip boundary on one surface produces a lower hydrodynamic pressure in a lubricated sliding contact at various conditions (slope incline, and slip length), resulting in a reduced load support which reduces the positive effect of slip on friction. However, if the surface is designed with an optimal artificial slip pattern (the slip zone is applied on 0.65 of contact length), even

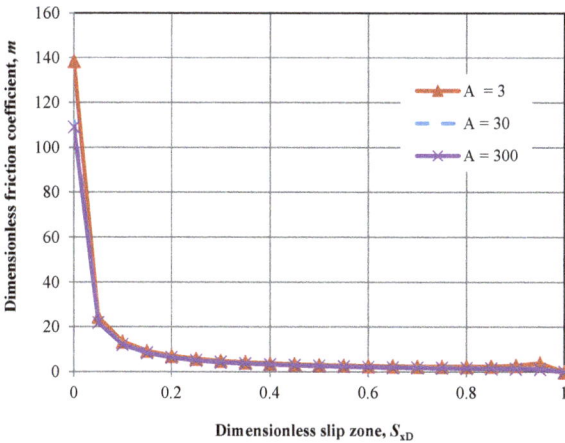

Figure 11. Effect of the dimensionless slip zone of the artificial slip surface S_{xD} on the dimensionless friction coefficient m for several values of dimensionless slip length A. All profiles are calculated for parallel sliding surfaces ($h^* = 1$).

when there is no wedge effect, the load support has a maximum value. In addition, the friction force can decrease significantly. Therefore, it is very beneficial to make one of the contacting surfaces in lubricated-MEMS with an artificial slip surface for achieving ideal lubrication performance, i.e. reduced friction coefficient and increased load support.

Author details

M. Tauviqirrahman[1,2], R. Ismail[1,2], J. Jamari[1] and D.J. Schipper[2*]

*Address all correspondence to: d.j.schipper@utwente.nl

1 Laboratory for Engineering Design and Tribology, Mechanical Engineering Department, University of Diponegoro, Jl. Prof. H. Sudharto, Kampus UNDIP Tembalang, Semarang, Indonesia

2 Laboratory for Surface Technology and Tribology, University of Twente Drienerlolaan, Enschede, The Netherlands

References

[1] Houston MR, Howe RT, Maboudian R. Effect of Hydrogen Termination on The Work of Adhesion Between Rough Polycrystalline Silicon Surfaces. Journal of Applied Physics 1997;81(8): 3474–3483.

[2] Maboudian R, Ashurst WR, Carraro C. Self-Assembled Monolayers as Anti-Stiction Coatings For MEMS: Characteristics and Recent Developments. Sensors and Actuators A: Physical 2000;82(1–3): 219–223.

[3] Tagawa M, Ikemura M, Nakayama Y, Ohmae N. Effect of Water Adsorption on Microtribological Properties of Hydrogenated Diamond-Like Carbon Films. Tribology Letters 2004;17(3): 575–580.

[4] Smallwood SA, Eapen KC, Patton ST, Zabinski JS. Performance Results of MEMS Coated with a Conformal DLC. Wear 2006;260: 1179–1189.

[5] van Spengen WM, Puers R, De wolf I. On The Physics of Stiction and Its Impact on The Reliability of Microstructures. Journal of Adhesion Science and Technology 2003;17(4): 563–582.

[6] Pit R, Hervet H, Leger L. Direct Experimental Evidence of Slip in Hexadecane: Solid Interfaces. Physical Review Letters 2000;85: 980–983.

[7] Craig VSJ, Neto C, Williams DRM. Shear-Dependent Boundary Slip in an Aqueous Newtonian Liquid. Physical Review Letters 2001;87(054504).

[8] Bonaccurso E, Kappl M, Butt HJ. Hydrodynamic Force Measurements: Boundary Slip of Hydrophilic Surfaces and Electrokinetic Effects. Physical Review Letters 2002;88(076103).

[9] Bonaccurso E, Butt HJ, Craig VSJ. Surface Roughness and Hydrodynamic Boundary Slip of A Newtonian Fluid in A Completely Wetting System. Physical Review Letters 2003;90(144501).

[10] Zhu YX, Granick S. Rate-Dependent Slip of Newtonian Liquid at Smooth Surfaces. Physical Review Letters 2001;87(096105).

[11] Priezjev NV, Darhuber AA, Troian SM. Slip Behaviour in Liquid Films on Surfaces of Patterned Wettability: Comparison Between Continuum and Molecular Dynamics Simulations. Physical Review E 2005;71(041608).

[12] Cottin-Bizonne C, Barentin C, Charlaix E, Bocquet L, Barrat JL. Dynamics of Simple Liquids at Heterogeneous Surfaces: Molecular Dynamics Simulations and Hydrodynamic Description. European Physical Journal E 2004;15: 427– 438.

[13] Harting J, Kunert C, Herrmann HJ. Lattice Boltzmann Simulations of Apparent Slip in Hydrophobic Channels. Europhysic Letters 2006; 75: 328–334.

[14] Li BM, Kwok DY. Discrete Boltzmann Equation for Microfluidics. Physical Review Letters 2003;90(124502).

[15] Spikes HA. The Half-Wetted Bearing. Part 1: Extended Reynolds Equation. Proceedings of the Institution of Mechanical Engineers, Part J: Journal of Engineering Tribology 2003; 217: 1–14.

[16] Spikes HA. The Half-Wetted Bearing. Part 2: Potential Application in Low Load Contacts. Proceedings of the Institution of Mechanical Engineers, Part J: Journal of Engineering Tribology 2003;217: 15–26.

[17] Salant RF, Fortier AE. Numerical Analysis of A Slider Bearing with A Heterogeneous Slip/No-Slip Surface. Tribology Transactions 2004; 47: 328–334.

[18] Fortier AE, Salant RF. Numerical Analysis of A Journal Bearing with A Heterogeneous Slip/No-Slip Surface. ASME Journal of Tribology 2005;127: 820–825.

[19] Wu CW, Ma GJ, Zhou P. Low Friction and High Load Support Capacity of Slider Bearing with A Mixed Slip Surface. ASME Journal of Tribology 2006;128: 904–907.

[20] Choo JH, Glovnea RP, Forrest AK, Spikes HA. A Low Friction Bearing Based on Liquid Slip at The Wall. ASME Journal of Tribology 2007;129: 611–620.

[21] Tauviqirrahman M, Ismail R, Jamari, Schipper DJ. Effect of Boundary Slip on The Load Support in A Lubricated Sliding Contact. AIP Conference Proceedings 2011;1415: 51-54. doi:10.1063/1.3667218.

[22] Aurelian F, Patrick M, Mohamed H. Wall Slip Effects in (Elasto) Hydrodynamic Journal Bearing. Tribology International 2011;44: 868–877.

[23] Tauviqirrahman M, Ismail R, Jamari, Schipper DJ. Wall Slip Effects in a Lubricated MEMS. International Journal of Energy Machinery 2011;4(1): 13–22.

[24] Navier CLMH. Mémoire Sur Les Lois Du Mouvement Des Fluides. Mémoires de l'Académie Royale des Sciences de l'Institut de France 1823:6: 389–440.

[25] Israelachvili J. Intermolecular and Surface Force, vol. 1, second ed. Academic Press, London; 1995.

[26] Reynolds O. On The Theory of Lubrication and Its Application to Mr. Beauchamp Tower's Experiments, Including An Experimental Determination of The Viscosity of Olive Oil. Philosophical Transactions of the Royal Society of London, Part I 1886;177: 157–234.

[27] Patankar SV. Numerical Heat Transfer and Fluid Flow. Taylor & Francis, Levittown 1980:30–58.

[28] Cameron A. The Principles of Lubrication. Longman Green and Co., ltd; 1966.

[29] Kompvopoulos K. Surface Engineering and Microtribology for Microelectromechanical System. Wear 1996;200: 305–327.

[30] Tretheway DC, Meinhart, CD. Apparent Fluid Slip at Hydrophobic Microchannel Walls. Physics of Fluids 2002;14: L9-12.

[31] Watanabe K, Yanuar, Udagawa H. Drag Reduction of Newtonian Fluid in a Circular Pipe with a Highly Water-Repellant Wall. Journal of Fluid Mechanics 1999;381: 225–238.

[32] Ou J, Perot B, Rothstein JP. Laminar Drag Reduction in Microchannels using Ultrahydrophobic Surfaces. Physics of Fluids 2004;16: 4635.

[33] Patankar NA. On the Modelling of Hydrophobic Contact Angles on Rough Surfaces. Langmuir 2003;19: 1249–1253.

[34] Lafuma A, Quere D. Superhydrophobic States. Nature Materials 2003;2: 457–460.

Chapter 9

Friction and Wear of a Grease Lubricated Contact — An Energetic Approach

Erik Kuhn

Additional information is available at the end of the chapter

1. Introduction

Lubricating greases are colloid disperse systems with visco-elastical properties. These special lubricants have a wide range of application. More than 90% of the ball bearings are grease lubricated but also gears and journal bearings are application examples for greases.

The composition of greases consists of a base oil and a thickener (of course some additives can be found in commercial lubricants). The thickener forms a network which leads to a complex rheological and tribological behaviour.

Tribological contacts lubricated by greases are often working under mixed friction conditions.

The situation inside the grease film determines the content of fluid friction. The situation inside the tribological gap influences the content of solid friction.

A special phenomenon is the structural degradation due to the effects of friction. It leads to a dependence on time of the grease behaviour.

Aims of this chapter are the description of the liquid friction process inside the grease film and of the procedure of structural degradation. Both phenomena have a strong influence on the friction and wear behaviour of the whole tribo-system.

2. General definitions and terminology

The presented work is based on a conception of friction and wear that investigates the process from an energy point of view. Some ideas differ from well-known definitions so that they need to be introduced here for better understanding.

A *tribological contact* describes the geometrical situation of rubbing bodies with interactions in the sense of the tribological process.

Friction is (only) an energy expenditure.

Wear is a production of irreversibility due to affected friction energy. It covers all elements of a tribo-system [1].

This chapter distinguishes between wear volume and volume of the removed material.

Wear volume is presented by the material area where irreversible friction effects lead to an excess of a critical energetic level. (in contrast to the removed material volume (loss of material)).

Mixed friction describes an energy expenditure in several states of friction that exists simultaneously inside the same tribo-system (see also [2]). A direct contact of the solid bodies is not necessary for mixed friction of a lubricated couple (mix of solid and liquid friction).

Lubricating greases are colloid disperse systems with visco-elastical properties.

The grease structure is characterised by the geometry and the distribution of the thickener, the interactions between the thickener and the tribo-system and the ability of storing energy.

3. Energy balance for a grease lubricated contact

Grease lubricated contacts are often working in mixed friction. That means liquid friction (grease film) and solid friction (asperity deformation) have to be considered. An idea of the situation inside the lubricated contact is presented in the next figures.

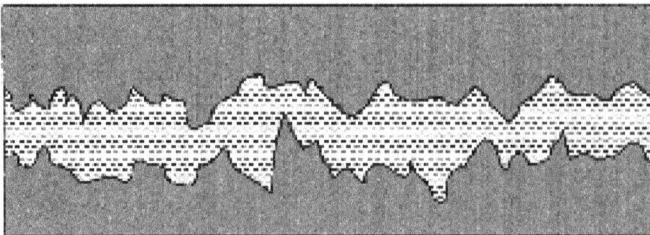

Figure 1. Grease lubricated contact with two rough surfaces

The surface profile is modelled with a spherical shape of the asperities and the gap configuration can be illustrated with Fig.2.

The situation of a single contact in consideration of the simple idea of lubricant layers inside the grease film is presented in Fig.3.

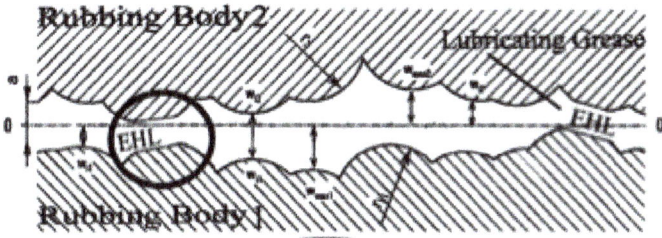

Figure 2. Modelled mixed friction contact of a grease lubricated gap

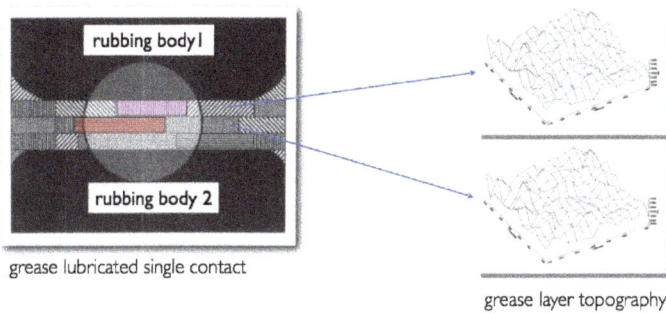

grease lubricated single contact

grease layer topography

Figure 3. Single contact and modelled contact situation inside the grease film

The grease topography comes from IR-microscopy and presents the density distribution. In Fig.3 the contact of two density areas inside the greases film is highlighted.

Investigations of the surface profile and of the density distribution of the grease lead to a discrete contact model. Random variables are asperity height $\xi_{1,2}$, radius of the modelled asperity shape $\rho_{1,2}$ and the observed density $\delta_{1,2}$. All random parameters can be described with the Gaussian distribution.

For the contact probability F illustrated in Fig.3 (left) can be obtained [3]

$$F(z,u,\rho) = \int_0^z \int_0^{z-\xi_2} f(\xi_1) f(\xi_2) d\xi_1 d\xi_2 \cdot \int_{uk1}^{ug1} f(\rho_1) d\rho_1 \cdot \int_{uk2}^{ug2} f(\rho_2) \cdot \int_{\rho k1}^{\rho g2} f(\delta_1) d\delta_1 \cdot \int_{\rho k2}^{\rho g2} f(\delta_2) d\delta_2 \tag{1}$$

An energy balance is created by investigation of a single grease lubricated contact. The expended energy is formed as a sum of different contents

$$W_{friction} = \sum_i W_{i=1..n} \tag{2}$$

The assumption of mixed friction leads to the consideration of two main contents of the friction energy.

$$W_{mixed-friction} = W_{solid} + W_{fluid} \qquad (3)$$

A proposal is made with

$$W_{solid} = e_{elast/plast} \cdot V_{elast/plast} \cdot n_{elast/plast} \qquad (4)$$

$$W_{liquid} = e_{rheo1,2,3} \cdot V_{rheo1,2,3} \cdot n_{rheo1,2,3} + e_{solidif} \cdot n_{solidif} \cdot V_{solidif} + e_{tensile} \cdot V_{tensile} \cdot n_{tensile} + e_{wave} \cdot V_{wave} \cdot n_{wave} \qquad (5)$$

In Eq.(4) and (5) the notation *energy density* for the selected mechanism multiplied by *stressed volume* and by the number of contacts with the same mechanism is used. For a single contact n=1 holds. The solid friction from Eq.(4) considers the deformation (elastically/plastically) of the stressed asperities. Information about this is given in [4].

The summands in Eq.(5) describe the shearing process in different gap situations (index *rheo*), solidification effects in the center of the observed gap (index *solidif*), tensile stress in the outlet of the contact (index *tensile*) and the formation of a friction wave (index *wave*).

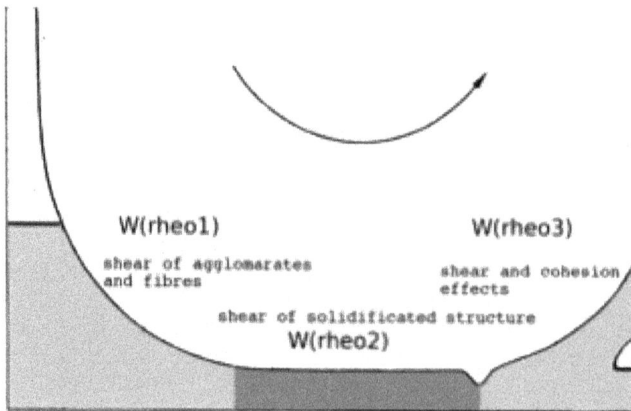

Figure 4. Different stress situations for the grease inside the contact geometry $(e_{rheo1,2,3})$

4. Liquid friction inside the grease film

4.1. Stress situation inside the gap geometry

As mentioned before a micro single contact of a grease lubricated couple is observed and the stress situation is analysed. The idea of the process model developed here is illustrated in Fig.5.

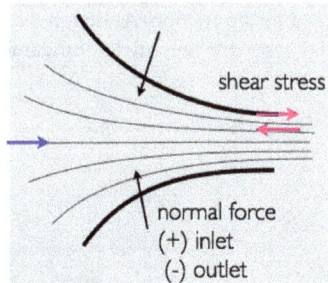

Figure 5. Grease lubricated gap between two asperities (micro single contact). Shear stress is the most important stress mechanism of the grease. Normal force (+) leads to solidification effects. Normal force (-) leads to tensile stress.

In addition the possibility of a friction wave inside the grease film is observed. The idea is clarified in the Fig.6.

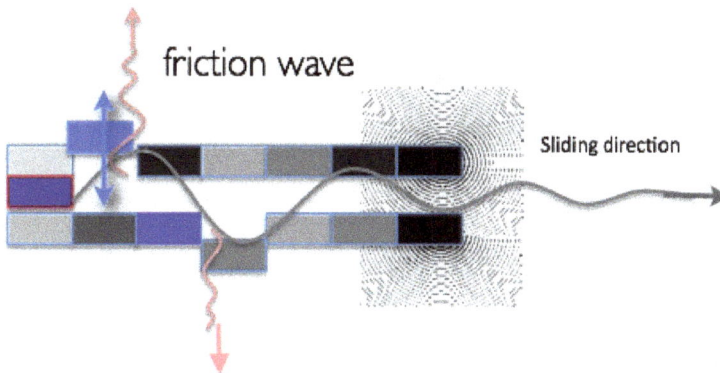

Figure 6. Schematic representation of a friction wave caused by the contact of volume elements of the grease film with different properties

4.2. Empirical proposals to quantify the friction energy

The experimental work is focused on the quantification of t the energy expenditure during the shear process of a lubricant. This requires an experimental procedure that simulates the liquid friction inside the grease film.

A proposal is made by the use of a rheometer [5], [6]. Shearing a grease sample in a rotating or oscillating test can be interpreted as a fluid friction experiment. All reaction measured by the rheometer is caused by the grease behaviour. The plate and cone in a rheometer configuration does not lead to the state of mixed friction. Although the test conditions are fare from real contact situation rheometer tests are helpful for fundamental investigations of fluid friction.

A picture of a cone-plate configuration is shown in Fig.7.

Figure 7. Evolution of the observed shear stress (left) during a rheometer experiment with a cone-plate system (right) in rotational modus

The energy per volume that is necessary to shear the grease during the experiment can be obtained from a shear test (Fig.7). The experimental conditions are constant shear rate and constant test temperature. To compare different grease samples the test time for each experiment has to be held constant.

An empirical proposal [7] is made with

$$e_{rheo} = \dot{\gamma}_{const} \cdot \int_0^t \tau(\varsigma) d\varsigma \qquad (6)$$

with e_{rheo} the energy density for the shear process [J/m³], t the test time [s], $\dot{\gamma}$ the shear rate [1/s] and ς the current time [s].

A more interesting experimental procedure is an oscillating rheometer test [Fig.8]. The grease behaviour can be observed in a wide range of the oscillating amplitude. For the investigation of fluid friction inside the grease film experiments within the linear visco-elastical range can be analysed.

To quantify the friction behaviour the following proposal [3] can be used

$$e_{rheo-elast} = \frac{G' \cdot \gamma_{elast}^{2}}{\cos \delta} \tag{7}$$

G' is the storage modulus [Pa], γ is the deformation [-] and δ is the phase different angle. Eq. (7) describes the energy per volume $[J/m^{3}]$ to deform the sample elastically and is related to the thickener.

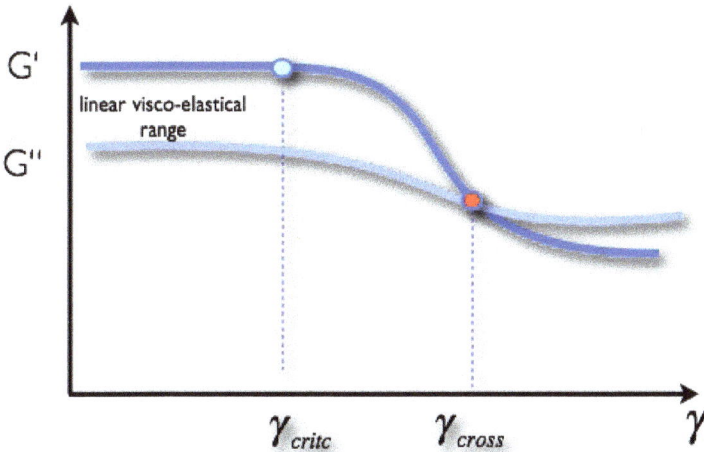

Figure 8. Typical evolution of the storage modulus and the loss modulus during an amplitude sweep (oscillating measurement)

4.3. Results from experimental work

The energy expenditure expressed by the rheological energy density $e_{rheo-elast}$ presents the liquid friction behaviour of the investigated grease samples. Equation (7) observes only the shear mechanism inside the tribological gap. To compare the greases the same deformation (oscillating amplitude) has to be used. Some greases with the same thickener type (Li-soap) and the same base oil (mineral oil) were observed by a variation of soap content and test temperature.

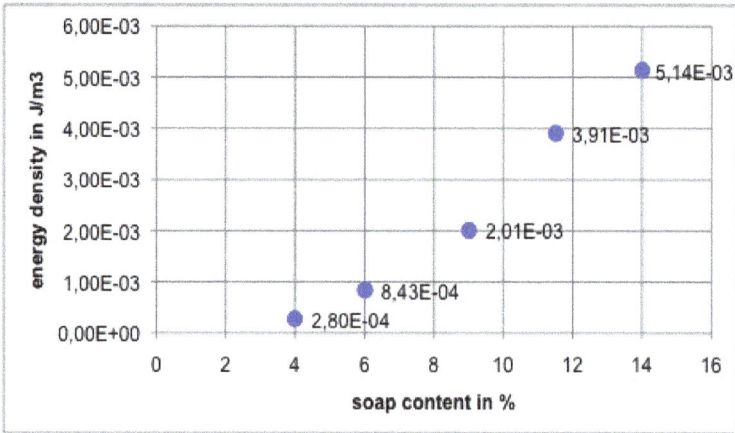

Figure 9. Energy densities from the linear visco-elastical range = 20° C for different soap contents

An increase of the soap content leads to an increase of the liquid friction. Experiments with a temperature $\vartheta = 50°C$ and the same grease samples deliver the results in Fig.10.

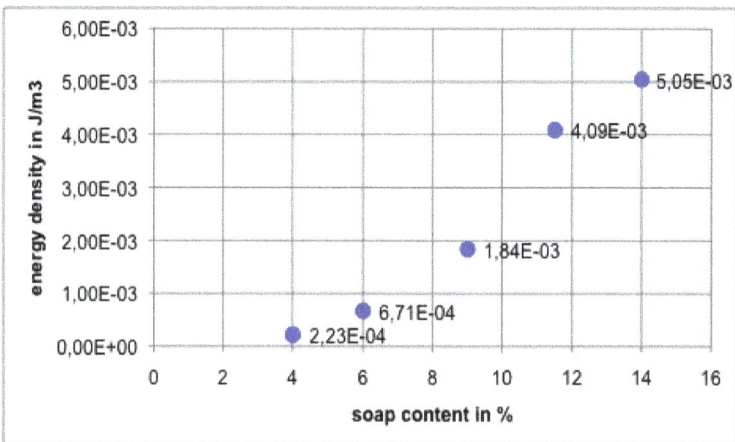

Figure 10. Energy densities for a test temperature = 50°C (same conditions as Fig.9)

Compared with the test temperature of $\vartheta = 20°C$ lower values of the energy densities are obtained. This behaviour is in accordance with the experience that fluid friction is decreasing with an increasing temperature. A more or less linear correlation of energy density and soap content can be observed.

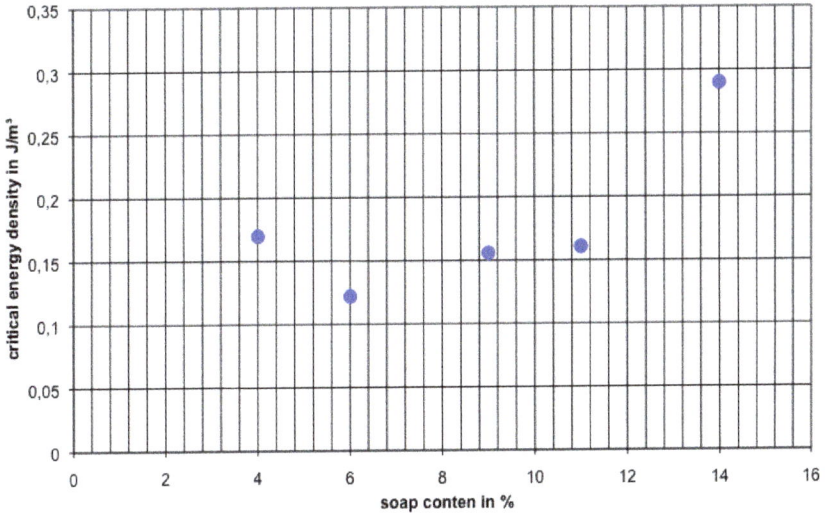

Figure 11. Critical energy level vs. soap content

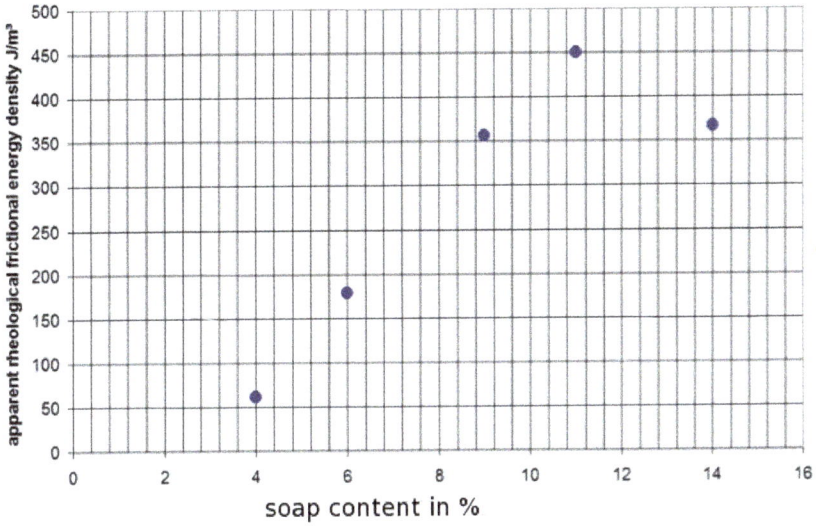

Figure 12. Energy densities for the crossing point vs. soap content

5. Irreversible effects due to friction

5.1. Idea of the structural degradation of lubricating greases

A typical curve obtained from rheometer experiments for constant shear rate and temperature (rotational mode) shows a strong dependence on time. The drop of shear stress versus stress time is an indirect expression of the structural degradation and well known from many papers [8],[9],[10].

To illustrate the friction effects AFM-investigations made by [11] are presented below (Fig. 13). The change of thickener structure caused by the liquid friction is evident. The geometry and distribution of the thickener is completely different to the initial situation and it can be assumed that the new grease structure shows a different tribological behaviour.

Volume elements inside the grease film are modelled to observe the tribological process. Because of the thickener distribution these volume elements have different properties as elasticity, density, level of accumulated energy, level of critical energy etc.. The consequence of the property distribution is a different tribological behaviour of the observed volume elements forming a lubricant layer. The contact situation of two assumed grease layer composed of different volume elements is presented in Fig. 14.

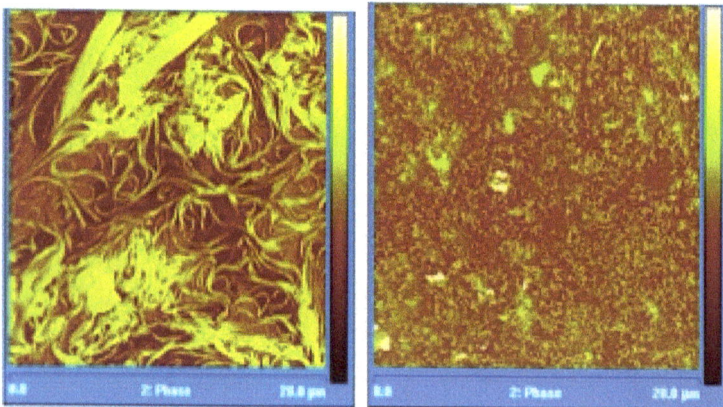

Figure 13. Left- fresh grease sample, right- grease stressed in a rheometer [11]

Liquid friction between two modelled grease layer. Different volume elements have different rheological and tribological properties.

An energy stress is applied to the control volume if liquid friction take place (Fig.15). Due to this energy stress an energy accumulation process, a dissipation process and a transition process (overstep of a critical energy level) starts.

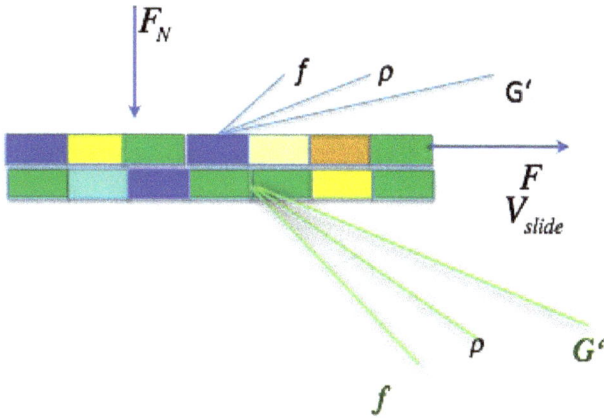

Figure 14. Liquid friction between two modelled grease layer. Different volume elements have different rheological and tribological properties.

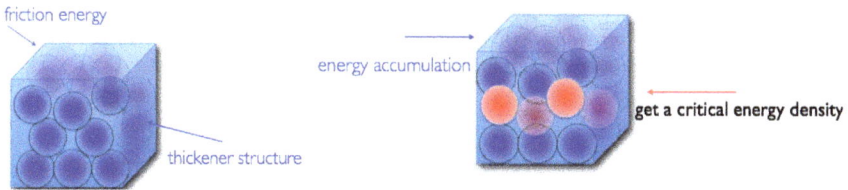

Figure 15. Modelled volume element inside the grease film. Left – unstressed, right – stressed by liquid friction

The degradation of the grease structure begins with a transition process within a volume element. Overstepping a critical energy density initiates an irreversible change of the structure.

A contact model is developed to describe the energetic situation of critical exceedance.

To quantify the number of exceedance of critical energy level a stationary Gaussian process is used. Information about the conditions and definitions can obtained in [1] and [12].

$$E\big[N_0(u)\big] = \lim_{\varepsilon \to 0} \frac{1}{\varepsilon \cdot L} \cdot \int_0^L E[A(x,\varepsilon) \cdot h'(x)]dx \tag{8}$$

$$E\big[N_0(u)\big] = \frac{1}{2\pi} \cdot \sqrt{\frac{m_2}{m_0}} \cdot e^{\frac{u^2}{2m_0}} \tag{9}$$

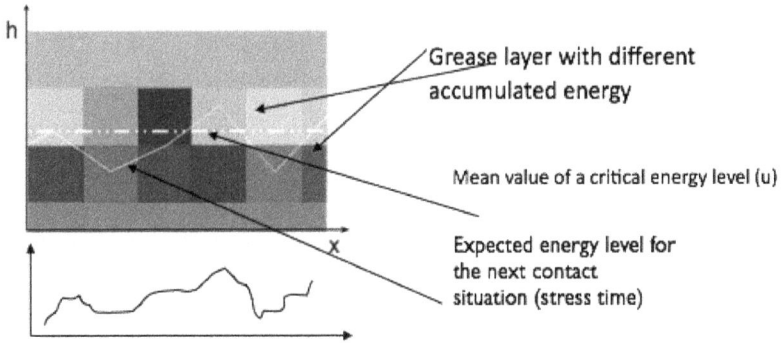

Figure 16. New model to describe the energetic situation fort he transition process inside the grease film [1]

Initial situation is described with Eq. (8) and the expectation of the number of overstepping N_0 can be determined with Eq. (9). This proposal uses spectral moments m_2; m_0.

An example [1] was quantified by using density distribution from IR-microscopy to determine the parameter m_2; m_0. A mean value of the critical energy level was assumed.

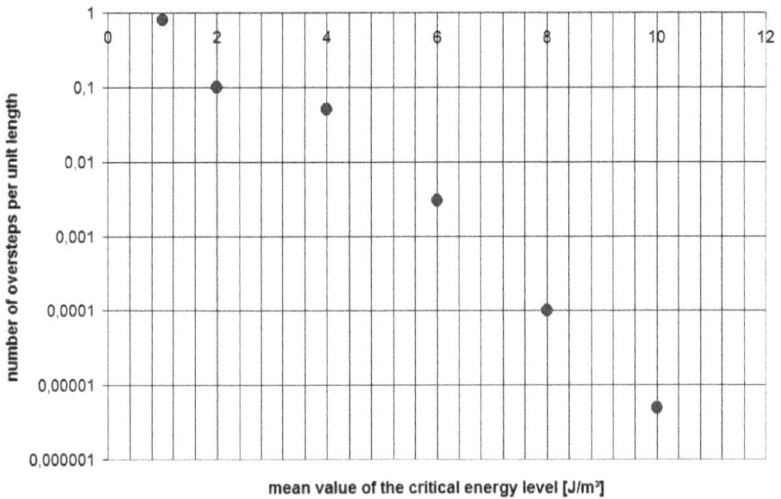

Figure 17. Influence of the critical energy level of the number of oversteps [1]

5.2. Thermodynamic investiagtions

5.2.1. General aspects

Friction process within a tribo-system is an irreversible process. It means that input of friction energy leads to irreversible effects. This approach interpreted friction and wear process as an cause-effect-chain.

Different authors tried to find relations between the system behaviour expressed by mass loss (wear) and entropy flow/production [15]-[19]. *Abdel-Aal* [18] expressed the conjecture that relation between frictional heat generation and heat dissipation is related to wear transition. This author pointed out [19] that there exists a one to one relationship between entropy generation and mass loss. *Doelling et al.* [17] give the argumentation that there exists a strong correlation between components wear and entropy flow.

Ling et al. [13] give an experimental description of a correlation between wear and entropy flow in lubricated sliding systems.

„Sliding wear is an irreversible degradation of surfaces induced by friction. On a microscopic scale irreversible physical interactions between the sliding surfaces – including plastic deformation of asperities, fracture, delamination, abrasive plowing, and corrosive wear, among others – creates friction resistance forces, dissipates power, and generates irreversible entropy. Since the physical interactions responsible for friction and wear monotonically produce entropy, entropy becomes a time base for wear" [13].

"The entropy production, in fact, enable one to bridge the atomic scale phenomena with the macro scale response" [14] in [13]. "In addition friction and wear, from the vantage point of thermodynamics irreversible transform mechanical energy into other forms through dissipative processes. Therefore, entropy production is believed to be a propitious measure for a systematic study of wear and friction" [15].

All these investigations pointed out that the application of irreversible thermodynamics is a promising tool to analyse the tribological processes.

5.2.2. Entropy and structural degradation

The aim of this chapter is a description of relation between energetic situation of the tribo-system and the degradation of grease structure. Tribo-sytems as solid surface 1, solid surface 2 and lubricant are investigated but also subsystems as lubricant layer 1 against lubricant layer 2.

Source of irreversible processes are thermodynamic forces $X_i (i=1, 2...)$ (gradient of temperature, gradient of concentration...). These forces Xi evoke corresponding flows $I_i (i=1, 2, ...)$ (heat flow, diffusion flow...) The generalisation of classical thermodynamic to describe irreversible effects leads to an investigation of local equilibrium. That means there exist macroscopic small system areas that are provided in equilibrium while the whole system is out of equilibrium. Two principles of thermodynamic are used: the linear dependence of flows and thermodynamic forces, and the *Onsager-reciprocity* [20].

For the entropy generation can be written

$$\frac{dS}{dt} = \sum_i I_i \cdot X_i$$

The variation of entropy is influenced by two terms

$$dS = dS_{out} + dS_{in} \tag{10}$$

Heat transfer across the boundary of the modelled system (subsystem) delivers the entropy dS_{out}. Entropy related to mechanisms taking place inside the system is described with dS_{in} (entropy production). The intention is to relate the entropy production inside the system with different irreversible effects caused by the friction process within the grease layer. It can be written

$$\rho \cdot \frac{dS}{dt} + div\sigma = \Theta \tag{11}$$

with σ - entropy flow density ($\sigma = \Im / T$; $\Im$ =heat flow density) and Θ - local entropy increase per unit time ($\Theta = -(\Im, gradT)/T^2$) [20]

Written in differential form and related to time

$$\frac{dS}{dt} = \frac{dS_{in}}{dt} - \frac{dQ_{1-2}}{dt} + s_e \cdot \frac{dm_e}{dt} - s_a \cdot \frac{dm_a}{dt} \tag{12}$$

with $s_e \cdot \dfrac{dm_e}{dt}$ - entropy transported into the system with mass transport; $s_a \cdot \dfrac{dm_a}{dt}$ - entropy transported out of the system by mass loss. S_{in} describes the entropy production inside the system and $(\pm)S_{Q1-2}$ leads to a change of system entropy by heat transfer across the system boundaries.

Eq. (12) describes the entropy balance for an open thermodynamic system (with exchange of mass). As Abdel-Aal [18] pointed out only the entropy source strength, namely entropy created in the system, should be used as a basis for systematic description of the irreversible process (degradation of materials).

The process of structural degradation can be described as a process of energy accumulation, energy dissipation and transition of critical energy levels. Each of these mechanisms delivers a contribution to an entropy balance and will change the production term in Eq.(10) and (12). Entropy production (for the observed volume element) is determined by fluid friction and its effects. The process of solid friction (for a mixed friction contact) delivers only a heat portion into the modelled system. The tribo-subsystem can be illustrated with Fig.18.

The transport processes modeled from the investigated tribo-system are presented in Fig. 19, 20 and 21.

As a consequence of the modelled mechanisms for the entropy production can be written as

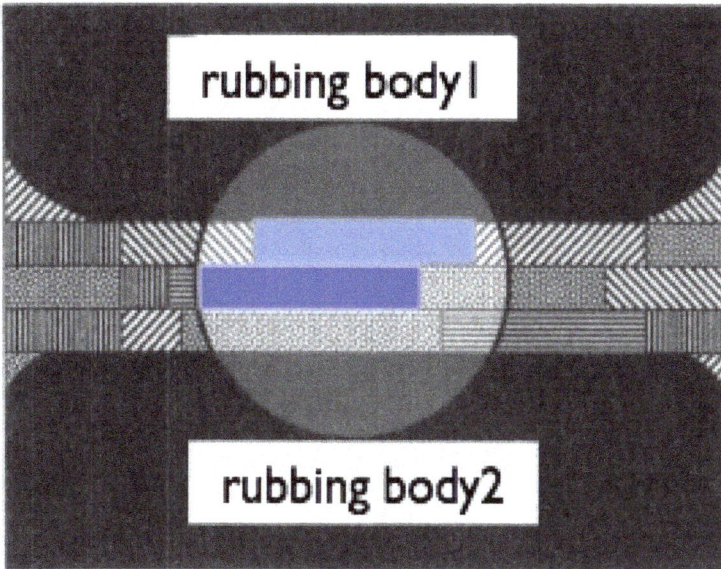

Figure 18. Tribo-subsystem (grease layer against grease layer) inside the general tribo-system

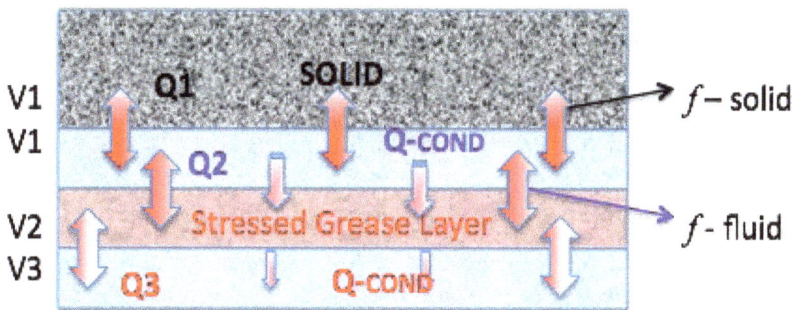

Figure 19. Observed contact situation (section of a complete tribological system). Solid rubbing body and some modelled grease layers. Heat transport: Q1 - heat amount from solid friction transported into the solid material and into the grease layer; Q2 - heat amount from liquid friction flows into layer 1 and 2; Q3 - same as Q2; Q_{COND} presents the heat flow by conduction from layer to layer; V - velocities of the observed layers

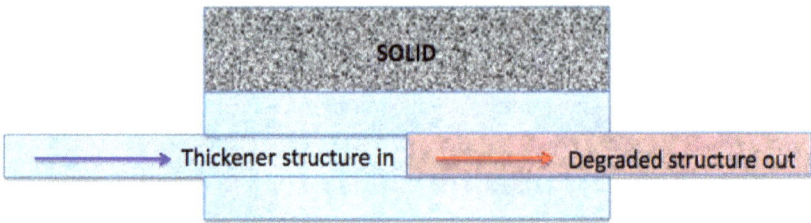

Figure 20. Imagine of transport of thickener structure in and out of the observed volume element (mass transport)

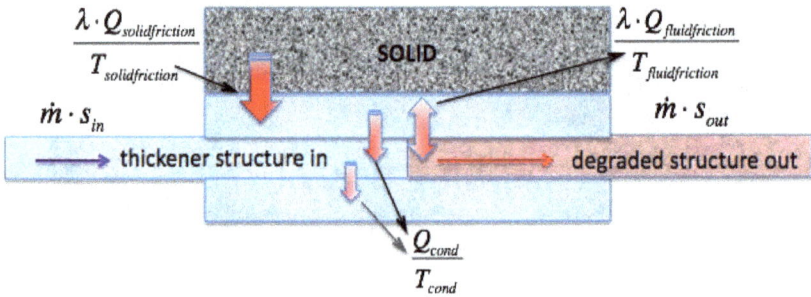

Figure 21. Modelled entropy transport processes inside the observed system (entropy flow) [22]

As a consequence of the modelled mechanisms for the entropy production can be

Written with

$$\frac{dS_{in}}{dt} = \frac{dS_{acc}}{dt} + \frac{dS_{diss}}{dt} + \frac{dS_{trans}}{dt} \tag{13}$$

It means the process of energy accumulation the process of energy dissipation and the transition of an critical energy level produce entropy. Any chemical potential is disregarded in this investigation. We may rewrite Eq.(12)

$$\frac{dS}{dt} = \left(\frac{dS_{acc}}{dt} + \frac{dS_{diss}}{dt} + \frac{dS_{trans}}{dt} \right) - \frac{dQ_{1-2}}{dt} + s_e \cdot \frac{dm_e}{dt} - s_a \cdot \frac{dm_a}{dt} \tag{14}$$

An assumption is made that heat flow gets a (−) that means for a balance that heat leaves the observed volume element.

It can be proposed

$$S_{acc} = \frac{e_{def} \cdot \varsigma_R \cdot V_{acc}}{T_{acc}} \tag{15}$$

with e_{def} the energy density used for deformation process, ς_R describes the part of friction energy which is accumulated, V_{acc} the accumulation volume and T_{acc} the temperature of accumulation process.

$$S_{trans} = \frac{G' \cdot \gamma_{critic}^2}{\cos \delta} \tag{16}$$

from oscillating rheometer measurements (see Eq.(7)). The temperature situation is assumed as $T_{solid} > T_{layer1} > T_{layer2}$. Furthermore it is assumed that different temperature appears for different mechanism. It means

$T_{acc} \neq T_{diss} \neq T_{trans}$.

In general it is conceivable that part of heat generated by solid friction (asperity

deformation) enters the first modelled grease layer. Part of this thermal load will be conducted into the next lubricant layer. Liquid friction between the modelled layers leads to a transport of heat into the layers.

An interesting description of the entropy production term for different processes at sliding interfaces comes from [21].

To link the energetic situation with the structural degradation the entropy production term has to be simplified. The friction energy W_f is observed with the temperature T_f. It can be obtained

$$\overset{*}{e}_{Rrheo} = T_f \cdot (\rho_a \cdot s_a) - \frac{T_f}{V_a}(S_e - S_{Q1-2}) \tag{17}$$

An interpretation delivers $(\rho_a \cdot s_a)$ as an entropy density leaving the system with the mass exchange. It means an increasing energetic release by entropy flow out of the system with the degraded structure leads to an increasing capacity to withstand stresses expressed by $\overset{*}{e}_{Rheo}$.

To use some experimental results some assumption were made. The specific entropy can be determined by

$$s = c \cdot \ln\left(\frac{T_{str}}{T_{unstr}}\right) \tag{18}$$

with index str for the stressed layer and $unstr$ for the unstressed layer. Test temperature in the rheometer was used for T_{unstr}. An increasing temperature with an increasing soap content during the friction process was assumed because all measurements show a direct correlation between soap content and fluid friction (see Fig. 10). To link the specific entropy leaving the system with the degradation process the parameter e^*_{Rheo} for the crossing point (rheometer tests) is used.

Fig. 22 presents the tendency of e^*_{Rheo} vs. s.

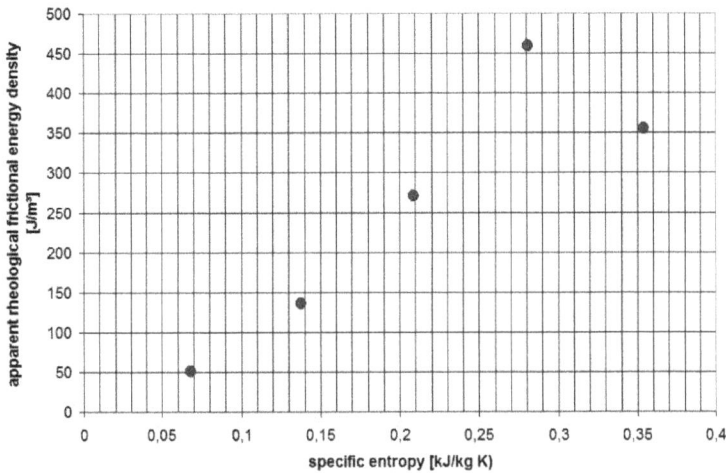

Figure 22. Correlation between e^*_{Rheo} (the crossing point) and specific entropy

Fig. 22 presents the assumed correlation between energetic stress and energetic release.

Finally some conclusions were made

$$\frac{dS}{dt} > \dot{S}_{prod} + \dot{S}_{Q1-2} + \dot{m} \cdot s_e \tag{19}$$

- A significant energetic release by the transport of degraded structure out of the system can be observed.

$$\frac{dS}{dt} = \dot{S}_{prod} + \dot{S}_{Q1-2} + \dot{m} \cdot s_e \tag{20}$$

- No significant energetic release by the transport of degraded structure out of the system can be observed.

$$\frac{dS}{dt} < \dot{S}_{prod} + \dot{S}_{Q1-2} + \dot{m} \cdot s_e$$

- An additional entropy source happens (for example the mentioned friction wave).

6. Conclusions

Some new definitions of general tribological subjects are made. A proposal for an energy balance of a grease lubricated contact is given and empirical proposals for quantification are presented. With the help of rheometer tests the fluid friction was investigated and results are illustrated. The degradation process of the grease structure is described with energetic parameters. An open thermodynamic system is created and described. The influence of the energy flow on the degradation process is presented too.

Author details

Erik Kuhn

Address all correspondence to: erik.kuhn@haw-hamburg.de

Department of Mechanical Engineering and Production, Hamburg University of Applied Sciences, Germany

References

[1] Kuhn,E.: Tribology of lubricating greases. An energetical approach of the tribological process (in German). expert verlag, 2009

[2] Fleischer,G.: Criterions of mixed friction (in German). Tribologie und Schmierung-stechnik, Berlin Technik Verlag 1985

[3] Kuhn,E.: Investigation of the Structural Degradation of Lubricating Greases due to Tribological Stress. Intern. Colloquium Tribology Esslingen, 2012.

[4] Fleischer,G.: Scientific findingy by Tross from a contemprorary point of view. (in German) 3rd Arnold Tross Colloquium, Hamburg 2007

[5] Kuhn,E.: Inherent tribo-system response to optimise the process conditions. 8th Arnold Tross Colloquium, Hamburg 2012

[6] Kuhn,E.: Tribological stress and structural behaviour of lubricating greases. ECO-TRIB 7.-9.6.2011 Vienna

[7] Kuhn,E.: Energetics of the time dependent flow behaviour of greases. Applied Rheology June 1997, p.118-122

[8] Czarny,R.: Lubricating greases. *WNT Publisher,* Warsaw, 2004 (in Polish)

[9] Delgado,M.A.: Manufacture and flow process of lubricating greases. PhD thesis.*University of Huelva,* Spain, 2005

[10] Åström, H.: Grease in elastohydrodynamic lubrication. PhD thesis, *Lulea University,* 1993

[11] Franco, J.M. ; Delgado,M.A. and Valencia,C.: Combined oxidative-shear resistance of castor oil-based lubricating greases. *3rd Arnold Tross Colloquium. Hamburg 2007.* Proc. pp.18-59

[12] Dierich,P.: Modelling the influence of roughnes on the wear prediction. Habil-thesis. (in German), Zittai 1986

[13] Ling,F.F.; Bryant,M.D. ang Doelling,K.L.: On irreversible thermodynamics for wear prediction. Wear 253(2002)1165-1172

[14] Buldum,A.; Ciraci,S.: Atomic-scale study of dry sliding friction. Phys. Rev. B55(4) (1997) 2606-2611.

[15] Aghdam,A.B., Khonsari,M.M.: On the relation between wear and entropy in dry sliding contact. Wear 270(2011) 781-790

[16] Doelling, K.L.; Ling, L.L.; Bryant, M.D. and Heilman, B.P.: An experimental study of the correlation between wear and entropy flow in machinery components. *Journal of Applies Physics* . 88(2000)5

[17] Abdel-Aal, H.A.: On the interdependence between kinetics of friction- released thermal energy and the transition in wear mechanisms during sliding of metallic pairs. *Wear* 253, (2002), pp. 11-12

[18] Abdel-Aal, H.A.: Wear and irreversible entropy generation in dry sliding.*Annals Dunarea de Jos of Galati, Fascicle,*VIII,pp.34-44

[19] Basarov, I.P.: Thermodynamic. *D.V.W.,* Berlin 1964 (in German)

[20] Bryant, M.D.: Entropy and dissipative processes of friction and wear. SERBIATRIB 09, *11th Intern. Conf. on Tribology* 13.5.-15.5.2009, pp.3-8.

[21] Kuhn,E.: Experimental investigations of the structural degradation of lubricating greases. GfT-conference Göttingen 2012

Sustainability of Tribosystems

Chapter 10

A Sensor System for Online Oil Condition Monitoring of Operating Components

Manfred R. Mauntz, Jürgen Gegner,
Ulrich Kuipers and Stefan Klingau

Additional information is available at the end of the chapter

1. Introduction

A web-based oil diagnosis system for continuous online lubricant condition monitoring is presented. The new approach utilizes sensor detection of chemical aging of engineering oils and their additives or first traces of wear debris by precision measurement of the electrical properties. The basic concept and physical background are introduced in detail.

The application potential of the sensor system is discussed on the example of the early identification of critical operating conditions for premature white etching cracks failures of rolling bearings in industrial gearboxes. Causative vibration loading is revealed prior to any component damage. Large roller bearings in wind energy gearboxes unusually often fail prematurely, i.e. clearly before the nominal L_{10} life. The failure is characterized by axial raceway cracks, from which branching and spreading crack systems, partly decorated by white etching microstructure, develop into the depth by corrosion fatigue. High localized friction coefficients, resulting from the specific vibration caused mixed friction operating conditions, initiate tensile stress induced cleavage-like brittle spontaneous surface cracking. The basic idea of the novel failure detection condition monitoring system is the early identification of chemical aging of the lubricant and its additives under the influence of vibration loading.

The sensor effectively controls the proper operation conditions of, e.g., bearings and gears in gearboxes. The online diagnostics system measures components of the specific complex impedance of oils. For instance, metal abrasion due to wear debris, broken oil molecules, forming acids or oil soaps, result in an increase of the electrical conductivity, which directly correlates with the degree of contamination of the oil. For additivated lubricants, the stage of degradation of the additives can also be derived from changes in the dielectric constant. The determination of the reduction in the oil quality by contaminations and the quasi continuous evaluation of wear and chemical aging follow the holistic approach of a real-time monitoring of an alteration in the condition of the oil-machine system. The measuring signals can be transmitted online to a web-based monitoring system via LAN, WLAN or serial interfaces of the sensor. Control of the relevant damage mechanisms, e.g. tribiological wear or oil aging, during proper operation below certain tolerance limits then allows preventive, condition-oriented maintenance to be carried out, if necessary, long before regular overhauling. Outage durations are reduced and the life of components and machines is increased.

2. Basic sensor concept and physical principles

2.1. Basic sensor concept

The basic sensor concept of the novel engineering oil monitoring system is based on the measurement of complex oil impedance components X, particularly the specific electrical conductivity κ and the relative permittivity ε_r. Due to their temperature dependence, the oil temperature T is also recorded [1-3]. Two or more electrodes, between which the oil flows, serve as a basic sensor. Resistance and capacity are measured independently of each other. Zero-mean periodic quantities are used to prevent polarization effects. Figure 1 shows the sensor in its triple plate design.

Figure 1. Sensor in triple plate design.

Oils are electrical non-conductors. The electrical residual conductivity of pure oils lies in the range below 1 pS/m. For comparison, the electrical conductivity of the electrical non-conductor distilled water is larger by six orders of magnitude.

Broken oil molecules, acids, abrasive (metallic) wear, ions, oil soaps, etc., cause an increase of the oil conductivity κ. It rises with increasing ion concentration and mobility. The electrical conductivity of almost all impurities is high compared to the extremely low corresponding property of original pure oils.

The basic sensor represents an electrode arrangement, in which the measured oil is used as electrical conductor and as dielectric material for conductivity and relative permittivit

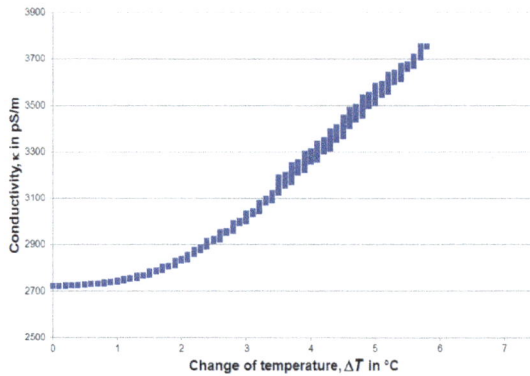

Figure 2. Temperature dependence of the electrical conductivity of sample oil.

measurement, respectively. Oil is an electrical non-conductor. High resistance of the basic sensor and resulting low measurement currents provide best interference sensitivity to interspersed electromagnetic fields. Due to the very small currents, moreover, sufficient interference suppression is achieved. To prevent polarization effects, zero-mean alternating current voltages are measured as test signals. However, no capacitive current components may be measured simultaneously during a conductivity measurement because the capacitive current is much higher than its ohmic components. Thus, rather high requirements are set on analog sensor electronic systems, which are met with the reported measurement procedure.

The conductivities of the insulating construction elements and insulation of electrical feed-through are about the same size as for the pure oils to be analyzed. The developed basic sensors and precise sensor electronic system ensure that the conductivity of feedthroughs and substrates may not be included into the test results. The active basic sensor unit consists of two or several basic sensor plates which are fixed to metal pins of a glass/metal feedthrough in a constant distant from each other. The plates of the basic sensor are arranged in the middle of the measuring chamber, allowing for an adequate incident flow of the flowing medium. A special alignment of the sensor housing parts is thus not necessary in this design. The extension characteristics of the sensor housing materials and the glass/metal feedthrough pins are exactly adjusted to the material characteristics of the used feedthrough glass. The compression strength is above 10 MPa

2.2. Temperature compensation

The ion mobility and thus the electrical conductivity κ depend upon the internal friction of the oil and therefore also on its temperature. The oil conductivity increases with temperature. Figure 2 shows the dependence of the conductivity κ on the temperature change ΔT.

Already for about 3 °C alteration in temperature, the conductivity changes by about 25%. The electrical conductivity κ is a temperature function that depends on oil impurities rather than on the oil itself. The type of pollution and its temperature dependence cannot be assumed to be known. To improve the comparability of measurements, a self-learning adaptive temperature compensation algorithm is implemented. An integral alteration of the oil quality can then be assessed by the temperature compensated conductivity value, whereas the type of contamination is not determinable. The relative permittivity is measured with the same basic sensor arrangement as used for the electrical conductivity.

The electrical conductivity and relative permittivity are to be measured with respect to a reference temperature T_R as close as possible to the operating temperature of the oil. These parameters can be evaluated by means of temperature-dependent approximating polynomials, as demonstrated below exemplarily for the electrical conductivity:

$$\kappa_R = \kappa_{R,0} + \left(a\Delta T_C + b\Delta T_C^2 + c\Delta T_C^3\right) \times \kappa_M \tag{1}$$

Here, κ_R and $\kappa_{R,0}$ denote the approximate and previously calculated (old) electrical conductivity of the oil at the reference temperature T_R, respectively. T_C stands for the current temperature of the oil and κ_M is the electrical conductivity measured without temperature compensation. Moreover, a, b, und c are the coefficients of the approximating polynomial to be adaptively determined. The temperature difference is defined as follows:

$$\Delta T = T_R - T_C \tag{2}$$

The oil temperature T_C is measured for this temperature compensation. The use of a polynomial of the third order in Eq. (1) ensures good approximation while keeping the computational effort for the applied microcomputer reasonably low. Figure 3 shows the measured values of the electrical conductivity κ after temperature compensation.

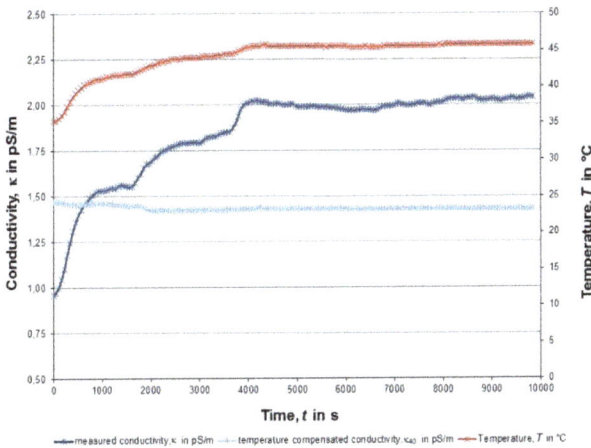

Figure 3. Measured conductivity values after temperature compensation.

2.3. Calculation and linear approximation of relative permittivity and conductivity

In a series of experiments on the non-additivated lubricating oil FVA03, fresh demineralized water was added to a volume of 3.01%. The oil conductivity data measured as a function of the water content are found to follow a linear relationship in good approximation. The theoretical course of the relative permittivity is calculated for dilute solutions according to different mixing rules by truncating a Taylor series expansion of the model equations after the linear term. The model of Lichtenecker is evaluated in Figure 4. Lichtenecker developed the formula of Eq. (3) for calculating the dielectric constant of a homogeneous mixture ε_r [4]:

$$\varepsilon_r = \varepsilon_{r,add}^{f} \times \varepsilon_{r,oil}^{1-f} \tag{3}$$

The permittivity of the addition and the oil, respectively, is denoted $\varepsilon_{r,add}$ and $\varepsilon_{r,oil}$. With the volume fraction f of the addition, $1-f$ becomes the volume fraction of the oil.

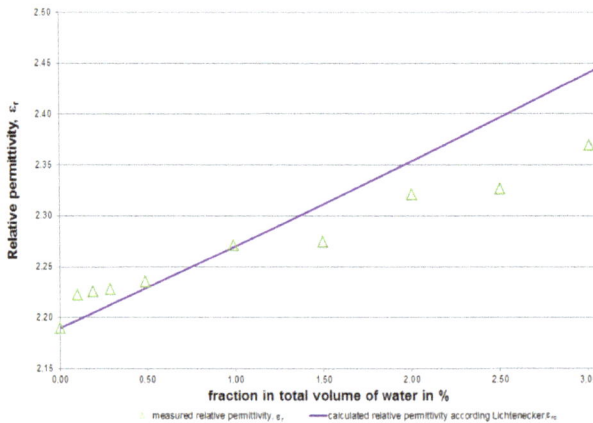

Figure 4. Electrical permittivity ε_r measured as a function of the water content and model fit according to Eq. (3).

3. Premature failures of rolling bearings and correlation with oil aging

Bearings in industrial, e.g. wind turbine, gearboxes unusually often suffer from a significantly shorter life than calculated by white etching cracks [5, 6]. Figure 5 shows the light-optical micrograph of a typical metallographic microsection [5]. The overrolling direction from left to right indicates surface initiation and top-down propagation of the extended crack system.

These early failures are characterized by mostly axial raceway cracks, revealing vertical semi to fully circular cleavage-like lenticular brittle spontaneous incipient cracks in preparatively opened original fracture faces [5, 6]. Occasionally, pock-like spallings are associated with the

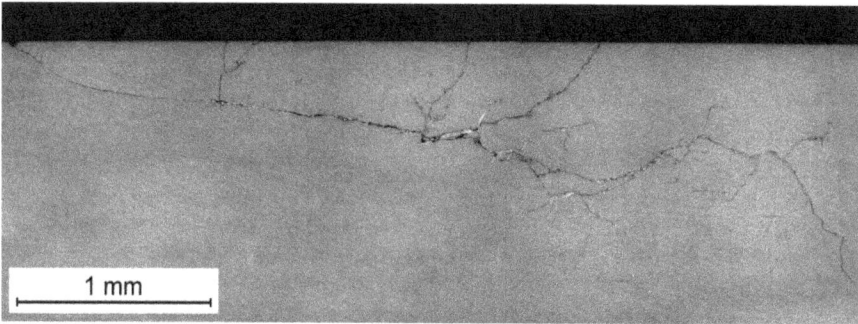

Figure 5. Radial microsection of a branching and spreading white etching crack system.

surface cracks, as shown exemplarily in Figure 6 [6]. The developing deep crack systems are open to the raceway, from which oil penetrates and promotes further corrosion fatigue crack growth as well as local secondary microstructural changes in the form of crack path decorating white etching constituents. It is evident from fractography and X-ray diffraction (XRD) residual stress analyses that the cleavage-like incipient cracks are caused by frictional tangential tensile stresses [5, 6], which occur in subregions of the contact area in specific, vibrationally induced mixed friction operating conditions [5–8].

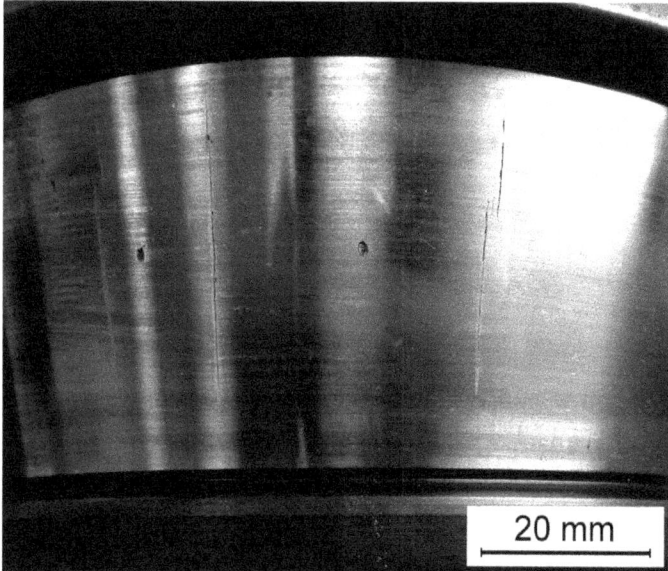

Figure 6. Inner ring raceway with typical axial cracks and few associated pock-like spallings.

XRD based material response analyses of run rolling bearings, suffering from white etching cracks on still largely undamaged raceways, reveal the causative vibration loading [5]. It is further reported that lubricant aging occurs under the influence of vibrations [7, 8]. An example of an infrared spectrum of used oil from rig test run of a roller bearing is provided in Figure 7 [7]. The verified O–H and C=O oxidation bands indicate operational acidification of the oil, also reflected in the dissolution of MnS inclusion lines on the raceway (cf. Figure 8), as a result of polycondensation reactions towards resinification and beginning lacquer formation. It is this aging of the lubricating oil and its additives, which can be detected at an early stage by the new sensor so that a gearbox operating at critical conditions is identified.

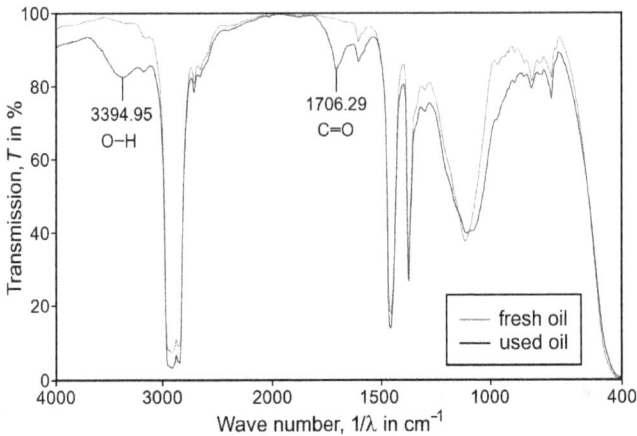

Figure 7. Oxidation peaks in the infrared spectrum of a used non-additivated aliphatic lubricating oil run under vibration loading in a rolling bearing rig test (λ is the wavelength).

The mentioned crack initiation by tribochemical reactions is also found on lateral surfaces of rollers. In Figure 8 [5], a scanning electron microscope (SEM) image, taken in the secondary electron imaging mode, is shown on the left. Residues of manganese and sulphur, detected in the crack-like defect by energy dispersive X-ray spectroscopy (Figure 8, on the right), indicate the causative tribo chemical dissolution of nonmetallic MnS inclusions [5, 7, 8].

4. Trial of the oil sensor system on a bearing and gear test rig

On a bearing and gear test rig, the new sensor based oil quality monitoring system is applied. Various load cycles are run and speeds and torques are measured. The results of the trial are described, evaluated and discussed in the following sections.

4.1. Loss of power and trial run characteristics

The speed-related power $P(n)$ of the test rig is given as follows:

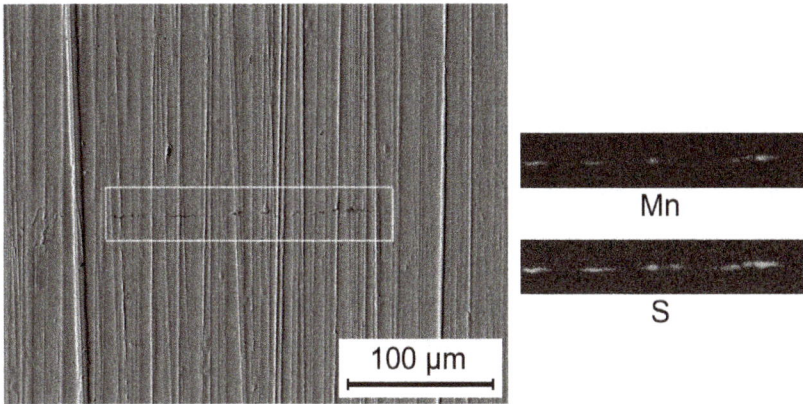

Figure 8. SEM image of a crack on a bearing roller with elemental mapping of Mn and S.

$$P = M \cdot \omega \ with \ \omega = 2\pi \cdot n \tag{4}$$

Here, M denotes the torque, ω and n respectively stand for the angular velocity and rotational speed. The implemented power loss ΔP is derived from the transmission ratio $N_{\ddot{U}}$:

$$\Delta P = \omega_1 \cdot \left(M_1 - N_{\ddot{U}} \cdot M_2 \right) \tag{5}$$

M_1 and M_2 indicate the torque of the drive and the load, respectively. In the first trial on the test rig, the rolling bearing is intentionally damaged. The time-dependent power loss in the gear, as derived from the measuring signal characteristics, is represented in Figure 9.

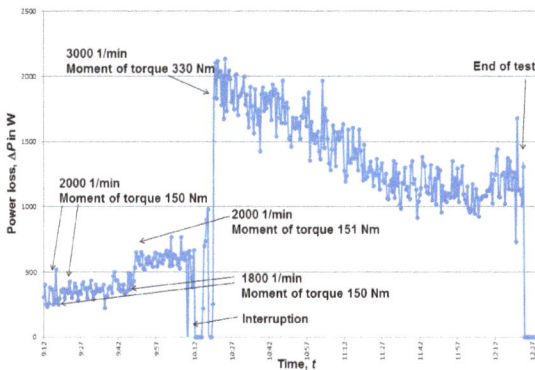

Figure 9. Calculated power loss ΔP.

After switching to a higher load, the power loss increases abruptly before the bearings run in. Towards the end of the trial, the bearings reveal indication of advanced deterioration. A wide lubrication gap and vibrations when overrolling spalling results in higher oscillation amplitudes, which leads to the automatic shutdown of the test rig eventually. The measuring results obtained with the oil sensor system are presented in the following.

4.2. Conductivity of the lubricating oil

Figure 10 shows the test readings of the conductivity measurement of the lubricating oil. The current bearing wear and the deteriorating oil condition in the conducted trial are reflected in the change of the electrical conductivity plotted vs. running time in the diagram.

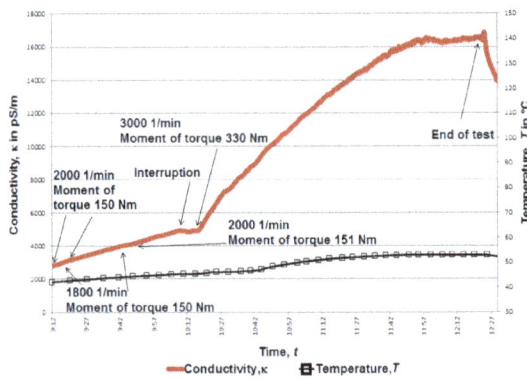

Figure 10. Measurement of the electrical conductivity κ vs. running time t.

New oil from the storage container exhibits a conductivity κ of 2312 pS/m. After filling into the trial gear and before the start-up of the test rig, a conductivity of 2791 pS/m is measured. This increase can be attributed to existing residual impurities in the gear. During the trial run, the conductivity κ of the gear oil increases to 16868 pS/m. Besides changes in temperature, conductivity increase is caused, e.g., by wear debris and removed material from spalling, impurities, broken oil molecules or forming oil soap. As described above, the temperature dependence of the electrical conductivity of the used gear oil is compensated and the oil conductivity measured in the gear trial is converted into the relevant conductivity value at 40 °C. Figure 11 shows the development of the temperature compensated oil conductivity with running time during the gear trial.

In the case of an initially low load, the electrical conductivity increases linearly with running time. It is to be assumed that the low bearing wear in this area also increases proportional to the time.

During the necessary intermediate shutdown (interruption) and run-up of the drive machine to 330 Nm, the conductivity is virtually constant. After switching over to the higher load, the

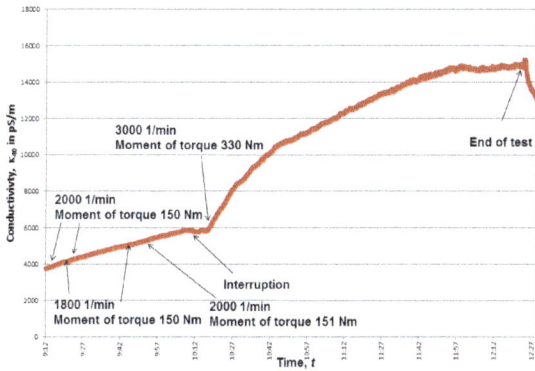

Figure 11. Time curve of the temperature compensated oil conductivity derived from Figure 10.

oil conductivity increases strongly. Here, the bearing run-in (shakedown) is shown as reduction in the conductivity increase. More than about 30 minutes prior to the final forced shutdown of the trial run by an oscillation sensor, the conductivity remains almost constant followed by a temporary rise directly before disconnection. After switching off the test gear, the oil conductivity decreases strongly. This clearly emphasizes the influence of the additives. During the loading stages, more impurities per time unit are produced than bound to additives. After shutting down the test rig, no further oil contamination occurs while the effect of additives still continues.

The variations in electrical conductivity are depicted in Figure 12. In this diagram, the curve follows averages respectively calculated over 3 minutes.

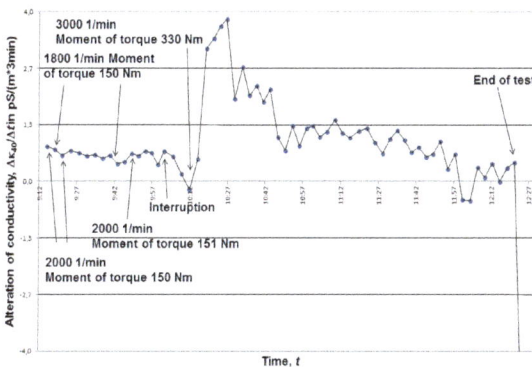

Figure 12. Alteration of the electrical conductivity, expressed as $\Delta \kappa_{40}/\Delta t$, vs. running time t.

When starting up at 2000 revolutions per minute and a torque of 150 Nm, a relatively constant change in conductivity from 0.6 to 0.8 pS/(m×3 min), equivalent to 3.3 to 4.4 fS/(m×s), occurs. In the case of higher load (330 Nm, 3000 min^{-1}), the change in conductivity rises up to 3.8 pS/(m×3 min), i.e. 21.1 fS/(m×s). After the intermediate load increase, the effect on the change of the oil conductivity appears stronger. This may be attributed to the time-dependent formation of impurities and changes in bearing stressing as can be expected during the development of spalling. Figure 13 shows the inner ring of the failed planet bearing with massive damage of the raceway at the end of the trial.

Figure 13. Heavily spalled inner ring raceway of the tested cylindrical roller bearing.

The connection between the change in conductivity κ and the loss of power ΔP in the gear is also evaluated. Figure 14 represents this progress graphically. Both the increasing change in oil conductivity and the gear power loss correlate with the bearing wear.

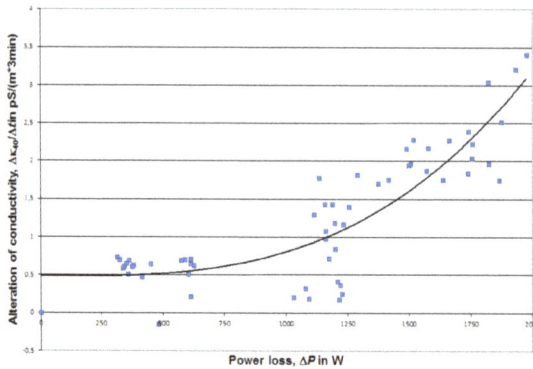

Figure 14. Alteration of the electrical conductivity as function of power loss of the gearbox.

In the diagram of Figure 14, a trend line is drawn as polynomial of the third order. The higher the increases in conductivity and loss of power in the gear, the stronger the bearing wear occurs. Exceptional changes to the system, e.g. switching of the load conditions, are not taken into account.

4.3. Relative permittivity of the lubricating oil

In addition to the electrical conductivity, the relative permittivity ε_r of the oil is measured. In the case of oils not enriched with additives, the water content can be determined that way. There are good prospects that the dwindling efficacy of the additives can be detected by means of the dielectric constant measurement. Figure 15 shows the time development of the relative permittivity during the trial run. Due to the dependence on temperature, this development is also depicted in the diagram.

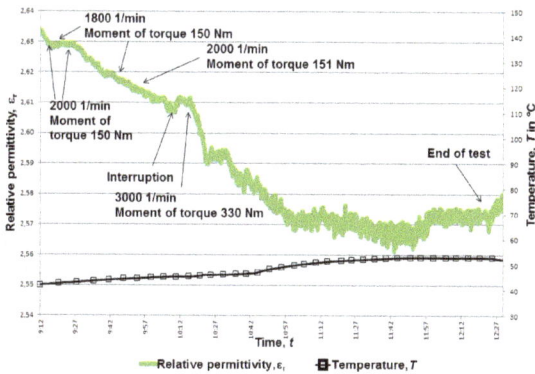

Figure 15. Time curves of the relative permittivity ε_r and the temperature T as a function of the running time t.

The change of the relative permittivity could be caused by a combination of the effects of a chemical reaction of additives, water evaporation from the oil and the temperature dependence of the cell constants as well as the relative permittivity itself. During the trial run, the temperature increases from 42 to 52.9 °C. The temperature dependence justifies the developed adaptive, self-learning temperature compensation technique. Figure 16 shows the temperature compensated time development of the relative permittivity during the gear trial.

5. Approach for condition monitoring of additivated lubricating oils

A direct connection between the electrical conductivity and the degree of contamination of oils is found. An increase of the electrical conductivity of the oil in operation can thus be interpreted as increasing wear or contamination of the lubricant. The aging of the oil is also evident in the degradation of additives. The used additives reveal high conductivity compared with the oil.

The consumption of the additives is reflected in a reduction of the electrical conductivity and permittivity of the oil. The gradient, i.e. the time derivative, of the conductivity or the dielectric constant progression respectively represents a measure of the additive degradation and consumption. The full additive degradation is indicated by the slope of zero (bathtub curve).

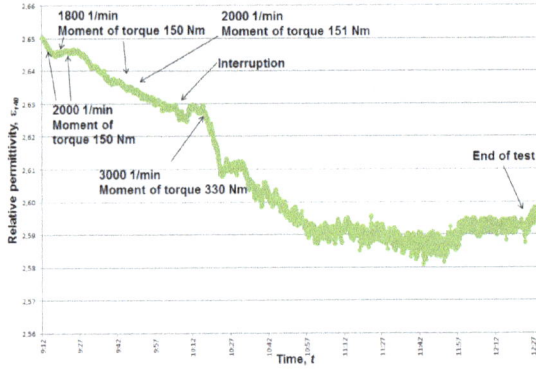

Figure 16. Time curve of the temperature compensated relative permittivity ε_r.

Then the measurement signal increases further with increasing pollution, water entry, etc. Figure 17 schematically shows the temperature compensated time curve of the permittivity of additivated oil continuously contaminated by the addition of wear debris, water or oil acids from chemical aging. Once the additives are consumed, the vanishing shielding effect results in a characteristic re-increase.

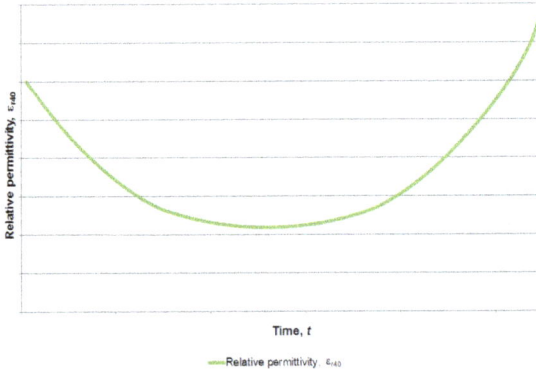

Figure 17. Temperature compensated permittivity.

The most commercially available particle counters only detect particles as small as 4 μm. In a very early stage of wear of bearings, gears, hydraulic cylinders, etc., however, particularly smaller particles are produced. A preventive maintenance lowing, rather than rigid inspection intervals, therefore requires recognition of even the smallest particles. These particles are far more common in the oils of functioning machines than larger ones. Oil aging can be involved in the failure, for instance, of rolling bearings [7].

6. Web-based decentralized lubricant quality monitoring system

The integration into a suitable communication structure and the realization of an online monitoring system offers an interesting practice-oriented utilization of the oil sensor system. This is briefly discussed below.

Preferred areas of application of the sensor system are energy production and automated technical plants that are operated locally, like e.g. wind turbines, generators, hydraulic systems or gearboxes. Plant employers are interested in continuous automated in vivo examination of the oil quality rather than interrupting the operation for regular sampling. Online oil status monitoring significantly improves the economic and ecological efficiency by increasing operating safety, reducing down times or adjusting oil change intervals to actual requirements. Once the oil condition monitoring sensors are installed on the plants, the measuring data can be displayed and evaluated elsewhere. A flexible decentralized monitoring system also enables the analysis of measuring signals and monitoring of the plants by external providers. A user-orientated service ensuring the quantitative evaluation of changes in the oil-machine system, including the recommendation of resulting preventive maintenance measures, relieves plant operators, increases reliability and saves costs.

In a web-based decentralized online oil condition monitoring system, the sensor signals are preferably transferred through the Internet to a database server and recorded on an HTML page as user interface [8]. Figure 18 shows the displayed measured data.

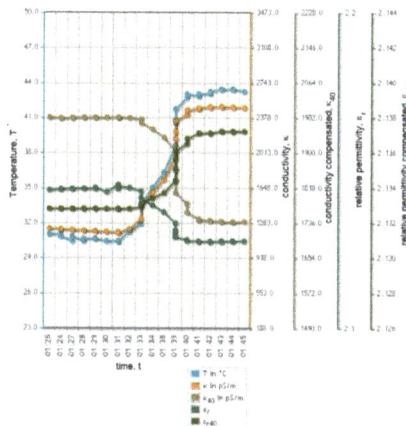

Figure 18. Displayed measured data.

Following authentication, a simple web browser permits access via the wired or wireless LAN. In case of alarm signals, an immediate automated generation of warning messages, for instance by e-mail or SMS, is possible from any computer with Internet connection. Figure 19 shows the new sensor system with communication unit [9].

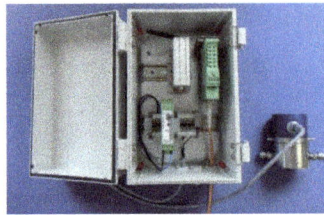

Figure 19. WearSens® sensor system with communication unit.

7. Conclusions

The online diagnostics system measures components of the specific complex impedance of oils. For instance, metal abrasion due to bearing wear at the tribological contact, broken oil molecules, acids or oil soap cause an increase in electrical conductivity that directly correlates with the degree of pollution of the oil. The dielectrical properties of the oils are especially determined by the water content, which, in the case of products that are not enriched with additives, becomes accessible by an additional accurate measurement of the relative permittivity. In the case of oils enriched with additives, statements on the degradation of additives can also be deduced from recorded changes in the relative permittivity.

Indication of damage and wear is measured as an integral factor of, e.g., the degree of pollution, oil aging and acidification, water content and the decomposition state of additives or abrasion of the bearings. It provides informative data on lubricant aging and material loading as well as the wear of the bearings and gears for the online operative monitoring of components of machines. Additional loading, for instance, by vibration induced mixed friction in rolling-sliding contact (rolling bearings, gears, cams, etc.) causes specific faster oil aging, e.g., in the course of premature component failures. Verified in roller bearing vibration rig tests, the oil suffers from significant acidification by polycondensation reactions and incipient resinification, as proven by infrared spectroscopy of used lubricant. The application potential of the sensor is discussed on the example of the prevention of early rolling bearing failures in industrial gearboxes, of which vibrational contact loading is the root cause.

For an efficient machine utilization and targeted damage prevention, the new electrical online oil condition monitoring system offers the prospect to carry out timely preventative maintenance on demand rather than in rigid inspection intervals. The determination of impurities or reduction in the quality of the lubricants and the quasi continuous evaluation of the bearing and gear wear and oil aging meet the holistic approach of a real-time monitoring of a change in the condition of the oil-machine system.

The measuring signals can be transmitted to a web-based condition monitoring system via LAN, WLAN or serial interfaces of the sensor system. The monitoring of the tribological wear mechanisms during proper operation below the tolerance limits of the components then allows preventive, condition-oriented maintenance to be carried out, if necessary, long before regular overhauling, thus reducing outages caused by wear while simultaneously increasing the overall lifetime of the oil-machine system.

On a bearing and gear rig test, various load cycles are run and the functionality of the introduced electric online condition monitoring sensor system is tested successfully. The evaluation of the experiment is presented.

Author details

Manfred R. Mauntz[1*], Jürgen Gegner[2], Ulrich Kuipers[3] and Stefan Klingau[1]

*Address all correspondence to: mrm@cmc-instruments.de

1 cmc Instruments GmbH, Eschborn, Germany

2 University of Siegen, Siegen, Germany

3 South Westphalia University of Applied Sciences, Hagen, Germany

References

[1] Gegner, J., Kuipers, U., Mauntz, M. Ölsensorsystem zur Echtzeit-Zustandsüberwachung von technischen Anlagen und Maschinen, Technisches Messen 77; 2010. pp. 283-292.

[2] Kuipers, U., Mauntz, M. Ölsensorsystem – Sensorsystem zur Messung von Komponenten der komplexen Impedanz elektrisch gering leitender und nichtleitender Fluide, dessen Realisierung und Anwendung, German Patent Application N° 10 2008 047 366.9, Applicant: cmc Instruments GmbH, German Patent Office, Munich, Filing date: 15.09.2008, in German.

[3] Kuipers, U., Mauntz, M. Verfahren, Schaltungsanordnung, Sensor zur Messung physikalischer Größen in Fluiden sowie deren Verwendung, European Patent Application N° EP 09000244, European Patent Office, Munich; 09.01.2009, in German.

[4] Lichtenecker, K., Rother, K. Die Herleitung der logarithmischen Mischungsgesetzes aus allgemeinen Prinzipien der stationären Strömung, Physikalische Zeitschrift, 1931, 32, pp. 255-260.

[5] Gegner, J. Tribological Aspects of Rolling Bearing Failures, In: C.-H. Kuo (ed.), Tribology – Lubricants and Lubrication, Rijeka: InTech; 2011. Chap. 2, pp. 33-94.

[6] Nierlich, W., Gegner, J. Einführung der Normalspannungshypothese für Mischreibung im Wälz-Gleitkontakt, Düsseldorf: VDI Reports 2147, VDI Wissensforum; 2011. pp. 277-290, in German.

[7] Gegner, J., Nierlich, W. Operational Residual Stress Formation in Vibration-Loaded Rolling Contact, Advances in X-ray Analysis, Vol. 52; 2008, pp. 722-731.

[8] Nierlich, W., Gegner, J. Material Response Bearing Testing under Vibration Loading, In: J. M. Beswick (ed.), Advances in Rolling Contact Fatigue Strength Testing and Related Substitute Technologies, STP 1548, ASTM International, West Conshohocken, Pennsylvania, USA, 2012.

[9] Gegner, J., Kuipers U., Mauntz, M. New Electric Online Oil Condition Monitoring Sensor – an Innovation in Early Failure Detection of Industrial Gears, 4th International Multi-Conference on Engineering and Technological Innovation, 19.-22-07.2011, Orlando, Florida, USA 2011, Proceedings Volume I, International Institute of Informatics and Systemics, Winter Garten, Florida, USA, 2011, pp. 238-242.

[10] Gegner, J., Kuipers U., Mauntz, M. High-precision online sensor condition monitoring of industrial oils in service for the early detection of contamination and chemical aging, Sensor+Test Conferences 07.-09.06.2011, Nürnberg, AMA Service GmbH, Wunstorf, 2011, pp. 702-709.

Chapter 11

Innovative "Green" Tribological Solutions for Clean Small Engines

Xana Fernández-Pérez, Amaya Igartua,
Roman Nevshupa, Patricio Zabala, Borja Zabala,
Rolf Luther, Flavia Gili and Claudio Genovesio

Additional information is available at the end of the chapter

1. Introduction

Since its invention in the last quarter of the nineteenth century and during all the twentieth century, two-stroke engines penetrated in many industrial, automotive and handheld applications where engines with high specific power, simple design, light overall weight and low cost are required. Presently, two-stroke engines are commonly used in motorcycles, scooters, chainsaw, agricultural machinery, railways grinding machines, outboard applications, etc.

Usually, the moving parts of a two-stroke motor are lubricated either by using mixture of oil with fuel or by pumping oil from a separate tank. Both designs use total-loss lubrication method, with the oil being burnt in the combustion chamber. Therefore, the lubricating oil must meet specific requirements: it must have an optimal balance of light and heavy oil components to lubricate at high temperature, it must produce no deposits (carbon sooty and other) on moving parts, and it should be ash-less. In addition, the oil should provide good protection of moving parts at high speed under deceleration of engine with the throttle closed, when the engine usually suffers from oil-starvation.

Also, two-stroke engine produce more contaminants than four-stroke engines, due to oil burning in the combustion chamber. Therefore, it is very important to reduce these contaminations to meet ecological requirements.

Most challenging issue of the European technological strategy resides in complete substitution of fossil-based fuels and lubricating oils with renewable eco-friendly and high performance materials. Esters and polyglycols were identified as alternative base oils because of their high biodegradability, low toxicity; low ash formation and absence of polymer components, in [1]. Synthetic esters are characterized by their polar structure, high wear resistance, good viscosity-temperature behaviour, miscibility with non-fossil fuels. Esther-base oils can be blended with various components like antifoam agents, oxidation inhibitors, pourpoint depressants, antirust agents, detergents, anti-wear agents, friction reducers, viscosity index improvers, etc., to create environmental friendly prototype engine oils and to meet the changing environmental requirements in low sulphur fuels and other alternative fuels and their application to engine oils.

Low metal additives content and clean-burnt characteristics result in less engine fouling with much reduced ring stick and lower levels of dirt built-up on ring grooves, skirts and under crowns. Owing to the presence of polar ester groups in the molecule which have higher adhesion to metal surface, esters have much better lubricity than hydrocarbons. The performance of the ester-based lubricating oils can be further improved by selecting a proper base oil and additive package.

Another important problem is related with performance of fuel injector system when bio fuels are used. Diesel injection nozzles consist of a body (usually in Ni-Cr steel) and needle valve (High speed steel, HSS), fitted together with very strict tolerances. The design of the nozzle, i.e., the number of orifices, their diameters, positions and drilling angles depend on specific engine application. The current trend is to use multi-hole nozzles with very small holes with diameter of only 0.10 - 0.14 mm in order to improve the fuel atomization and flow pattern.

Heat treatments are applied to the body and the needle to obtain the necessary hardness both on the surface and in the core of the parts and to face the following problems:

• fatigue failures at high stress areas due to repeated pulses of very high injection pressures;

• thermal shocks.

Adequate finishing of the orifice surfaces is very important also to optimize the erosion resistance.

The usage of new diesel blends characterized by different physical and chemical properties as compared with the traditional fuels could lead to modifications both in the choice of materials, geometry and positioning of orifices or their surface finishing to ensure the correct spray pattern. This work describes the results of our recent studies aimed at solving the problems related to the introduction of new eco-friendly oils and lubricants.

2. New prototype engine oils

2.1. Oil characterization

Three different synthetic ester base oils have been selected to formulate three prototype engine oils with the same additive composition. These oils are different mixtures of fully saturated polyglicol-ester and mono-ester types and designated as SEMO 4, SEMO 5 and SEMO 10. Same additive package has then been added to the three bases. After comparative characterization of these prototype oils and selection of the oil with the best tribological performance (SEMO 10), a new improved formulation was developed based on the selected lubricating oil, designated SEMO 36. In addition, conventional mineral oil for two-stroke engines was used as reference oil. The additive package composition of the reference oil is different but it is ash-free as well as the other SEMO oils.

Oil viscosity was characterized according to ASTM D-445-06 standard procedure in [3], and viscosity index was determined using ASTM D-2270-04 in [4].

Deposit forming tendency of the oils was characterized by the Coker test at 250 °C during 12 h. Some physical and rheological properties of the lubricating oils are shown in Table 1. Among the prototype lubricating oils, SEMO 10 has the lowest viscosity both at 40 and 100 °C, the highest flash point and the lowest deposit forming tendency.

Unleaded petrol (E228) and bioethanol E85 (mixture of 85% of ethanol with 15% of gasoline) were selected to test miscibility of the lubricating oils with standard and alternative fuels. For this purpose two different lubricant/fuel ratios were used. Regarding to the miscibility method A (90% lubricant in fuel), SEMO 10 as well as SEMO 5 demonstrated good miscibility both with unleaded petrol and E85. Compared to this, the results for the 2% mixtures according (method B) differed. All tested lubricants proved to be perfectly miscible with EN228 fuel, whereas only SEMO 36 demonstrated to be fully miscible with E85. According to both miscibility methods the reference oil was only miscible with EN228. SEMO 36, when compared to its original prototype SEMO 10, has a much higher viscosity. Flash point for this lubricant is lower than for SEMO 10 but still higher than 200 °C.

		Ref. Oil	SEMO 4	SEMO 5	SEMO 10	SEMO 36
Density, g/ml	0.877	0.915	0.917	0.935	0.999	
Viscosity @ 40°C, mm²/s	59.5	84.9	94	45.8	113.3	
Viscosity @ 100°C, mm²/s	8.6	12.5	13.2	8.0	18.3	
Viscosity index	117	144	140	147	181	
Flash point, °C	120	204	190	260	218	
Pour Point, °C	-21	-39	-33	-39	not tested	
Deposit forming *	9	4	3	9	not tested	
Miscibility Method A (90% lubricant in fuel)	EN228	Good	Good	Good	Good	not tested
	E85	Poor	Poor	Good	Good	Good
Miscibility Method B (2% lubricant in fuel)	EN228	Good	Good	Good	Good	Good
	E85	Poor	Poor	Poor	Poor	Good

*Rating on base 10

Table 1. Properties of the engine oils

Wettability of the surface of the cylinder liner by lubricating oil is important for corrosion- and wear-protection of the piston rings and cylinder liner at the start-up when the temperature of the components is low. In this work, the wetting characteristic of the tested oils was determined using the Sessile Drop method. The resulting contact angles of the drops of various oils on the honed surface of the cylinder liner are shown in Table 2. Same method could not be used to determine wettability of the piston ring because of the small width of the ring. Therefore, the following procedure for qualitative comparison of the wettability of the piston ring by different oils in [12] was applied: 1 μl of oil was placed on the circular flat surface of the phosphate cast iron piston ring and then, after 30 s, the extension of the oil drop along this surface was measured.

Oil	Contact angle on honed cast iron, (°)	Spread distance of the oil on the piston ring, (mm)
SEMO 4	46.1±3.1	5.33±0.04
SEMO 5	43.4±0.2	5.39±0.05
SEMO 10	33.1±1.1	7.01±0.12
SEMO 36	50.8±0.5	3.78±0.17

Table 2. Contact angle and oil spread distance

The contact angle for SEMO 36 oil on the honed cast iron was the highest among all the tested lubricating oils. The contact angles of SEMO 5 and SEMO 4 were very similar one to each other and only slightly lower than for SEMO 36. SEMO 10 had the lowest contact angle and the largest drop spread for all tested oils. The behaviour of the drop spread of the tested lubricating oils over the piston ring surface is similar to that of the contact angle, bearing in mind that large contact angle values correspond to small spread distances.

Biodegradability and toxicity of the lubricating oils were examined according to the recommendations of the Organization for Economic Co-operation and Development (OECD) in [5]. Biodegradability of lubricating oils was tested using OECD 301F Manometric Respirometry Method consisting of the measurement of oxygen uptake by a stirred solution of the test substance in a mineral medium, inoculated with micro-organisms in [6]. Toxicity of the lubricating oils was studied using "Alga, Growth Inhibition Test" OECD 201 in [7] and "Daphnia Magna" 24 h Acute Immobilisation Test OECD 202 in [8]. In the "Alga, Growth Inhibition Test", selected green algae were exposed to various concentrations of the test oils over several generations under defined conditions. Results of biodegradability test are shown in Table 3. As expected, all synthetic ester base oils successfully passed the biodegradability test, while the reference mineral oil was not biodegradable according to the standard procedure OECD 301 Biodegradation of SEMO 5 and SEMO 10 exceeded 70%. In toxicity tests both with Alga and Daphnia Magna, the oils were classified as not harmful for aquatic organisms according to the standard procedures OECD 201 and 202 (see Table 4).

Time (days)	SEMO 4	SEMO 5	SEMO 10	REF
7	29.4	41.3	25.2	27.6
14	36.2	68.4	53.5	38.7
21	52.0	79.6	69.6	33.4
28	61.1	81.2	75.7	51.2
Ultimate	> 60%	> 60%	> 60%	< 60%

Table 3. Biodegradability of oils (% of biodegraded oil) in [12].

Oil	EC50/EL50 with Alga, mg/l	Classification OECD 201	EC50/EL50 with Daphnia Magna, mg/l	Classification OECD 202
SEMO 4	>100	not harmful *	>1000	not harmful *
SEMO 5	>100	not harmful *	>1000	not harmful *
SEMO 10	>100	not harmful *	>1000	not harmful *
SEMO 36	-	-	>1000	not harmful *

*With respect to aquatic organisms.

EC50/EL50 is that concentration of test substance which results in a 50% reduction in either growth or growth rate relative to the control.

Table 4. Results of the toxicity tests in [12].

2.2. Tribological evaluation according to DIN 51834-2

Tribological evaluation of lubricating oils was done using ball-on-disk configuration with reciprocating motion according to the standard procedure DIN 51834-2 in [9]. Ball and disk were made of 100Cr6 steel. The ball, 10 mm in diameter, performed reciprocating motion with a stroke of 1 mm and a friction frequency 50 Hz. Normal load was 50 N during short run-in period 45 s and 300 N during the test 60 min. The ball and the disk were immersed in the lubricating oil, which temperature during the test was constant and 50 °C. Friction force was measured as function of time. Friction coefficient was calculated as the ratio of the tangential force to the normal force.

After test completion, diameter of the wear scar on the ball was measured using optical microscope, and, from this data, volume wear of the ball was calculated for each lubricating oils tested.

Evolution of the friction coefficient in friction evaluation tests is shown in Figure 1. Oils with low additive content: SEMO 4, SEMO 5 and SEMO 10 showed an interval of frictional insta-bility after the run-in period. In the instability period, which lasted from 400 up to 800 s, there are some sharp peaks indicating damage of surface and seizure, probably due to micro-welding. The reference lubricating oil had a less pronounced instability period without sharp peaks, while SEMO 36 did not present any instability. Final values of friction coefficient after 60 min and the diameters of the wear scar on the ball are shown in Table 5.

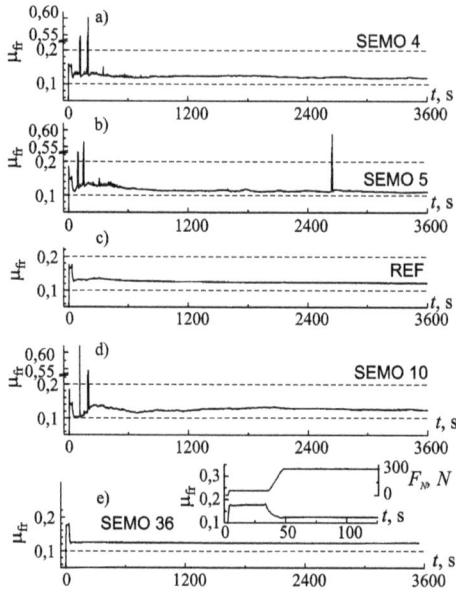

Figure 1. Evolution of friction coefficient in time during tribological evaluation tests of the following oils: a) SEMO 4, b) SEMO 5, c) reference oil, d) SEMO 10, e) SEMO 36. Inset in graph e) shows the initial part of the plot together with the curve of the normal load in [12].

Oil	Final μ_{fr}	Diameter of wear scar, μm	Estimation of worn volume, mm³
SEMO 4	$0.119 \pm 3.40*10^{-5}$	884	$6.01*10^{-3}$
SEMO 5	$0.111 \pm 4.34*10^{-5}$	885	$6.02*10^{-3}$
REF	$0.122 \pm 2.73*10^{-5}$	517	$6.99*10^{-4}$
SEMO 10	$0.125 \pm 3.81*10^{-5}$	929	$7.32*10^{-3}$
SEMO 36	$0.123 \pm 1.54*10^{-5}$	459	$4.36*10^{-4}$

Table 5. Friction coefficient and wear of the ball in tribological evaluation test of oils (DIN 51834-2) in [12].

The volume of worn material of the ball was estimated geometrically on the basis of the diameter of the wear scar using the following equation (1):

$$V = \frac{\pi}{3}\left(R^3 - (R^2 + a^2)\sqrt{R^2 - a^2}\right) \tag{1}$$

where $R = 5$ *mm* is the radius of the ball and a is the radius of the circular wear scar.

Wear specific energy, E_w, that is, the ratio of the dissipated energy, E, during friction per unit mass of worn material Δm, is an important characteristic which shows the ability of a material to resist wearing. This is a complex parameter taking into account both friction, which characterizes energy supply to the material in the friction zone, and wear intensity. This parameter is considered a very useful tool to compare standard tribological evaluation and simulated tests in [10]. Wear specific energy was determined using the following equation in [12]:

$$E_w = \frac{E}{\Delta m} = \frac{v_m F_N \int_{t_i}^{t_f} \mu_{fr}(t)dt}{\Delta m} \tag{2}$$

Where v_m is mean sliding velocity obtained with a reciprocating frequency of 50 Hz and 1 mm stroke, F_N is the normal load, μ_{fr} is the friction coefficient, t_i and t_f are respectively the initial and final time points of friction test time interval.

In this study, only ball wear was determined as specified by DIN 51834-2. So, the absolute value of wear specific energy could not be determined; since wear of the disk was not measured. However, by using the ball mass loss in the denominator of eq. (2), the upper bound estimation of the wear specific energy can be determined. This upper bound can be used for qualitative comparison of anti-wear properties of the lubricating oils under constant friction conditions. These values, determined using eq. (2), are shown in Figure 2. SEMO 36 and the reference oil have much higher values of the wear specific energy, than other oils. Therefore, these lubricating oils improve contacting surfaces wear protection since much larger energy should be dissipated to produce the same wear as compared to SEMO 4, SEMO 5 and SEMO 10 lubricants.

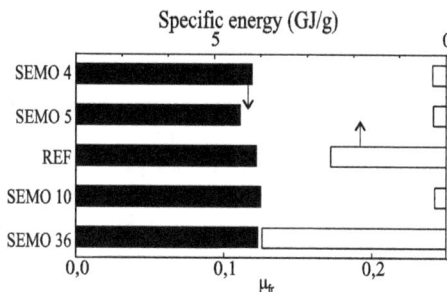

Figure 2. Friction coefficient and wear specific energy in [12].

2.3. Piston ring/cylinder liner simulation

Tribological simulation was performed using cast iron phosphated piston ring and cast iron cylinder liner using reciprocating motion configuration. The samples for the tests were cut from real engine parts (Minsel M165 two-stroke engine manufactured by Abamotor Energía) keeping original curved surfaces and surface finishing. The conformal contact between the piston ring and the cylinder counterpart was reproduced by placing a piston ring on a suitable frame, A, and fixing it by means of a clamp, B (Figure 3). Wear of the components was determined by weighting and geometry measurements.

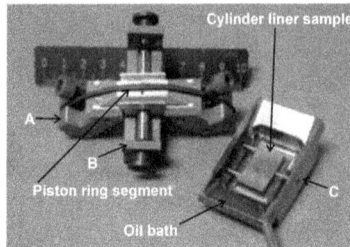

A - frame for fixing the piston ring segment, B – fixing clamp, C – base with oil bath for fixing cylinder liner sample.

Figure 3. Experimental set-up for piston ring/cylinder liner simulation.

The piston ring segments performed a reciprocating motion with a stroke of 1 mm and a friction frequency 40 Hz. Normal load was 50 N during short run-in period 45 s and 300 N during the test 90 min. During the test, the piston ring segment and the cylinder liner sample were immersed in the oil, which temperature was constant at 200 °C.

The mass change of the piston ring segments and cylinder liner sample was determined from weighting the components before and after friction tests. Since the mass change can be due to two competitive processes: (i) wear out and (ii) deposit formation from the oil at elevated temperature, estimation of wear out by weighting can give erroneous results. Indeed, after the tests the surface colour became yellowish and remained after dissolvent cleaning indicating some sparingly soluble deposits formed on the surface due to some chemical reaction. Therefore, in addition to the determination of the mass change, worn volume was calculated from surface geometry. Surface morphology of the friction zone was studied using white light confocal microscopy at three different zones along the wear track on the cylinder liner sample. The acquired 3D surface images were 0.5 mm wide in the direction of friction and each image contained 138 cross-section profiles of the wear track yielding totally 414 profiles for each sample. Firstly, the cross-section profiles were averaged for each sample and then among different samples tested using the same lubricating oil. Worn volume of the samples of cylinder liner was calculated as a product of a mean cross-section area of the groove and the total length of the groove. The cross-section area was determined by numerical integration of the cross-section profiles and then worn mass was calculated from the worn volume using the density of cast iron.

Surface chemical composition of the friction zone of cylinder liner samples was characterized using Energy Dispersion X-Ray Spectroscopy (EDS).

Evolution of friction coefficient in time during friction between piston ring segment and a piece of the cylinder liner is shown in Figure 4. It is possible to highlight the increment of the coefficient of friction μ_{fr} for lubricants SEMO 4 and SEMO 5 overtaking the constant value reached by the lubricant of reference. In fact, for SEMO 4 and SEMO 5, friction coefficient gradually rose during the experiment (90 min) and did not stabilize. The growth behaviour was almost linear in time.

Initial friction coefficient was about 0.2 and the final one about 0.33 in both cases. SEMO 10 and SEMO 36 showed different behaviour. The initial values were 0.2 and 0.14 for SEMO 10 and SEMO 36, correspondingly.

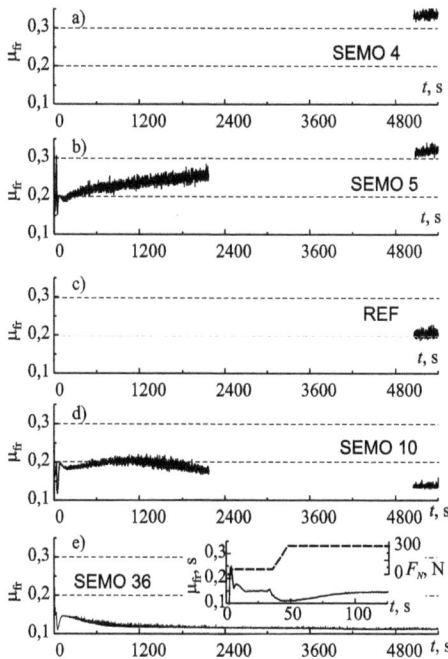

Figure 4. Evolution of friction coefficient in time during piston ring/cylinder liner simulation test. a) SEMO 4, b) SEMO 5, c) reference oil, d) SEMO 10, e) SEMO 36. Inset in the graph e) shows the initial part of the plot together with the curve of the normal load in [12].

At the beginning, after a run-in period, friction coefficient increased and reached maximum. For SEMO 36 the maximum was reached usually between 100 and 200 s from the beginning of the test, while for SEMO 10 the period of increase was longer and the maximum was reached after 700 to 1700 s from the beginning of the test. After reaching the maximum, friction coefficient decreased slowly and stabilized at 0.14 and 0.11 for SEMO 10 and SEMO 36, correspondingly. The friction coefficient of lubricant SEMO 10 showed a slow decline until reaching a constant value lower than the reference one. Friction coefficient for the improved lubricant SEMO 36 levelled out rapidly at a very low value and showed less scatter, probably due to some sort of surface deposition on the contact surfaces.

The averaged cross-section profiles of each liner sample tested are reported in Figure 5. Different scales of magnitude are used for better visualization of the mean contact surface profile. It is possible to notice very good performance of the lubricant SEMO 10 and its improvement in the lubricant SEMO 36. Samples tested using SEMO 4 and SEMO 5 had deep grooves with the maximum depth 22 to 25μm. The samples tested using the reference oil and SEMO 10 had less deep grooves with a maximum depth of 4 to 5μm. Surface of the samples tested using SEMO 36 oil had some thin scratches in the direction of friction while grooves had not been formed.

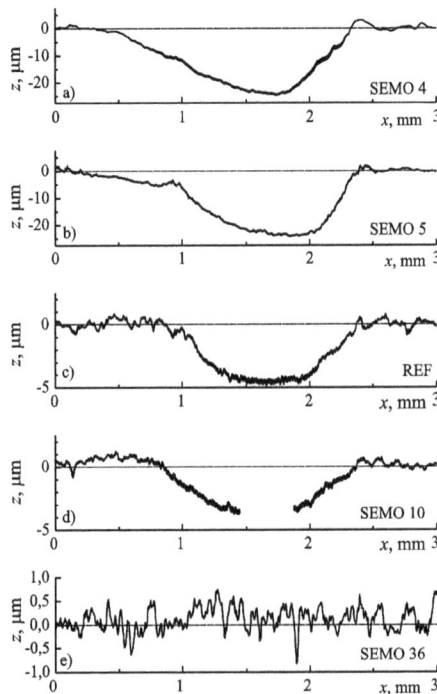

Figure 5. Average cross-section profile of the friction zone of cylinder liner samples tested using different lubricants in [12].

Figure 6 shows images of the friction zone of the piston ring segments after friction simulation tests with different lubricating oils. Wear and damage of the surface as function of the oil used was similar to that in the cylinder liner. In tests with SEMO 4 and SEMO 5, the material in the friction zone was heavily damaged. The wear can be classified to be of the adhesive type with intensive plastic deformation and edging. When the reference oil and SEMO 10 oil were used in the tests, the damage of the material was less pronounced than for SEMO 4 and SEMO 5, but the wear in all cases was of the adhesive type. Only small damage was observed on the piston ring segments when using SEMO 36. In this case, only summits of the circular grooves of the piston ring presented some wear and deformation. From the point of view of hydrodynamic lubrication these results may seem to be surprising, since, with the same additive composition, higher wear rate occurs for thinner oil (SEMO 10 in our case) than more viscous oils (SEMO 4 and SEMO 5). Therefore, these results lead to the following conclusions: 1) the lubrication regime should be of a boundary type and 2) surface protection against wear for SEMO 10 and SEMO 36 oils seems to be resulting from the formation of surface layer as a result of adsorption of oil components or tribochemical reactions between the oil components and the base material.

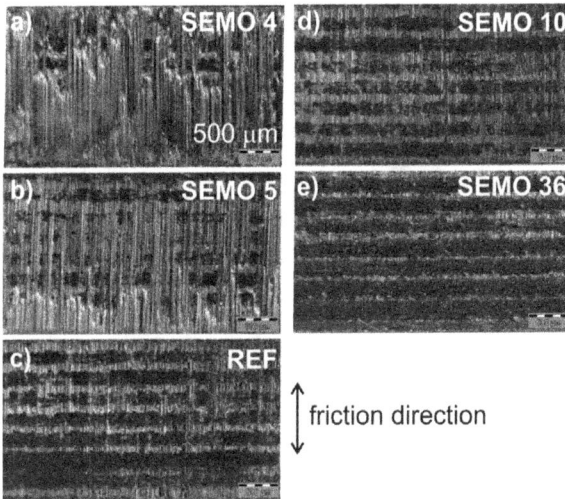

Figure 6. Optical images of the friction zone of the piston ring segments after friction simulation tests using different lubricants. The scale of each image is the same and shown by a scale bar in [12].

Results of the mass change measurements of the components are shown in Table 6. Worn mass calculated from the worn volume is plotted vs. measured mass change in Figure 7 (dots). The experimental data are fitted by linear function with two adjusted parameters: slope and intersect (dashed line in Figure 7). The solid line is a linear fit with a fixed slope 1 and adjusted intersect. Coefficients of determination for these linear regressions are 0.983 and 0.949, correspondingly, indicating statistically significant linear relationship between the mass change and worn mass determined from the geometry of the groove. Therefore, the deposit formation has not much influence on the mass change and the last can be used as a measure of the components wear out in these tests. The upper bound of the wear specific energy was determined in accordance with eq. (2), using the cylinder liner mass change in the denominator of eq. (2).

Oil	Final μ_{fr}	Mass change, mg		Worn volume (cylinder.) mm³
		cylinder	segment	
SEMO 4	0.34	-3.95	-1.41	-1.12*10⁻²
SEMO 5	0.32	-3.68	-3.22	-9.8*10⁻³
REF	0.20	-1.1	-0.64	-1.74*10⁻³
SEMO 10	0.14	-0.94	-1.25	-2.3*10⁻³
SEMO 36	0.11	-0.09	1.23	0

Table 6. Results of friction simulation tests in [12].

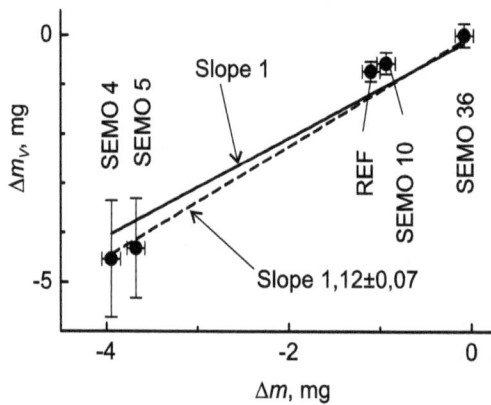

Figure 7. Mass wear determined from the geometry of the groove vs. mass change of the cylinder liner samples. The dashed line is a linear regression of experimental data with two adjusted parameters: slope and intercept. The solid line is a linear regression with a fixed slope 1 in [12].

Final friction coefficient and wear specific energy are shown in Figure 8. SEMO 36 oil showed the best antifriction and wear resistance characteristics among all tested lubricants. Friction coefficient was almost a half of that for the reference oil, while specific wear energy was 7.8 times higher than for the reference oil. In comparison with the ball-on-disk tests, wear specific energy for SEMO 36 lubricant was much lower in the tribological simulation test; however, oil temperature in these two tests was different. When the ball-on-disk evaluation tests were performed at the same temperature as in the simulation test (200 °C), the value of wear specific energy was similar to that in the simulation test: 0.14 GJ/g in the ball-on-disk at 200 °C vs. 0.18 GJ/g in the piston ring/cylinder liner simulation test. Although these values are only upper bound estimations of the real values, they are close to one another. According to the structural-energetic approach in [10], this means that the dominating wear mechanism in both cases is the same. Then, a significant decrease in the wear specific energy from 3.97 to 0.14 GJ/g with temperature increase from 50 to 200 °C implies changing in dominating wear mechanism at higher temperature. It can be stated that, under the applied experimental conditions, the chemical compositions of the base oil and the additives had greater influence on the tribological performance of the lubricants than their rheological properties.

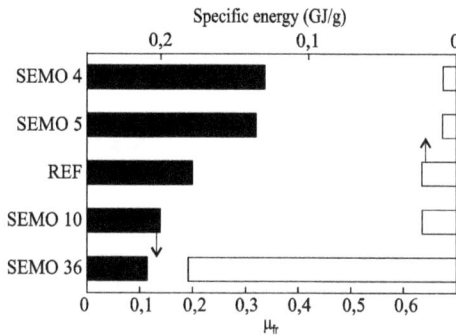

Figure 8. Friction coefficient and wear specific energy in friction simulation tests in [12].

2.4. Surface characterization

Surface chemical composition of the friction zone of cylinder liner samples was characterized using Energy Dispersive X-Ray Spectroscopy (EDS). Table 7 shows surface chemical composition for three different surfaces:

• friction zone of the cylinder tested using SEMO 36 lubricant,

• untouched surface of the same cylinder, and

• reference cylinder not immersed neither heated in lubricating oil.

	Fe	O	C	Si	Mn
Friction zone	61.5±1.9	11.5±0.7	23.5±3.2	2.83±0.68	0.615±0.085
Untouched zone	84.0±3.6	7.36±0.46	3.41±3.31	4.19±0.07	1.11±0.17
Reference sample	92.5±0.3	2.95±0.46	0.1	3.59±0.52	0.935±0.255

Table 7. Surface chemical composition (at.%) of the cylinder liner samples tested with SEMO 36 lubricating oil

Silicon and manganese were alloying elements of the base material and did not show important variations in their concentration, whereas the most important variation was in the carbon and oxygen content. There was no significant difference for other elements since the oils had no metal-containing additives. Figure 9 shows surface concentration of four elements relative to iron. After the test, during which a cylinder was immersed in the SEMO 36 oil and heated at 200 °C, carbon and oxygen concentrations on untouched surfaces were slightly higher than on the reference sample, e.g., the sample not immersed into the oil. However, carbon and oxygen concentrations drastically increased on the surface of the friction zone, on which carbon was each forth atom. Also, in contrast to the untouched surface and the reference surface, on the surface of the friction zone, carbon concentration was higher, than the oxygen one. One can infer from these data that friction induced tribochemical reactions between oil components and base material to form surface layer enriched with carbon and oxygen. This surface layer or sliding lacquer may protect mating surfaces from adhesion and/or damage yielding lower friction and wear in [10].

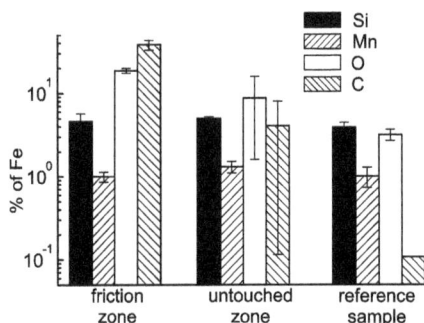

Figure 9. Surface concentration of elements relative to iron in [12].

2.5. Experimental evaluation in real two-stroke engines (Minsel M165)

After previous simulation tribological test the performance of the oils was evaluated in real two-stroke engines (Minsel M165) with a swept volume of 158 cm³, a stroke of 54 mm, compression ratio 7,1:1, power (ISO 1585) 3.53/4.8 kW/HP, maximum torque 120 Nm and 4500 rpm rotation speed. Scuffing tests were performed using various lubricating oil – petrol mixtures in order to evaluate the lubricating performance of the lubricants under extreme load conditions. The test conditions applied are shown in Table 8, and the tested oil-fuel compositions are shown in Table 9.

Figure 10 shows the photographs of the engine components after scuffing tests, in which the reference mineral oil was used in a mixture with pure petrol and bioethanol. Increase in the bioethanol content in the fuel led to decrease in carbon soot deposition on the engine cylinder and piston. Also, when bioethanol was used, the surface was less damaged under extreme working conditions.

Test step	Speed, rpm	Time, min	Power	
			%	HP
1	2000	5	0	0
2	4000	20	50	2.4
3	4000	20	75	3.2
4	2000	5	0	0
5	4500	90	100	full load
6	2000	5	0	0

Table 8. Experimental conditions for scuffing tests of real two-stroke engines

Fuel	Reference oil	New developed oils	
		SEMO 10	SEMO 36
Petrol	2%	2%	2%
E10	2%	-	-
E20	2%	-	-
E85	2%	-	2%

Table 9. Oil – petrol combinations tested in a real two-stroke engine scuffing test

Figure 10. Macro images of two-stroke engine components after scuffing test using a mixture of mineral oil with petrol (a,d), bioethanol E10 (b,e) and bioethanol E20 (c,f) in [12].

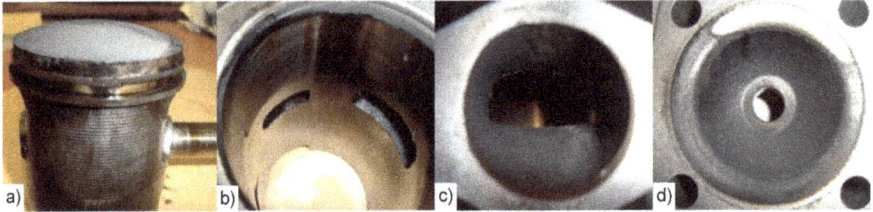

Figure 11. Macro images of two-stroke engine components after scuffing test using mixture of SEMO 10 lubricating oil with petrol: a) piston, b) cylinder, c) exhaust side, d) intake in [12].

Figures 11 show the photographs of the engine components after scuffing tests using a SEMO 10 – petrol mixture. Some seizure between compression piston ring and cylinder was observed when using a mixture of SEMO 10 with petrol. Several vertical abrasion marks were formed in the exhaust zone of the cylinder, where the temperature was higher. However, the piston and cylinder were quite clean with only some carbon soot deposits in the exhaust zone. The state of the cylinder head was quite healthy and clean in the intake zone, the carbon residues were considered normal.

Figure 12 shows the pictures of the engine components after scuffing test using SEMO 36 lubricating oil with petrol and bioethanol fuels. When using a mixture of SEMO 36 with bioethanol E85 or petrol, no scuffing or seizure was observed. Only light scratches were found on the cylinder surface, which were more pronounced when using petrol. In this case, carbon soot deposits formed intensively on the top part of the piston. The piston and cylinder were very clean, when using bioethanol.

In addition, gaseous emissions from the engine were analyzed for various fuel-oil mixtures with different proportions of bioethanol to petrol: 20%, 30% and 85%. The gas emissions were measured using the Directive CE 2002/88, Portable, SH3 modality as reference limits. The differences in power and consumption were negligible when using bioethanol E10 and E20. When compared with the petrol, the NO_x emissions showed an increasing trend and the emissions of CO and CH diminished in tests with bioethanol and reference oil. When using E85, the reference mineral oil was not miscible, but the new developed oil SEMO 36 was totally miscible. When using bioethanol E85, a considerable reduction in engine power was observed yielding value 13% to 22% less than in the tests with petrol. At the same time fuel consumption increased slightly between 7% and 20%, and gaseous emissions were considerably reduced (see Table 10). When using SEMO 36 the reduction in NO_x emission was the most significant as compared with other gases and was probably due to the lower temperature generated.

Figure 12. Macro images of two-stroke engine components after scuffing test using mixture of SEMO 36 lubricating oil with bioethanol E85 (a, b, c) and petrol (d, e, f): a), b), d), e) piston, c), f) cylinder in [12].

Oil Type and %	Power (kW)	Consumption (g/kWh)	NO$_x$ (g/kWh)	CH$_x$ (g/kWh)	CO (g/kWh)
SH3 Limit Normative			5.36	161	603
Petrol/Ref. Oil 2%	5.46	397	1.469	139.8	333.2
E10 + 2% Ref. oil	5.44	385	1.573	124.1	314.8
E20 + 2% Ref. oil	5.5	382	2.29	128	251.5
E85 + 2% Ref. oil (not miscible)	4.8	427	2.29	109.8	43.11
E85 + 2% SEMO 36 (miscible)	4.3	478	0.689	119.5	32.93

Table 10. Emission of gases from two-stroke engine tested with different lubricating oil– petrol combinations in [12].

2.6. Life cycle

The lifecycle analysis for a 2-stroke engine fed by petrol and E85 was carried out using the model M 165 Minsel engine running in a tiller during 1000 h, which characteristics are shown in Table 11.

Model of machine	Tiller 3002
Machine weight	90-110 kg
Engine model	M165 Minsel 2-stroke
Engine weight	12.8 kg
Engine life	1000 h
Scuffing test results	OK
Engine power	3 kW
Emissions	Directive 97/68/CE and later 2002/88/CE and 2004/26/CE

Table 11. Characteristics of the engine used in life-cycle analysis

Two fuel + oil pairs named as "Cleanengine systems" were compared with the Conventional system for the same engine working in the same application. In the alternative Cleanengine system I the engine was fed by a mixture of bioethanol E20 and mineral oil. In the alternative Cleanengine system II, the engine was fed by bioethanol and newly developed advanced and biodegradable lubricating oil SEMO 36. The fuel and oil consumption for the conventional and two alternative systems is shown in Table 12.

	Conventional	Cleanengine system (I)	Cleanengine system (II)
Fuel consumption per functional unit	Petrol 900 kg	BioE20 1123 kg	BioE85 1405 kg
Oil consumption per functional unit	Mineral oil 36 kg	Mineral oil 23 kg	SEMO 36 29 kg

Table 12. Parameters of the conventional and alternative systems used in the life-cycle analysis

The Eco-indicator 99 Methodology was used for the Impact Assessment method. The components of the environmental impact are shown in Figure 13 a), while the total environmental impact is shown in Figure 13 b). Almost all components of the environmental impact as well as the total environmental impact were higher for fossil fuel. However, the climate change was more affected by the renewable system.

The global environmental impact evaluated by Lifecycle Assessment tools for the Cleanengine system I and II using bioethanol was lower than for the reference system using petrol. The comparison between two alternative systems Cleanengine I and Cleanengine II showed that the last one had slightly higher environmental impact due to higher fuel and lubricant consumption that can be related to the lower calorific value of the ethanol compared to the petrol. While the reduction of the environmental impact is attributed to the reduction in emissions, the use of a biodegradable nontoxic lubricant will further reduce the environmental impact of the Cleanengine II system.

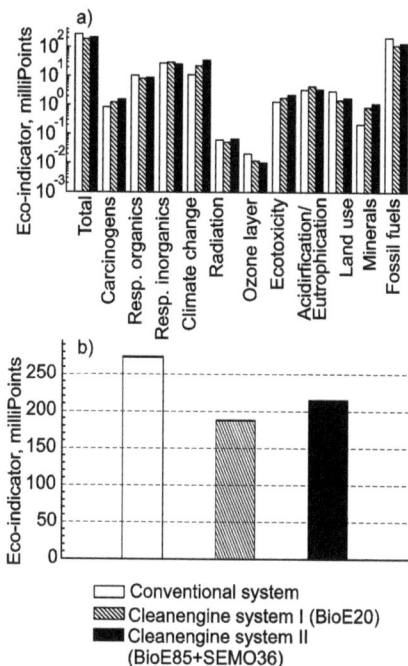

Figure 13. Results of the life-cycle and environmental impact analysis for the conventional and two alternative systems: a) components of the environment impact, b) total environmental impact in [12].

3. Nozzles for future engines

Compared to conventional liquid hydrocarbon fuels, bio-fuels exhibit considerable differences in their physical properties which significantly influence on the injector flow as well as on primary and secondary spray break-up processes. As a consequence, spray mixture formation of bio-fuels is considered to be largely different compared to conventional fuels under engine operating conditions with severe consequences on the combustion and emission characteristics. Hence, injection and combustion system optimization as well as optimization of the injector configuration (number of nozzle holes, diameter, spray targeting, etc.) for bio-fuels requires a detailed knowledge of how the fuel properties influence the injector flow and spray atomization characteristics. Optimization of the nozzles materials and design is an important task which will open new markets and enlarge the number of potential customers for eco-friendly applications.

3.1. Tribological evaluation

Different metal-doped DLC coatings were developed by Physical Vapour Deposition method (PVD). Friction and wear tests were carried out using SRV tribometer with "cylinder-on-disc" configuration in lubricated conditions. The coatings were deposited on steel cylinders and disks. The cylinder, 15 mm in diameter, performed reciprocating motion with a stroke of 2 mm and a friction frequency 50 Hz. Normal load was 50 N during short run-in period 30 s and 200 N during the test 60 min. The cylinder and the disk were immersed in fluids, which temperature during the test was constant and 25 °C.

Both Cr- and Ti DLC coatings had good friction and wear behaviour and they could be a good alternative to improve tribological properties of the actual uncoated nozzles.

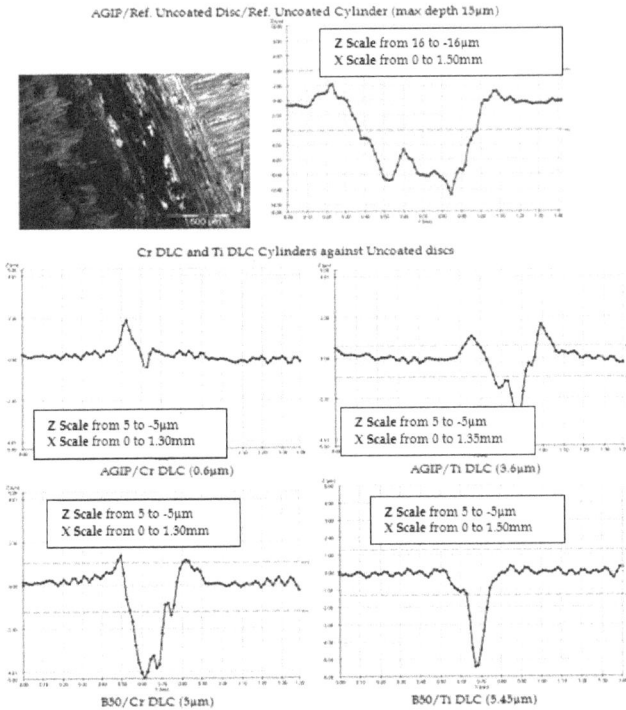

Figure 14. Average cross-section profile of the friction zone of discs samples (uncoated reference, Cr and Ti DLC) tested using different fuels, AGIP and B50. Different scales of magnitude are used for better visualization of the mean contact surface profile.

Surface morphology of the friction zone was studied using white light confocal microscopy. The averaged cross-section profiles for each sample tested are shown in Figure 14. It is possible to notice very good performance of the coatings, them had deeper grooves with a maximum depth of 5.45 μm. Cr DLC tested against AGIP fuel had better performance than Ti DLC. Two different scales of magnitude, Z, are used for better visualization of the mean contact surface profile. From 5 to -5μm for coated discs (Cr DLC and Ti DLC lubricated with AGIP ref and B50) and from 16 to -16μm for uncoated ref samples.

3.2. Corrosion characterization

Corrosion resistance of different materials and coatings used for nozzles fabrication (Cr and Ti DLC) was characterized using electrochemical impedance spectroscopy and potentiodynamic polarization techniques in order to determine the kinetics parameters and the corrosion mechanisms of these materials in NaCl 0.5M or K_2SO_4 0.2M in [12].

Base nozzles material, uncoated steel X82WMo, was also characterized under corrosion conditions and compared with DLC coated samples of the same material. The electrolyte used in these tests was K_2SO_4 0.2M. Cr DLC coating offered excellent corrosion protection. The coating did not exhibit any pores or defects, protecting effectively the substrate during immersion.

Open-circuit potential (OCP) was measured during 2200 s in order to analyze the samples tendency with the exposure time. After that, an electrochemical impedance spectroscopy was performed in a frequencies range from 10 k to 10 mHz. Once impedance measurements finished, a potentiodynamic potential swept was applied from OCP-0.2 V to OCP+0.6 V at a scan rate of 0.5 mV/s.

Coated nozzles had more positive potential than the reference ones. For all surfaces, OCP was stable after first 2200 s of immersion. The difference between three nozzles regarding impedance results was very notable. Cr and Ti DLC coated samples had a semicircle Nyquist diagrams implying that the electrolyte did not reach the substrate during the immersion in the dissolution. The coating acted as an effective protective barrier. Uncoated nozzle had lower corrosion resistance. Two time constants could be clearly distinguished from two maxima in the Nyquist plots.

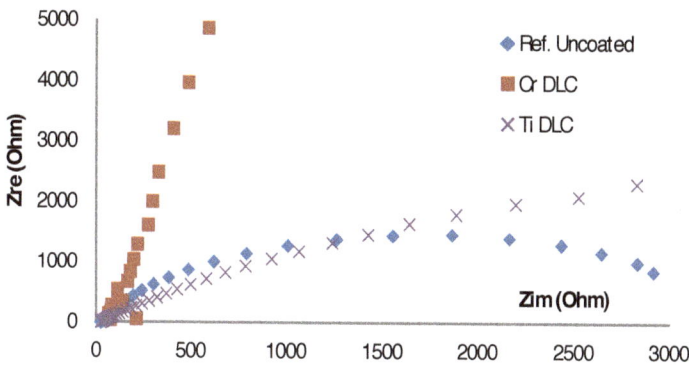

Figure 15. Nyquist diagrams. Impedance data of coated and uncoated nozzle in K_2SO_4.

Table 13 shows the parameters obtained from equivalent circuit simulation of the experimental data and Figure 16 shows the equivalent circuits used in the simulation process.

Samples (Nozzle appl.)	OCP (V)	R1 (kΩ/cm²)	CPE1 (μF/cm²)	R2 (kΩ/cm²)	CPE2 (μF/cm²)	ZO (μΩ−1·σ 1/2; σ 1/2)
Nozzle Uncoated	-0.556	0.035	Y_o=58.45 n=0.82	1.10	Y_o=140.7 n=0.94	
Nozzle Cr DLC	-0.022	12780	Y_o=0.75 n=0.94	-	-	-
Nozzle Ti-DLC	-0.354	10.38	Y_o=0.34 n=0.639	-	-	Y_o=0.02 B=3.23

Table 13. Equivalent circuit parameters of coated and uncoated nozzles

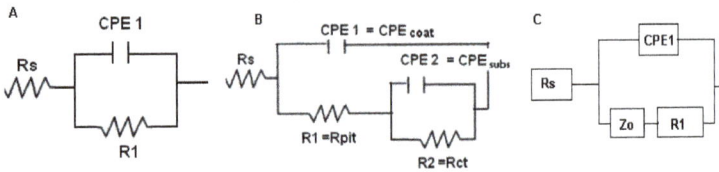

Figure 16. Equivalent circuits used for the experimental data simulation. Circuit A) for Nozzle Cr DLC; circuit B) for uncoated nozzle and circuit C) for Ti DLC coating.

Polarization curves for the coated nozzle are shown in Figure 17.

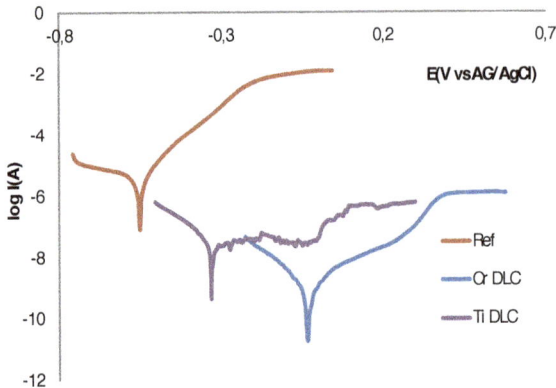

Figure 17. Polarization curves on coated and uncoated nozzles immersed in K_2SO_4

Cr DLC coating had passive behaviour and low corrosion current of the order of 10^{-9}A for potentials near to OCP. Coating Ti DLC also had passive behaviour in a wide zone of the anodic branch. Cr DLC and Ti DLC notably improved substrate corrosion behaviour reducing its corrosion current by several orders of magnitude (see table 14).

Samples	E_{corr} (V)	I_{corr} ($\mu A/cm^2$)
Nozzle Ref. Uncoated	-0.533	18.5
Nozzle Cr DLC	-0.039	0.003
Nozzle Ti DLC	-0.331	0.18

Table 14. Corrosion current of coated and uncoated nozzles calculated using Tafel approach

3.3. Experimental evaluation in real four-stroke engines (Minsel M430)

The injectors were tested in the Minsel M-430 engine manufactured by Abamotor Energía, SL. The parameters of the engine and test conditions are shown in Table 15.

Bore	85 mm
Stroke	75 mm
Displacement	426 c.c.
Compresion ratio	19,3: 1
Power NB	8,4 / 8,7 Cv
Rpm	3000
Torque	23Nm / 2000 RPM
Dry weight	45 Kg

Table 15. Characteristics of the engine used in engine tests to evaluate the different alternative nozzles (Cr DLC and Ti DLC).

During the test the engine worked for 50 hours at full load (3000 rpm). Biodiesel B30 was used as a fuel, which was a mixture of FAME (100% Biodiesel) with diesel B at a rate of 30%.

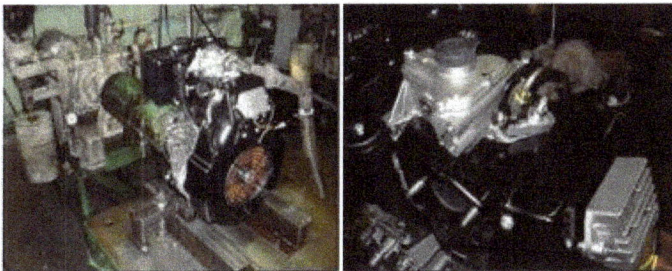

Figure 18. The engine on test bench and the tested nozzles installed on the engine

3.4. Nozzle characterization after test in the engine

Scanning electron microscope (SEM) and energy dispersion X-Ray spectroscopy (EDS) were used for characterization of the nozzles geometry after the engine tests. Cr DLC coating had better behaviour than Ti DLC.

The microanalysis showed that for the all coatings the deposited layer on the needle persisted after the test, with the exception of the tip where the Ti DLC layer has been detached

Additionally, the spray holes geometries of the nozzle body were analysed after endurance test with two different fluids: reference standard fuel and 30% biodiesel.

Figure 19 shows the scanning electron microscope images (EDS) of the nozzle body tip before the engine test (real part and its corresponding silicon model for orifices internal characteristics analysis), whereas Figures 20 and 21 show the nozzles after the tests with standard diesel fuel and B30 fuel, correspondingly. Though large quantities of carbonaceous deposits could be observed on free surfaces for both fuels, no deposits were found on internal surface of spray holes.

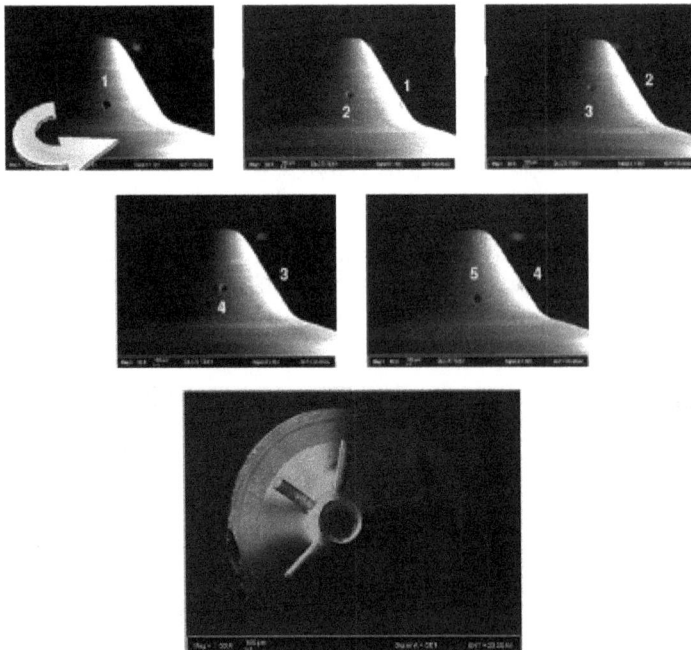

Figure 19. Nozzle body tip and silicon model with labelled holes

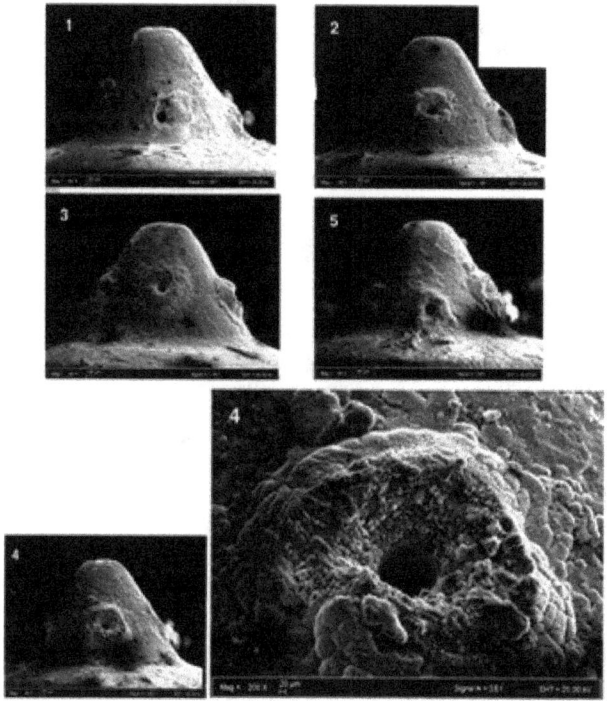

Figure 20. Images of the nozzle after endurance test with standard diesel fuel

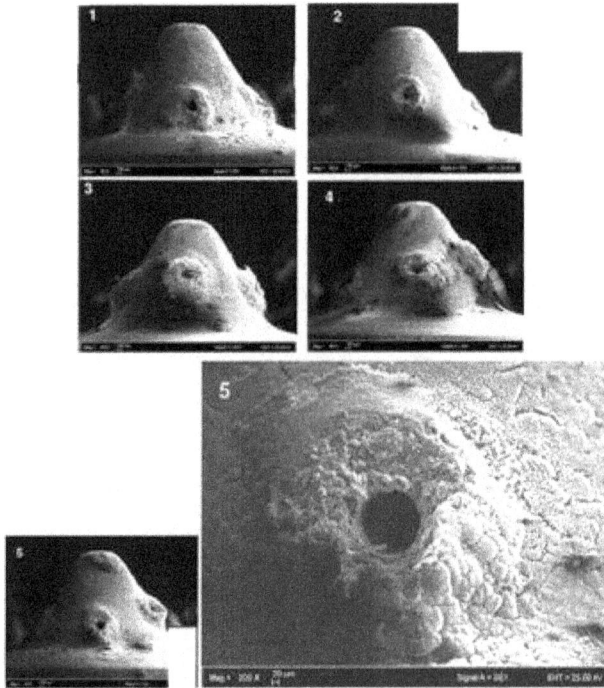

Figure 21. Images of the nozzle after endurance test with B30 fuel

Finally, nozzle deposits were analyzed by Thermal Gravimetric Analysis (TGA), which showed no big difference in deposits composition for the nozzles operated with standard diesel and B30 blend.

4. Conclusions

Fully formulated prototype lubricants based on synthetic esters had low toxicity for aqueous organisms (algae and Daphnia Magna) and high biodegradability evaluated by the Manometric Respirometry Method.

Among three developed prototype lubricating oils, SEMO 10 had the best tribological performance which was comparable with that of the reference mineral oil. Further improvement of the tribological properties of this lubricating oil was achieved by additive re-formulation. The developed lubricant, SEMO 36, exceeded the reference mineral oil in tribological performance.

Our findings indicated that, in addition to the rheological properties of the lubricating oil, deposit build-up was an important factor controlling the tribological performance of the oil both in simulation experiments and real two-stroke engines. Two kinds of deposits: carbon soot and transparent sliding lacquer were observed on the engine components after tests. Build-up of a transparent sliding lacquer was especially important in the case of SEMO 36 oil and it was related with considerable reduction both in wear rate and friction coefficient. For SEMO 36, surface chemical analysis of the friction zone showed important changes in surface chemical composition, which was especially marked by increase in carbon and oxygen content. It is evident that formation of the sliding deposits stemmed from tribochemical reactions between the oil components and base material (cast iron and steel). The chemical state of carbon and oxygen atoms on the surface of friction zone should be further investigated for better understanding of these mechanisms.

Tests in real two-stroke engines were performed using mixtures of the developed lubricant with petrol or bioethanol. In both cases, no seizure between piston ring and cylinder liner was observed. When using bioethanol, the engine components were clean without important carbon soot deposits.

Engine power slightly decreased and fuel consumption slightly increased - on a volumetric basis when bioethanol E85 was blended with the newly developed lubricating oil SEMO 36. However, these results might be related with lower calorific value of ethanol as compared with petrol. Besides, the new lubricating oil improved scuffing resistance in combination with miscible lubricants and significantly reduced the environmental impact. In addition to low toxicity and high biodegradability, emissions of CO, NO_x and hydrocarbons from engines lubricated with the newly developed lubricants were lower than with traditional mineral oil and much below the limits established for portable applications.

Concluding, a new generation of lubricating oils for two-stroke engines have been developed combining low friction, good protection against wear and scuffing, no ash residue, low carbon soot or other deposit formation. These lubricating oils are compatible with bioethanol E85.

Application of Cr DLC coating on injection nozzles significantly increased the corrosion resistance and improved behaviour in engine test.

Though Ti-DLC coating also improved substrate corrosion resistance, its performance in engine test was worse than for Cr DLC coating.

Deposit chemical composition and the nozzle performance did not significantly vary in endurance tests when standard diesel was substituted by B30 blend.

Acknowledgements

The authors acknowledge financial support of the European Commission, the project CleanEngine "Advanced technologies for highly efficient Clean Engines working with alternative fuels and lubes" under contract TST5-CT2006-031241, and the Spanish Minister of Science and

Innovation, for co-financing the project under contracts ENE 2008-00652-E/ALT "Tecnologías Avanzadas para motores limpios altamente eficientes, trabajando con combustibles y lubricantes renovables", RYC-2009-0412 and BIA2011-25653. Also, the authors acknowledge support received from other partners who participated in the projects: ARIZONA Chemical, OBR, GUASCOR Power, BAM, AVL and INSTITUTO MOTORI.

We appreciate the useful help of Olatz Areitioaurtena and Raquel Bayón on performing Biodegradability, toxicity and corrosion characterizations and tests.

Author details

Xana Fernández-Pérez[1], Amaya Igartua[1], Roman Nevshupa[2], Patricio Zabala[3], Borja Zabala[3], Rolf Luther[4], Flavia Gili[5] and Claudio Genovesio[6]

1 Fundación Tekniker, Avda, Otaola, Eibar, Spain

2 Instituto de Ciencias de la Construcción Eduardo Torroja (IETcc), c/ Serrano Galvache, Madrid, Spain

3 Abamotor Energía, SL, B. Astola, Abadiano, Spain

4 Fuchs Europe GmbH, Mannheim D, Germany

5 CRF StradaTorino Orbassano, Italy

6 FIRAD S.p.A, Fabbrica Italiana Ricambi Apparati Diesel, Bagnolo, Italy

References

[1] Igartua, A; Barriga, J; Aranzabe, A; (2005) *Biodegradable Lubricants*. Virtual Tribology Institute Edition, ISBN 83-70204-418-X.

[2] Woydt, M; Skopp, A; (2005) *Ash-free and bionotox engine oils*. In: Biodegradable lubricants, eds. A. Igartua, J. Barriga, A. Aranzabe, Radom: Virtual Tribology Institute, Institute of Terotechnology; p. IV.6-IV.9.

[3] ASTM D-445-06: Standard Test Method for Kinematic Viscosity of Transparent and Opaque Liquids (and Calculation of Dynamic Viscosity).

[4] ASTM D-2270-04: Standard Practice for Calculating Viscosity Index from Kinematic Viscosity at 40 and 100°C.

[5] OECD guidelines for testing of chemicals: Section 3; 2003. 12 p.

[6] OECD 301F: Manometric Respirometry Test. OECD guidelines for testing of chemicals; 1992.

[7] OECD 201: Alga, Growth Inhibition Test. OECD guidelines for testing of chemicals; 2006.

[8] OECD 202: Daphnia sp. Acute Immobilisation Test. OECD guidelines for testing of chemicals; 2004.

[9] DIN 51834-2. Tribological test in the translatory oscillation apparatus. Part 2: Standard Test Method for Measuring the Friction and Wear Properties of EP Lubricating Oils Using the SRV Test Machine; 2004.

[10] Kostetsky, BI; (1992) *The structural-energetic concept in the theory of friction and wear* (synergism and self-organization). Wear; 159:1-15.

[11] Nevshupa, RA ; (2009) The *role of athermal mechanisms in the activation of tribodesorption and triboluminisence in miniature and lightly loaded friction units.* Journal of Friction and Wear; 30:118-126.

[12] Igartua, A; Nevshupa, R; Fernández-Pérez, X; Conte, M; Zabala, R; Bernaola, J; Zabala, P; Luther, R; Rausch, R; (2011) *Alternative Eco-Friendly Lubes for Clean Two-Stroke Engines.* Tribology International, 44, 727-736.

[13] Martínez, L; Nevshupa, R; Álvarez, L; Huttel, Y; Méndez, J; Román, E; Mozas, E; Valdés, JR ; Jimenez, MA; Gachon, Y; Heau, C; Faverjon, F; (2009) *Application of diamond-like carbon coatings to elastomers frictional surfaces.*Tribology International, v. 42, pp. 584-590.

[14] Bayón, R; Nevshupa, R; Zubizarreta, C; Ruiz de Gopegui, U; Barriga, J; Igartua, A; (2010) *Characterisation of tribocorrosion behaviour of multilayer PVD coatings.* Analytical and Bioanalytical Chemistry.V. 396. P. 2855-2862.

[15] Bayón, R; Zubizarreta, C; Nevshupa, R; Rodriguez, JC; Fernández-Pérez, X; Ruiz de Gopegui, U; Igartua, A; (2011) *Rolling-sliding, scuffing and tribocorrosion behaviour of PVD multilayer coatings for gears application.* Industrial Lubrication and Tribology. V. 63/1. P. 17–26.

[16] Alajbegovic, A; Meister, G; Greif, D; Basara, B; (2001) *Three Phase Cavitating Flows in High Pressure Swirl Injectors.* 4th Int. Conf. on Multiphase Flow – ICMF'01, May 27 – June 1, 2001, New Orleans, Louisiana, U.S.A.

[17] Von Berg, E; Alajbegovic, A; Tatschl R; Krüger, C; Michels, U; (2001) *Multiphase Modeling of Diesel Sprays with the Eulerian/*Eulerian Approach (DaimlerChrysler AG), ILASS-Europe 2001, Sept. 2-6, 2001, Zürich, Switzerland

[18] Von Berg, E; Alajbegovic, A; Greif, D; Poredos, A; Tatschl, R; Winklhofer, E (2002); *Break-up Model for Diesel Jets based on Locally Resolved Flow Field in the Injection Hole,* ILASS-Europe 2002, Sept. 9-11, 2002, Zaragoza, Spain

Permissions

The contributors of this book come from diverse backgrounds, making this book a truly international effort. This book will bring forth new frontiers with its revolutionizing research information and detailed analysis of the nascent developments around the world.

We would like to thank Jürgen Gegner, for lending his expertise to make the book truly unique. He has played a crucial role in the development of this book. Without his invaluable contribution this book wouldn't have been possible. He has made vital efforts to compile up to date information on the varied aspects of this subject to make this book a valuable addition to the collection of many professionals and students.

This book was conceptualized with the vision of imparting up-to-date information and advanced data in this field. To ensure the same, a matchless editorial board was set up. Every individual on the board went through rigorous rounds of assessment to prove their worth. After which they invested a large part of their time researching and compiling the most relevant data for our readers. Conferences and sessions were held from time to time between the editorial board and the contributing authors to present the data in the most comprehensible form. The editorial team has worked tirelessly to provide valuable and valid information to help people across the globe.

Every chapter published in this book has been scrutinized by our experts. Their significance has been extensively debated. The topics covered herein carry significant findings which will fuel the growth of the discipline. They may even be implemented as practical applications or may be referred to as a beginning point for another development. Chapters in this book were first published by InTech; hereby published with permission under the Creative Commons Attribution License or equivalent.

The editorial board has been involved in producing this book since its inception. They have spent rigorous hours researching and exploring the diverse topics which have resulted in the successful publishing of this book. They have passed on their knowledge of decades through this book. To expedite this challenging task, the publisher supported the team at every step. A small team of assistant editors was also appointed to further simplify the editing procedure and attain best results for the readers.

Our editorial team has been hand-picked from every corner of the world. Their multi-ethnicity adds dynamic inputs to the discussions which result in innovative outcomes. These outcomes are then further discussed with the researchers and contributors who give their valuable feedback and opinion regarding the same. The feedback is then collaborated with the researches and they are edited in a comprehensive manner to aid the understanding of the subject.

Apart from the editorial board, the designing team has also invested a significant amount of their time in understanding the subject and creating the most relevant covers. They scrutinized every image to scout for the most suitable representation of the subject and create an appropriate cover for the book.

The publishing team has been involved in this book since its early stages. They were actively engaged in every process, be it collecting the data, connecting with the contributors or procuring relevant information. The team has been an ardent support to the editorial, designing and production team. Their endless efforts to recruit the best for this project, has resulted in the accomplishment of this book. They are a veteran in the field of academics and their pool of knowledge is as vast as their experience in printing. Their expertise and guidance has proved useful at every step. Their uncompromising quality standards have made this book an exceptional effort. Their encouragement from time to time has been an inspiration for everyone.

The publisher and the editorial board hope that this book will prove to be a valuable piece of knowledge for researchers, students, practitioners and scholars across the globe.

List of Contributors

Nehal S. Ahmed and Amal M. Nassar
Additives Lab., Department of Petroleum Applications, Egyptian Petroleum Research Institute, Nasr City, Cairo, Egypt

Walter Holweger
Schaeffler Technologies AG & Co. KG, R&D Central Materials, Germany

Maciej Paszkowski
Wroclaw University of Technology, Institute of Machines Design and Operation, Wroclaw, Poland

Virginia Sáenz de Viteri and Elena Fuentes
IK4-Tekniker, Eibar, Spain

Masataka Nosaka and Takahisa Kato
Department of Mechanical Engineering, University of Tokyo, Tokyo, Japan

Giuseppe Pintaude
Mechanical Academic Department, Federal University of Technology – Paraná, Curitiba, Brazil

Remigiusz Michalczewski, Marek Kalbarczyk, Michal Michalak, Witold Piekoszewski, Marian Szczerek, Waldemar Tuszynski and Jan Wulczynski
Institute for Sustainable Technologies - National Research Institute (ITeE-PIB), Tribology Department,
Radom, Poland

M. Tauviqirrahman and R. Ismail
Laboratory for Engineering Design and Tribology, Mechanical Engineering Department, University of Diponegoro, Jl. Prof. H. Sudharto, Kampus UNDIP Tembalang, Semarang, Indonesia
Laboratory for Surface Technology and Tribology, University of Twente Drienerlolaan, Enschede, The Netherlands

J. Jamari
Laboratory for Engineering Design and Tribology, Mechanical Engineering Department, University of Diponegoro, Jl. Prof. H. Sudharto, Kampus UNDIP Tembalang, Semarang, Indonesia

D.J. Schipper
Laboratory for Surface Technology and Tribology, University of Twente Drienerlolaan, Enschede, The Netherlands

Erik Kuhn
Department of Mechanical Engineering and Production, Hamburg University of Applied Sciences, Germany

Manfred R. Mauntz and Stefan Klingau
Cmc Instruments GmbH, Eschborn, Germany

Jürgen Gegner
University of Siegen, Siegen, Germany

Ulrich Kuipers
South Westphalia University of Applied Sciences, Hagen, Germany

Xana Fernández-Pérez and Amaya Igartua
Fundación Tekniker, Avda, Otaola, Eibar, Spain

Roman Nevshupa
Instituto de Ciencias de la Construcción Eduardo Torroja (IETcc), c/ Serrano Galvache, Madrid, Spain

Patricio Zabala and Borja Zabala
Abamotor Energía, SL, B. Astola, Abadiano, Spain

Rolf Luther
Fuchs Europe GmbH, Mannheim D, Germany

Flavia Gili
CRF StradaTorino Orbassano, Italy

Claudio Genovesio
FIRAD S.p.A, Fabbrica Italiana Ricambi Apparati Diesel, Bagnolo, Italy